Microwave
Solid-State Devices

SAMUEL Y. LIAO

Professor of Electrical Engineering
California State University, Fresno

Prentice-Hall, Inc., *Englewood Cliffs, New Jersey 07632*

Library of Congress Cataloging in Publication Data

Liao, Samuel Y.
    Microwave solid-state devices.

    Includes bibliographies and index.
    1. Microwave devices.   2. Solid state devices.
I. Title.
TK7876.L478   1984   621.381′33   84-2075
ISBN 0-13-580571-6

*The author dedicates this book
to the love of
his wife Lucia Hsiao-Chuang Lee Liao
and
their children
Grace, Kathy, Gary, and Jeannie*

Editorial/production supervision: Greg Hubit
Cover design: Judy Winthrop
Manufacturing buyer: Tony Caruso

Printed in the United States of America

10 9 8 7 6 5 4 3 2 1

ISBN 0-13-580571-6   01

Prentice-Hall International, Inc., *London*
Prentice-Hall of Australia Pty. Limited, *Sydney*
Editora Prentice-Hall do Brasil, Ltda., *Rio de Janeiro*
Prentice-Hall Canada Inc., *Toronto*
Prentice-Hall of India Private Limited, *New Delhi*
Prentice-Hall of Japan, Inc., *Tokyo*
Prentice-Hall of Southeast Asia Pte. Ltd., *Singapore*
Whitehall Books Limited, *Wellington, New Zealand*

Contents

Preface

This book is intended to serve as a text in a course on microwave solid-state devices at the senior or beginning graduate level in electrical engineering. Its primary purpose is to provide readers with an understanding of the microwave solid-state devices commonly used. It is assumed that readers have had previous courses in electromagnetics and solid-state electronics. Because the book is, to a large extent, self-contained, it can also be used as a textbook by physical science students and as a reference book by electronics engineers working in the microwave area.

This book is not the second edition of *Microwave Devices and Circuits*, but it is a continuation of that book.

The book is organized into twelve chapters:

1. Chapter 1 is introductory.
2. Chapter 2 studies wave propagation in different media.
3. Chapter 3 describes microwave solid-state junction-effect devices, including microwave transistors, tunnel diodes, MES diodes, and MIS diodes.
4. Chapter 4 treats microwave solid-state field-effect devices such as MESFETs, MOSFETs, VMOSFETs, UMOSFETs, and CCDs.
5. Chapter 5 deals with microwave solid-state Gunn-effect devices such as Gunn diodes, LSA diodes, InP diodes, and CdTe diodes.
6. Chapter 6 discusses microwave solid-state avalanche-effect devices, including IMPATT diodes, TRAPATT diodes, BARITT diodes, and parametric amplifiers.
7. Chapter 7 analyzes electrooptical-reflection-effect devices such as step-index fibers, graded-index fibers, and optical-fiber communication systems.

8. Chapter 8 describes solid-state lasers and laser modulators, including ruby laser, GaAs laser, Nd^{3+} : YAG laser, and the Pockels cell laser modulator.

9. Chapter 9 investigates invisible-heat-effect devices such as infrared devices and the FLIR system.

10. Chapter 10 treats the light-emitting devices such as LEDs, infrared LEDs, and visible LEDs; and liquid crystal displays (LCDs).

11. Chapter 11 deals with photon-absorption-effect devices including photon detectors, thermal detectors, and photothermic detectors.

12. Chapter 12 analyzes high-speed switching devices such as MISS diodes, *p-n-p-n* thyristor, and MESFET switches.

The arrangement of topics is flexible, and the instructor has a choice in the selection or order of the topics to suit either a one-semester or possibly a one-quarter course. Problems for each chapter are included to aid readers in further understanding the subjects discussed in the text. A solutions manual may be obtained from the publisher by instructors who have adopted the book for their courses.

The author is grateful to several anonymous reviewers for their many valuable comments and constructive suggestions, which helped to improve this book. The author would also like to acknowledge his appreciation for the publication of this book by Prentice-Hall, Inc.; and especially extend his thanks to Bernard M. Goodwin, Engineering Editor, for his constant guidance, and to Margaret McAbee, Production Editor, for her skillful editorial work.

SAMUEL Y. LIAO

Chapter 1

Introduction

The purpose of this book is to present the principles, operations, and applications of microwave solid-state devices. Microwave techniques have been increasingly adopted in such diverse areas as radio astronomy, long-distance communications, space exploration, satellite communications, radar systems, and missile electronic systems. As a result of the accelerating growth of microwave technology, research, and development in universities and industries, students preparing for and electronics engineers working in the microwave field need to understand the theory and characteristics of microwave solid-state devices for the design and production of microwave systems.

1-0 MICROWAVE FREQUENCIES

The term *microwave frequencies* traditionally refers to those frequencies from 1 GHz to 300 GHz or wavelengths measured from 30 centimeters to 1 millimeter. However, *microwaves* really indicate wavelengths in the micron ranges—that is, *microwave frequencies* up to infrared and visible light regions. In this book microwave frequencies mean those from 1 GHz to 10^6 GHz.

The microwave band designation that resulted from World War II radar security considerations was not officially recognized by any industrial, professional, or government organization until 1969. In August 1969 the U.S. Department of Defense, Office of Joint Chiefs of Staff, by a message to all services, directed the use of microwave frequency bands as listed in Table 1-0-1. On May 24, 1970, the Department of Defense adopted another band designation for microwave frequencies as shown in Table 1-0-2. In electronics industries and academic institutes,

1

however, IEEE (Institute of Electrical and Electronics Engineers, Inc.) microwave frequency bands are commonly used as shown in Table 1-0-3. These three band designations are given in Table 1-0-4 for comparison.

TABLE 1-0-1 U.S. MILITARY MICROWAVE BANDS

Designation	Frequency range (GHz)
P band	0.225– 0.390
L band	0.390– 1.550
S band	1.550– 3.900
C band	3.900– 6.200
X band	6.200– 10.900
K band	10.900– 36.000
Q band	36.000– 46.000
V band	46.000– 56.000
W band	56.000–100.000

TABLE 1-0-2 U.S. NEW MILITARY MICROWAVE BANDS

Designation	Frequency range (GHz)	Designation	Frequency range (GHz)
A band	0.100–0.250	H band	6.000– 8.000
B band	0.250–0.500	I band	8.000– 10.000
C band	0.500–1.000	J band	10.000– 20.000
D band	1.000–2.000	K band	20.000– 40.000
E band	2.000–3.000	L band	40.000– 60.000
F band	3.000–4.000	M band	60.000–100.000
G band	4.000–6.000		

TABLE 1-0-3 IEEE MICROWAVE FREQUENCY BANDS

Designation	Frequency range (GHz)	Wavelength (cm)
VHF	0.1– 0.3	300.00–100.00
UHF	0.3– 1.0	100.00– 30.00
L band	1.0– 2.0	30.00– 15.00
S band	2.0– 4.0	15.00– 7.50
C band	4.0– 8.0	7.50– 3.75
X band	8.0– 13.0	3.75– 2.31
Ku band	13.0– 18.0	2.31– 1.67
K band	18.0– 28.0	1.67– 1.07
Ka band	28.0– 40.0	1.07– 0.75
Millimeter	40.0– 300.0	0.75– 0.10
Submillimeter	300.0–3000.0	0.10– 0.01

TABLE 1-0-4 COMPARISON OF IEEE BANDS, OLD BANDS AND NEW BANDS

Frequency in GHz	0.1	0.15	0.2	0.3	0.4	0.5	0.6	0.75	1	1.5	2	3	4	5	6	7.5	10	15	20	30	40	50	60	75	100
Wavelength in cm	300	200	150	100	75	60	50	40	30	20	15	10	7.5	6	5	4	3	2	1.5	1	0.75	0.6	0.5	0.4	0.3

IEEE bands: VHF, UHF, L, S, C, X, Ku, K, Ka, MILLIMETER (W)

Old bands: P, L, S, C, X, K, Q, V, W

New bands	A	B	C	D	E	F	G	H	I	J	K	L	M

Wavelength in cm	300	200	150	100	75	60	50	40	30	20	15	10	7.5	6	5	4	3	2	1.5	1	0.75	0.6	0.5	0.4	0.3
Frequency in GHz	0.1	0.15	0.2	0.3	0.4	0.5	0.6	0.75	1	1.5	2	3	4	5	6	7.5	10	15	20	30	40	50	60	75	100

1-1 MICROWAVE SOLID-STATE DEVICES

After Shockley and his coworkers invented the transistor in 1948, the second half of the twentieth century became the solid-state electronic era. In 1954 and 1960 Townes and Maiman developed the maser and ruby laser, respectively. Esaki originated the tunnel diode in 1958, and in 1963 Gunn observed the Gunn effect in a GaAs diode. Read proposed the avalanche-effect diode in 1958 and Boyle created the charge-coupled device in 1969. In the 1970s optical-fiber devices gradually developed. All of these achievements in microwave solid-state devices were accomplished in less than 30 years. In the next century microwave electronic technology is expected to evolve into an electrooptical age. Table 1-1-1 lists the major microwave solid-state devices discussed in this book, along with their operational mechanisms and applications.

TABLE 1-1-1 MICROWAVE SOLID-STATE DEVICES AND THEIR APPLICATIONS

Operational mechanism	Devices	Applications
Junction effect	Microwave transistors	Transmitter, amplifier
	Tunnel diodes	Amplifier, switches
	Schottky-barrier diodes	Modulator, demodulator, detector, switches
	Microwave MIS diodes	CCDs
Field effect	MESFETs	Local oscillator, amplifier, modulator, driver stage
	MOSFETs	Oscillator, amplifier, switches, driver stage
	VMOSFETs, UMOSFETs, DMOSFETs, and VDMOSFETs	Oscillator, amplifier
	CCDs: SCCDs, BCCDs, and JCCDs	Infrared detection, signal processing
Gunn effect	Gunn diodes	Local oscillator, amplifier
	LSA diodes	Transmitter, power stage
	InP diodes	Oscillator
	CdTe diodes	Oscillator
Avalanche effect	IMPATT diodes	Driver stage, local oscillator, power stage
	TRAPATT diodes	Pulse generator, power stage
	BARITT and DOVETT diodes	Local oscillator
	Parametric amplifiers	Low-noise amplifier
Electrooptical-reflection effect	Optical fibers	Telecommunications and military command control
	Step-index fibers	Long-haul and high-bandwidth links
	Graded-index fibers	Medium-haul and high-bandwidth links

TABLE 1-1-1   MICROWAVE SOLID-STATE DEVICES AND THEIR
APPLICATIONS (CONTINUED)

Operational mechanism	Devices	Applications
Stimulated-emission effect	Ruby laser	Industrial and military uses
	Solid-state junction lasers	Light source for optical fibers
	GaAs laser	Industrial and military uses
	Nd^{3+}:YAG laser	Single-mode fibers
	Nd^{3+}:Glass laser	Industrial and military uses
Electrooptical effect	Pockels cells	Laser modulator
	Kerr cells	Laser modulator
Magnetooptical effect	Faraday-rotation modulator	Laser modulator
Invisible-heat effect	Blackbody	Infrared calibration
	FLIR system	Target identification
Light-emission effect	LEDs	Indicator, light source for optical fibers
	Infrared LEDs	Indicator, light source for optical fibers
	Visible LEDs	Indicator, light source for optical fibers
Light-illuminating effect	LCDs	Digital displays Alphameric displays Infrared display
Photon-absorption effect	Photon detectors	Radiation detection
	Thermal detectors	Radiation detection
	Photothermic detectors	Radiation detection
Switching effect	MISS diodes	High-speed switches
	p-n-p-n thyristors	Switches
	GaAs MESFET switches	High-speed switches

1-2 MICROWAVE SOLID-STATE ELECTRONIC SYSTEMS

A microwave solid-state electronic system usually consists of three major units: transmitter, transmission line or medium, and receiver. The transmission line or medium may be either a solid link or free space. The prime requirements of a transmitter are output power and conversion efficiency, whereas the prime requirement of a receiver is the noise figure. The characteristics of a microwave solid-state device determine the capability and reliability of that unit in the system.

Chapter 2

Wave Propagation in Materials, Ionosphere, and Radar Systems

2-0 INTRODUCTION

In microwave electronic systems it is often necessary to use a certain type of metallic-film coating on plastic substrate in order to attenuate light wave intensity at visible light frequencies or to use a certain type of crystal to change the phase angle of the light wave. The plastic dome window of a missile, for instance, is painted with a gold-film coating to attenuate the microwave radiation, and the potassium dideuterium phosphate $KD^*P(KD_2PO_4)$ crystal is commonly used in a microwave laser modulator to polarize the light wave. When a metallic film is very thin, its resistivity is much higher than the bulk metal. Commonly used dielectric crystals and magnetic ferrites for wave propagation are the anisotropic media types. In this chapter we analyze light wave propagation in certain metallic film coatings and anisotropic crystals, as well as in the ionosphere and radar systems.

2-1 PLANE WAVE PROPAGATION IN METALLIC-FILM COATING ON PLASTIC SUBSTRATE

In certain engineering applications it is often desirable to use a metallic-film coated glass to attenuate optimum electromagnetic radiation at microwave frequencies and also to transmit as much of light intensity at visible light frequencies as possible. The metallic-film coating normally must have a high melting point, high electrical conductivity, high adhesion to glass, high resistance to oxidation, and insensitivity to light and water, plus the capability of dissipating some power for de-icing, de-fogging, or maintaining certain temperature levels. The metallic-film coatings on a plastic substrate are used in such applications as windshields on airplanes or automobiles, medical equipment in hospitals, and on the dome windows of space vehicles, missiles, or other military devices.

2-1-1 Surface Resistances of Metallic Films

Very thin metallic films have a much higher resistivity than bulk metals because of the scattering of electrons from the surface of the film. If the film thickness is very large compared to the electron mean-free-path, the resistivity is expected to be nearly the same as that of a bulk metal. When film thickness is of the order of the electron mean-free-path, then the role of surface scattering becomes dominant. Fuchs [1] and Sondheimer [2] considered the general form of the solution of the Boltzmann equation for the case of a conducting film and found the film conductivity σ_f in terms of the bulk conductivity σ, the film thickness t, and the electron mean-free-path p:

$$\sigma_f = \frac{3t\sigma}{4p}\left[\ell n\left(\frac{p}{t}\right) + 0.4228\right] \qquad \text{for} \quad t \ll p \qquad (2\text{-}1\text{-}1)$$

The surface resistance of conducting films is normally quoted in units of ohms per square because in the equation for resistance

$$R = \frac{\text{specific resistivity} \times \text{length}}{\text{thickness} \times \text{width}} = \frac{\rho\ell}{tw} \qquad (2\text{-}1\text{-}2)$$

When the units of length ℓ and width w are chosen so as to have equal magnitude (i.e., resulting in a square), resistance R in ohms per square is independent of the dimensions of the square and equals

$$R_s = \frac{\rho_f}{t} = \frac{1}{t\sigma_f} \qquad \text{ohms/sq} \qquad (2\text{-}1\text{-}3)$$

According to the Fuchs–Sondheimer theory, the surface resistance of a metallic film is decreased as the thickness of the film is increased.

2-1-2 Optical Constants of Plastic Substrates and Metallic Films

The optical properties of materials are usually characterized by two constants, the refractive index n and the extinction index k. The refractive index is defined as the ratio of the phase velocities of light in vacuum and a medium. The extinction index is related to the exponential decay of the wave as it passes through a medium. Most optical plastics are suitable as substrate materials for a dome window and for application of a metallic film. Table 2-1-1 lists the values of the refractive index n of several commonly used nonabsorbing plastic substrate materials [3].

The measured values of the refractive index n and the extinction index k of thin metallic-film coatings deposited in a vacuum [3] are tabulated in Table 2-1-2.

TABLE 2-1-1 SUBSTRATE MATERIALS

Substrate material	Refractive index n
Corning Vycor	1.458
Crystal quartz	1.540
Fused silica	1.458
Plexiglass	1.490
Polycyclohexyl methacrylate	1.504
Polyester glass	1.500
Polymethyl methacrylate	1.491
Zinc crown glass	1.508

TABLE 2-1-2 REFRACTIVE INDEX n AND EXTINCTION INDEX k OF THIN METALLIC FILMS

Wavelength (Å)	Copper film		Gold film		Silver film	
	n	k	n	k	n	k
2000			1.427	1.215	1.13	1.23
2200					1.32	1.29
2300					1.38	1.31
2400					1.37	1.33
2500					1.39	1.34
2600					1.45	1.35
2700					1.51	1.33
2800					1.57	1.27
2900					1.60	1.17
3000					1.67	0.96
3100					1.54	0.54
3200					1.07	0.32
3300					0.30	0.55
3400					0.16	1.14
3500					0.12	1.35
3600					0.09	1.52
3700					0.06	1.70
3800						
4000						
4500	0.870	2.200	1.400	1.880		
4920						
5000	0.880	2.420	0.840	1.840		
5460						
5500	0.756	2.462	0.331	2.324		
6000	0.186	2.980	0.200	2.897		
6500	0.142	3.570	0.142	3.374		
7000	0.150	4.049	0.131	3.842		
7500	0.157	4.463	0.140	4.266		
8000	0.170	4.840	0.149	4.654		
8500	0.182	5.222	0.157	4.993		
9000	0.190	5.569	0.166	5.335		
9500	0.197	5.900	0.174	5.691		
10000	0.197	6.272	0.179	6.044		

2-1-3 Microwave Radiation Attenuation of Metallic-Film Coating on Plastic Substrate

A conductor with high conductivity and low permeability has low intrinsic impedance. When a radio wave propagates from a medium of high intrinsic impedance
into a medium of low intrinsic impedance, the reflection coefficient is high. From
electromagnetic plane-wave theory in the far field high attenuation occurs in a
medium made from material of high conductivity and low permeability. Good
conductors, such as gold, silver, and copper, have high conductivity and are often
used for attenuating electromagnetic energy. Microwave radiation attenuation by
metallic-film coating on a substrate consists of three parts [4]:

$$\text{Attenuation} = A + R + C \quad \text{dB} \tag{2-1-4}$$

where A = absorption or penetration loss in decibels inside the metallic-film coating
while the substrate is assumed to be nonabsorbing plastic glass

R = reflection loss in decibels from the multiple boundaries of a metallic-film
coating on substrate

C = correction term in decibels required to account for multiple internal
reflections when absorption loss A is much less than 10 dB for electrically
thin film.

Figure 2-1-1 shows the absorption and reflection of a metallic-film coating on a
plastic substrate.

(a) Absorption loss A.   As described in the usual way, the propagation
constant γ for a uniform plane wave in a good conducting material is given by

$$\gamma = \alpha + j\beta = (1+j)\sqrt{\pi f \mu \sigma_f} \quad \text{for} \quad \sigma_f \gg \omega\epsilon \tag{2-1-5}$$

If the plastic substrate is assumed to be a nonabsorbing material, the absorption loss
A of the metallic-film coating on a substrate is related only to the thickness t of the
coated film and the attenuation α:

$$A = 20\log_{10}e^{\alpha t} = 20(\alpha t)\log_{10}e = 20(0.4343)(\alpha t)$$
$$= 8.686t\sqrt{\pi f \mu \sigma_f} \quad \text{dB} \tag{2-1-6}$$

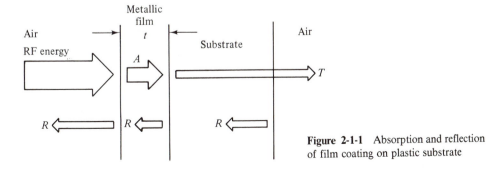

Figure 2-1-1 Absorption and reflection
of film coating on plastic substrate

where t = thickness of the film coating in meters

  μ = permeability of the film in henry per meter

  f = frequency in hertz

  σ_f = conductivity of the coated film in mhos per meter

Because the thickness of the coated film is very thin—for example, 100 Å (Å = 10^{-10} m) at most—the absorption loss A is very small and can be ignored.

(b) Reflection loss R.   The reflection loss R due to the multiple boundaries of the substrate glass coated with a metallic film can be analyzed by means of energy-transmission theory and it is expressed as

$$R = -20\log\frac{2|\eta_f|}{|\eta_a + \eta_f|} - 20\log\frac{2|\eta_g|}{|\eta_f + \eta_g|} - 20\log\frac{2|\eta_a|}{|\eta_g + \eta_a|}$$

$$= 20\log\frac{|\eta_a + \eta_f||\eta_f + \eta_g||\eta_g + \eta_a|}{8|\eta_f||\eta_g||\eta_a|} \qquad \text{dB} \qquad (2\text{-}1\text{-}7)$$

where η_f = intrinsic impedance of the coated metallic film

  η_g = intrinsic impedance of the glass substrate

  η_a = 377 ohms is the intrinsic impedance of air or free space

The intrinsic impedance of a metallic film is given by

$$|\eta_f| = \left|(1+j)\sqrt{\frac{\mu\omega}{2\sigma_f}}\right| = \sqrt{\frac{\mu\omega}{\sigma_f}} \qquad (2\text{-}1\text{-}8)$$

The intrinsic impedance of a glass substrate is expressed as

$$\eta_g \simeq \frac{\eta_a}{\sqrt{\epsilon_r}} = \frac{377}{\sqrt{3.78}} = 194 \text{ ohms} \qquad \text{for} \quad \sigma_g \ll \omega\epsilon_g \qquad (2\text{-}1\text{-}9)$$

where σ_g = about 10^{-12} mho per meter is the conductivity of the glass substrate

  ϵ_g = 4.77×10^{-11} farad per meter is the permittivity of the glass substrate

  ϵ_r = 3.78 is the relative permittivity of the glass substrate.

Substitution of the values of the intrinsic impedances η_f, η_g, and η_a in Eq. (2-1-7) yields the reflection loss as

$$R \simeq 20\log\left[28.33\sqrt{\frac{\sigma_f}{\mu f}}\right] = 88 + 10\log\left(\frac{\sigma_f}{f}\right) \qquad \text{dB} \qquad (2\text{-}1\text{-}10)$$

(c) Correction term C.   For very electrically thin film, the value of the absorption loss A is much less than 10 dB and the correction term is given by [5]

$$C = 20\log|1 - \rho 10^{-A/10}(\cos\theta - j\sin\theta)| \qquad (2\text{-}1\text{-}11)$$

where
$$\rho = \left(\frac{\eta_f - \eta_a}{\eta_f + \eta_a}\right)^2 \simeq 1 \qquad \text{for} \quad \eta_a \gg \eta_f$$

$$\theta = 3.59t\sqrt{f\mu\sigma_f}$$

Over the frequency range from 100 MHz to 40 GHz, the angle θ is much smaller than one degree so that $\cos\theta \simeq 1$ and $\sin\theta \simeq \theta$. Thus the correction term of Eq. (2-1-11) can be simplified to

$$C \simeq 20\log\left[3.59t\sqrt{f\mu\sigma_f}\right] = -48 + 20\log\left[t\sqrt{f\sigma_f}\right] \qquad \text{dB} \qquad (2\text{-}1\text{-}12)$$

Finally, the total microwave radiation attenuation by a metallic-film coating on a glass substrate defined in Eq. (2-1-4) in the far field becomes

$$\text{Attenuation} = 40 - 20\log(R_s) \qquad \text{dB} \qquad (2\text{-}1\text{-}13)$$

It is interesting to note that the microwave radiation attenuation due to the coated metallic film on a glass substrate in the far field is independent of frequency; it is related only to the surface resistance of the coated metallic film.

2-1-4 Light Transmittance of Metallic-Film Coating on Plastic Substrate

The complex refractive index of an optical material is given by [3] as

$$N = n - jK \qquad (see\ 2\text{-}1\text{-}2) \qquad (2\text{-}1\text{-}14)$$

It is assumed that light in air is normally incident on a thin absorbing film N_1 of thickness t_1, and that it is transmitted through an absorbing substrate of complex refractive index N_2 and then emerges into air. The incidence and the emergence media are dielectric of refractive index n_0. The reflection loss between the substrate and the air is small; for convenience, it is taken as zero. Figure 2-1-2 shows light transmittance, reflection, and absorption through a thin absorbing metallic film and a plastic substrate.

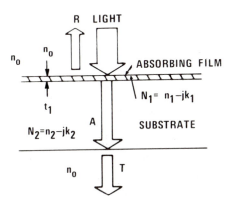

Figure 2-1-2 Light transmittance, reflection, and absorption through thin metallic-film coated on plastic substrate

Using the multireflection and transmission theory, the reflection loss is expressed by

$$R = \frac{a_1 e^\alpha + a_2 e^{-\alpha} + a_3 \cos \nu + a_4 \sin \nu}{b_1 e^\alpha + b_2 e^{-\alpha} + b_3 \cos \nu + b_4 \sin \nu} \tag{2-1-15}$$

where $a_1 = [(n_0 - n_1)^2 + k_1^2][(n_1 + n_2)^2 + (k_1 + k_2)^2]$

$a_2 = [(n_0 + n_1)^2 + k_1^2][(n_1 - n_2)^2 + (k_1 - k_2)^2]$

$a_3 = 2\{[n_0^2 - (n_1^2 + k_1^2)][(n_1^2 + k_1^2) - (n_2^2 + k_2^2)]$
$\quad\quad + 4n_0 k_1 (n_1 k_2 - n_2 k_1)\}$

$a_4 = 4\{[n_0^2 - (n_1^2 + k_1^2)](n_1 k_2 - n_2 k_1)$
$\quad\quad - n_0 k_1[(n_1^2 + k_1^2) - (n_2^2 + k_2^2)]\}$

$\alpha = \dfrac{4\pi k_1 t_1}{\lambda_0}$

$\lambda_0 = \dfrac{c}{f}$ is the wavelength in a vacuum

$c = 3 \times 10^8$ m/s is the velocity of light in vacuum; f is the frequency in hertz

$\nu = \dfrac{4\pi n_1 t_1}{\lambda_0}$

$b_1 = [(n_0 + n_1)^2 + k_1^2][(n_1 + n_2)^2 + (k_1 + k_2)^2]$

$b_2 = [(n_0 - n_1)^2 + k_1^2][(n_1 - n_2)^2 + (k_1 - k_2)^2]$

$b_3 = 2\{[n_0^2 - (n_1^2 + k_1^2)][(n_1^2 + k_1^2) - (n_2^2 + k_2^2)]$
$\quad\quad - 4n_0 k_1 (n_1 k_2 - n_2 k_1)\}$

$b_4 = 4\{[n_0^2 - (n_1^2 + k_1^2)](n_1 k_2 - n_2 k_1)$
$\quad\quad + n_0 k[(n_1^2 + k_1^2) - (n_2^2 + k_2^2)]\}$

Transmittance T is given by [3] as

$$T = \frac{16 n_0 n_2 (n_1^2 + k_1^2)}{b_1 e^\alpha + b_2 e^{-\alpha} + b_3 \cos \nu + b_4 \sin \nu} \tag{2-1-16}$$

Absorption loss A is given by

$$A = 1 - R - T \tag{2-1-17}$$

Total attenuation loss L is given by

$$L = A + R \tag{2-1-18}$$

When the concave surface of a plastic dome is uniformly coated with an electromagnetic interference shield of metallic film, however, the light is normally incident on the plastic substrate N_2, transmits through the thin metallic film N_1, and emerges into the air n_0. According to the electromagnetic theory of luminous transmission in transparent media, the light transmittance is the same whether the light is normally incident on the substrate medium N_2 or the absorbing film N_1. So the total attenuation loss is the same in both cases.

2-1-5 Plane Wave in Gold-Film Coating on Plastic Glass

Metallic-film coatings on plastic glasses have many engineering applications [6]. For example, a gold film is coated on the concave surface of a plastic-glass dome in a missile so that an optimum amount of microwave radiation is attenuated by the gold film while a sufficient light intensity is transmitted through the gold film at the same time.

(a) **Surface resistance.** The properties of bulk gold at room temperature are:

$$\text{Conductivity} \qquad \sigma = 4.10 \times 10^7 \quad \text{mhos-m}^{-1}$$
$$\text{Resistivity} \qquad \rho = 2.44 \times 10^{-8} \quad \text{ohm-m}$$
$$\text{Electron mean-free-path} \quad p = 570 \text{ Å}$$

It is assumed that the thickness t of the gold film varies from 10 to 100 Å. The surface resistances of the gold film are computed by using Eqs. (2-1-1) and (2-1-3) and are tabulated in Table 2-1-3. Figure 2-1-3 shows the surface resistances of the

TABLE 2-1-3 SURFACE RESISTANCE OF GOLD FILM

Thickness t (Å)	Conductivity σ_f (mho-m$^{-1} \times 10^7$)	Resistivity ρ_f (ohm-m $\times 10^{-7}$)	Surface resistance R_s (ohms/square)
100	1.17	0.86	8.60
90	1.11	0.90	10.00
80	1.03	0.97	12.13
70	0.96	1.04	14.86
60	0.86	1.17	19.50
50	0.77	1.30	26.00
40	0.68	1.48	37.00
30	0.54	1.86	62.00
20	0.41	2.41	120.50
10	0.24	4.15	415.00

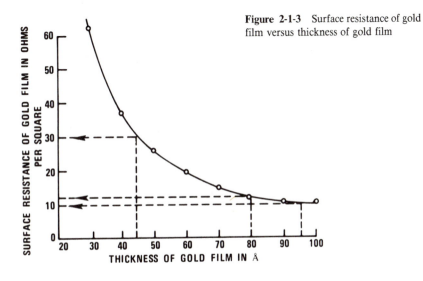

Figure 2-1-3 Surface resistance of gold film versus thickness of gold film

gold film in ohms per square against the thicknesses of the gold film from 10 to 100 Å. According to the Fuchs–Sondheimer theory, gold films have a typical surface resistance at about 10 to 30 ohms per square for a thickness of about 90 to 45 Å. Surface resistance is decreased as the thickness of the gold film is increased.

(b) Microwave radiation attenuation.   Substituting the values of the surface resistances for gold films in Eq. (2-1-13) yields the microwave radiation attenuation in decibels by the gold-film coating on a plastic glass. Figure 2-1-4 shows graphically the microwave radiation attenuation versus the surface resistance of the gold-film coating. For a coated gold film that has a surface resistance of 12 ohms per square, the microwave radiation attenuation is about 19 dB. The data agree with Hawthorne's conclusion [7].

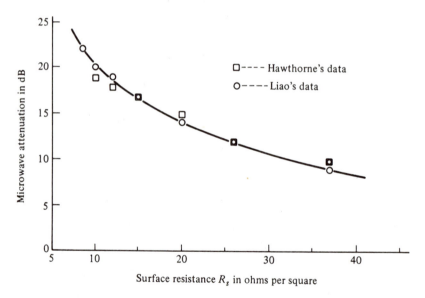

Figure 2-1-4   Microwave radiation attenuation versus surface resistance of gold film

(c) Light transmittance.   For the visible light region, the values of the refractive index n and the extinction index k of a gold-film coating on a plastic glass deposited in vacuum are taken from Table 2-1-2. The refractive index n_0 of air or vacuum is unity. The refractive index n_2 of the nonabsorbing plastic glass is taken as 1.50. Light transmittance T and the light reflection loss R of the gold-film coating on a plastic glass are computed by using Eqs. (2-1-16) and (2-1-15), respectively.

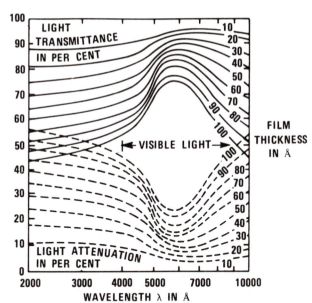

Figure 2-1-5 Light transmittance T and light attenuation loss L versus wavelength λ with film thickness t as a parameter for gold film

Absorption loss A and total attenuation L are calculated by using the values of T and R accordingly. The results are presented graphically in Fig. 2-1-5. It can be seen that, for a light transmittance of 80%, the thickness of the gold-film coating is about 80 Å. When absorption loss in the substrate material is considered, however, the light transmittance may be a little less than 80%.

(d) Optimum condition.   The surface resistance of a metallic film is decreased as the thickness of the film coating is increased. The luminous transmittance is decreased, however, as the surface resistance of the metallic film is decreased. This relationship for the visible light region is shown in Fig. 2-1-6.

Figure 2-1-7 illustrates the relationship of light transmittance versus wavelength for a given surface resistance of the gold film. If power dissipation of 5 watts per square is allowed for de-icing and de-fogging or keeping the temperature at a certain level by the gold-film coating on a plastic substrate and if the effective area of the coated film is 13 square inches, the surface resistance of the coated film must be 12 ohms per square. The power dissipation can be expressed as

$$P = \frac{V^2}{R} = \frac{(28)^2}{12 \times 13} = 5 \qquad \text{W/sq}$$

in which the voltage applied to the film coating terminations is 28 V. The optimum condition occurs at 18 dB of microwave radiation attenuation and 90% of light transmittance.

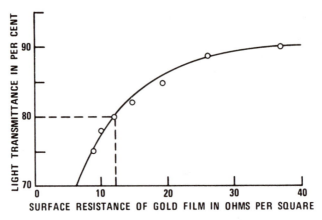

Figure 2-1-6 Light transmittance versus surface resistance of gold film

- LIGHT TRANSMITTANCE DECREASES, AS SURFACE RESISTANCE DECREASES
- FOR 80 PER CENT OF LIGHT TRANSMITTANCE, SURFACE RESISTANCE IS ABOUT 12 OHMS PER SQUARE

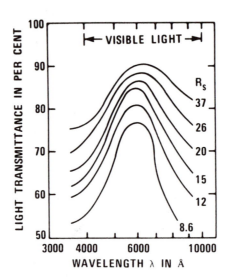

- LIGHT TRANSMITTANCE INCREASES AS SURFACE RESISTANCE INCREASES
- LIGHT TRANSMITTANCE OF 80 PER CENT OCCURS AT 12 OHMS PER SQUARE

Figure 2-1-7 Light transmittance versus wavelength with surface resistance R_s as a parameter for gold film

Example 2-1-5   Calculations of Gold-film Coating

A gold film of thickness 80 Å is coated on a plastic substrate with a refractive index of 1.50.

1. Calculate the gold-film surface resistance in ohms per square.
2. Compute the microwave attenuation in decibels.
3. Determine the light transmittance T at $\lambda = 0.6$ μm.
4. Find the light reflection loss R at $\lambda = 0.6$ μm.

Solution. 1. Gold-film surface resistance

(a) From Eq. (2-1-1) the gold-film conductivity is

$$\sigma_f = \frac{3 \times 80 \times 10^{-10} \times 4.1 \times 10^7}{4 \times 570 \times 10^{-10}} \left[\ell n \left(\frac{570}{80} \right) + 0.4228 \right]$$

$$= 1.03 \times 10^7 \text{ mhos-m}^{-1}$$

(b) The gold-film resistivity is

$$\rho_f = \frac{1}{\sigma_f} = \frac{1}{1.03 \times 10^7} = 0.97 \times 10^{-7} \text{ ohm-m}$$

(c) From Eq. (2-1-3) the gold-film surface resistance is

$$R_s = \frac{\rho_f}{t} = \frac{0.97 \times 10^{-7}}{80 \times 10^{-10}} = 12.12 \text{ ohms/sq}$$

2. From Eq. (2-1-13) the microwave attenuation is

$$\text{Micro-att} = 40 - 20 \log(12.12) = 18 \text{ dB}$$

3. The light transmittance T is computed with a program from Eq. (2-1-16) to be

$$T = 80\%$$

4. The light reflection loss R is computed with a program from Eq. (2-1-15) as

$$R = 20.0\%$$

2-1-6   Plane Wave in Silver- or Copper-Film Coating on Plastic Substrate

Silver- or copper-film coating on a plastic substrate is used in many engineering applications [8]. The surface resistance, microwave radiation attenuation, light transmittance, and optimum condition of both silver-film coating and copper-film coating can be described in the same way as for gold-film coating.

(a) Surface resistance.   At room temperature the properties of bulk silver and bulk copper are:

Silver: Conductivity $\sigma = 6.170 \times 10^7$ mhos-m^{-1}
           Resistivity $\rho = 1.620 \times 10^{-8}$ ohm-m
           Electron mean-free-path $p = 570$ Å

Copper: Conductivity $\sigma = 5.800 \times 10^7$ mhos-m^{-1}
           Resistivity $\rho = 1.724 \times 10^{-8}$ ohm-m
           Electron mean-free-path $p = 420$ A

It is assumed that the thickness t of the silver and copper films varies from 10 to 100 Å. The surface resistances of the silver and copper films are computed by using Eqs. (2-1-1) and (2-1-3) and are tabulated in Tables 2-1-4 and 2-1-5, respectively.

TABLE 2-1-4   SURFACE RESISTANCE R_s OF SILVER FILM

Thickness t (Å)	Conductivity σ_f (mho-m$^{-1} \times 10^7$)	Resistivity ρ_f (ohm-m $\times 10^{-7}$)	Surface resistance R_s (ohms per square)
100	1.75	0.571	5.71
90	1.66	0.602	6.69
80	1.55	0.645	8.06
70	1.44	0.695	9.93
60	1.31	0.763	12.72
50	1.17	0.855	17.10
40	0.99	1.010	25.25
30	0.81	1.230	41.00
20	0.60	1.670	83.33
10	0.36	2.760	276.00

TABLE 2-1-5   SURFACE RESISTANCE R_s OF COPPER FILM

Thickness t (Å)	Conductivity σ_f (mho-m$^{-1} \times 10^7$)	Resistivity ρ_f (ohm-m $\times 10^{-7}$)	Surface resistance R_s (ohms per square)
100	1.93	0.52	5.20
90	1.83	0.55	6.11
80	1.73	0.58	7.25
70	1.62	0.62	8.86
60	1.47	0.68	11.33
50	1.33	0.75	15.00
40	1.14	0.88	22.00
30	0.95	1.05	35.00
20	0.73	1.37	68.50
10	0.42	2.38	238.00

Figure 2-1-8 graphically plots the surface resistances of the silver and copper films in ohms per square versus the thickness from 10 to 100 Å.

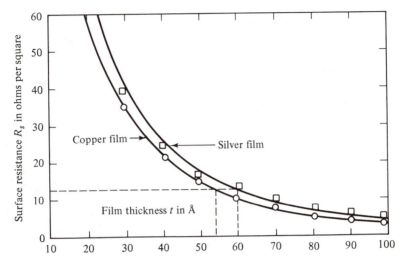

Figure 2-1-8   Surface resistance of silver and copper films versus thickness of the films

(b) RF radiation attenuation.   Substitution of the values of the surface resistances of the silver or copper films in Eq. (2-1-13) yields the microwave radiation attenuation in decibels by the silver- or copper-film coating on a plastic substrate. Figure 2-1-9 shows graphically the microwave radiation attenuation versus the surface resistance of the silver- or copper-film coating, respectively.

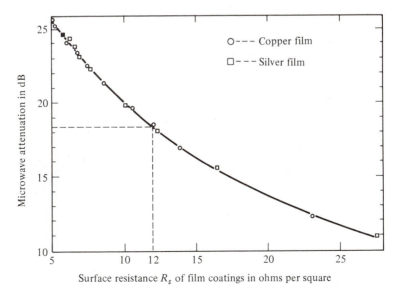

Figure 2-1-9   Microwave radiation attenuation versus surface resistance of silver and copper films

(c) Light transmittance. Light transmittance T and light reflection loss R of the silver- and copper-film coatings are computed by using Eqs. (2-1-16) and (2-1-15), respectively. The values of refractive index n and extinction index k of the silver- and copper-film coatings deposited in vacuum for the light-frequency range are taken from Table 2-1-2. The refractive index n_0 of air or vacuum is unity. The refractive index n_2 of the nonabsorbing plastic glass is taken as 1.50. The absorption loss A and the total attenuation L are calculated from the values of light transmittance T and light reflection loss R. The results are illustrated graphically in Figs. 2-1-10 and 2-1-11 for silver- and copper-film coatings, respectively.

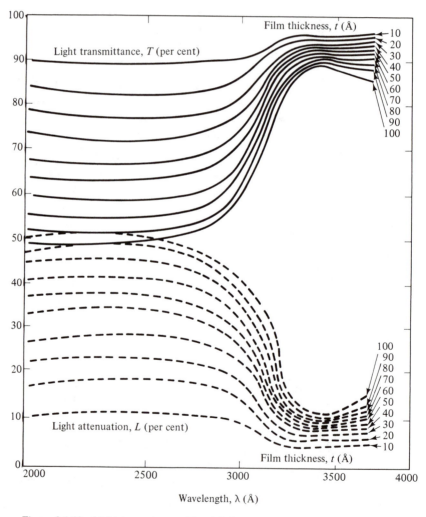

Figure 2-1-10   Light transmittance T and light attenuation loss L of silver film versus wavelength λ with film thickness t as a parameter

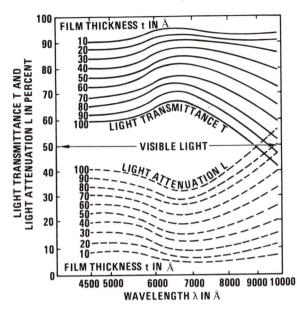

Figure 2-1-11 Light transmittance T and light attenuation loss L of copper film versus wavelength λ with film thickness t as a parameter

(d) Optimum condition.   The light transmittance increases as the surface resistance increases. This relationship is illustrated in Fig. 2-1-12 for silver and copper film, respectively. The optimum condition occurs at 18 dB of microwave radiation attenuation and 94% of light transmittance with a surface resistance of about 12 ohms per square.

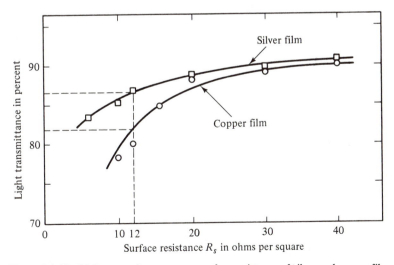

Figure 2-1-12 Light transmittance versus surface resistance of silver and copper films

Example 2-1-6   Computations of Copper-film Coating

A copper film of thickness 50 Å is coated on a plastic substrate with a refractive index of 1.50.

1. Calculate the copper-film surface resistance in ohms per square.
2. Compute the microwave attenuation in decibels.
3. Find the light transmittance T at $\lambda = 0.6\ \mu$m.
4. Determine the reflection loss R at $\lambda = 0.6\ \mu$m.

Solution.   1.   Copper-film surface resistance

(a)   From Eq. (2-1-1) the copper-film conductivity is

$$\sigma_f = \frac{3 \times 50 \times 10^{-10} \times 5.8 \times 10^7}{4 \times 420 \times 10^{-10}}\left[\ell n\left(\frac{420}{50}\right) + 0.4228\right]$$

$$= 1.33 \times 10^7\ \text{mhos-m}^{-1}$$

(b)   The copper-film resistivity is

$$\rho_f = \frac{1}{\sigma_f} = \frac{1}{1.33 \times 10^7} = 0.75 \times 10^{-7}\ \text{mho-m}$$

(c)   From Eq. (2-1-3) the copper-film surface resistance is

$$R_s = \frac{\rho_f}{t} = \frac{0.75 \times 10^{-7}}{50 \times 10^{-10}} = 15.00\ \text{ohms/sq}$$

2.   From Eq. (2-1-13) the microwave attenuation is

$$\text{Micro-att} = 40 - 20\log(15) = 16\ \text{dB}$$

3.   Light transmittance T is computed with a program from Eq. (2-1-16) as

$$T = 87.60\%$$

4.   Light reflection loss R is computed with a program from Eq. (2-1-15) as

$$R = 12.4\%$$

2-1-7   Design Example

A gold-film coating on a plastic is to be designed for a missile dome window so that the microwave radiation can be attenuated and the light intensity is transmitted to a certain level. Specifications are:

Minimum microwave attenuation	Att = 18 dB
Light transmittance	$T = 80\%$
Allowable power dissipation	$P = 5$ W/sq
Missile dome window	$A = 14$ sq in.
Refractive index of plastic substrate	$n = 1.50$

(a) Find the surface resistance in ohms per square.
(b) Determine the thickness of the gold-film coating.
(c) Find the microwave attenuation in decibels.
(d) Calculate the power dissipation in watts per square.

Solution.   (a) Using the graph in Fig. 2-1-6, we obtain the surface resistance for 80% of light transmittance as

$$R_s = 12 \text{ ohms/sq}$$

(b) Using Table 2-1-3, we find that the thickness of the gold-film coating is

$$t = 80 \text{ Å}$$

(c) Then from Fig. 2-1-4 the microwave attenuation is

$$\text{Micro-att} = 18 \text{ dB}$$

(d) The power dissipation is

$$P = \frac{V^2}{R} = \frac{(28)^2}{12 \times 14} = 4.67 \text{ W/sq}$$

where 28 V is the dc power supply in the missile.

2-2   LIGHT-WAVE PROPAGATION IN DIELECTRIC CRYSTALS

In discussing electromagnetic wave propagation, it has been assumed that the medium is homogeneous and isotropic. In other words, the dielectric permittivity ϵ and the magnetic permeability μ of the medium are constant. This assumption is not true for dielectric crystals and magnetic ferrites, however, because they are anisotropic media. The constitutive equations of an anisotropic medium are expressed as

$$\mathbf{D} = \hat{\epsilon}\mathbf{E} = \epsilon_0\mathbf{E} + \mathbf{P} \qquad (2\text{-}2\text{-}1)$$

$$\mathbf{B} = \hat{\mu}\mathbf{H} = \mu_0(\mathbf{H} + \mathbf{M}) \qquad (2\text{-}2\text{-}2)$$

where $\mathbf{P} = \epsilon_0\hat{\chi}_e\mathbf{E}$ is the electric polarization in dipole moment per unit volume
$\hat{\chi}_e =$ dielectric tensor susceptibility
$\hat{\epsilon} = \epsilon_0(1 + \hat{\chi}_e)$ is the dielectric tensor permittivity of the medium
$\mathbf{M} = \hat{\chi}_m\mathbf{H}$ is the magnetic polarization in dipole moment per unit volume, more commonly called magnetization
$\hat{\chi}_m =$ magnetic tensor susceptibility
$\hat{\mu} = \mu_0(1 + \hat{\chi}_m)$ is the magnetic tensor permeability of the medium

2-2-1   Polarization Tensor of Dielectric Crystals

The dielectric crystal consists of a regular periodic array of atoms (or ions) and is an anisotropic medium. Electric polarization will depend, both in magnitude and direction, on the direction of the applied field. In general, the polarization can be written

$$P_i = \epsilon_0 \sum_j \hat{\chi}_{ij} E_i \qquad (2\text{-}2\text{-}3)$$

where $\hat{\chi}_{ij}$ is called the electric susceptibility tensor, which is

$$\hat{\chi}_{ij} = \begin{bmatrix} \chi_{11} & \chi_{12} & \chi_{13} & \cdots & \chi_{1j} \\ \chi_{21} & \chi_{22} & \chi_{23} & \cdots & \chi_{2j} \\ \chi_{31} & \chi_{32} & \chi_{33} & \cdots & \chi_{3j} \\ \vdots & & & & \\ \chi_{i1} & \chi_{i2} & \chi_{i3} & \cdots & \chi_{ij} \end{bmatrix} \qquad (2\text{-}2\text{-}4)$$

The polarization tensor of a dielectric crystal for a 3×3 array is given by

$$\begin{bmatrix} P_x \\ P_y \\ P_z \end{bmatrix} = \epsilon_0 \begin{bmatrix} \chi_{11} & \chi_{12} & \chi_{13} \\ \chi_{21} & \chi_{22} & \chi_{23} \\ \chi_{31} & \chi_{32} & \chi_{33} \end{bmatrix} \begin{bmatrix} E_x \\ E_y \\ E_z \end{bmatrix} \qquad (2\text{-}2\text{-}5)$$

The polarization components can be written

$$P_x = \epsilon_0 (\chi_{11} E_x + \chi_{12} E_y + \chi_{13} E_z) \qquad (2\text{-}2\text{-}6)$$

$$P_y = \epsilon_0 (\chi_{21} E_x + \chi_{22} E_y + \chi_{23} E_z) \qquad (2\text{-}2\text{-}7)$$

$$P_z = \epsilon_0 (\chi_{31} E_x + \chi_{32} E_y + \chi_{33} E_z) \qquad (2\text{-}2\text{-}8)$$

It is possible to choose x, y, and z in such a way that the off-diagonal coefficients vanish and the diagonal elements are written

$$\epsilon_{11} = \epsilon_0 (1 + \chi_{11}) = \epsilon_0 \epsilon_{r11} \qquad (2\text{-}2\text{-}9)$$

$$\epsilon_{22} = \epsilon_0 (1 + \chi_{22}) = \epsilon_0 \epsilon_{r22} \qquad (2\text{-}2\text{-}10)$$

$$\epsilon_{33} = \epsilon_0 (1 + \chi_{33}) = \epsilon_0 \epsilon_{r33} \qquad (2\text{-}2\text{-}11)$$

The dielectric permittivities ϵ_{11}, ϵ_{22}, and ϵ_{33} are called the values of the principal axis of the permittivity tensor. If $\epsilon_{11} = \epsilon_{22} \neq \epsilon_{33}$, the crystal is known as a *uniaxial* crystal. If $\epsilon_{11} \neq \epsilon_{22} \neq \epsilon_{33}$, the crystal is known as a *biaxial* crystal. The anisotropy in a crystal is caused by nonsymmetric binding of the outermost electrons of the ions making up the crystal. As a result, the displacement of the bound electrons under the action of an applied field will depend on the direction of the field.

2-2-2   Birefringence

In an isotropic medium the wave polarization is independent of the field direction because $\epsilon_{11} = \epsilon_{22} = \epsilon_{33} = \epsilon_0$. But the situation is different in an anisotropic crystal. The wave velocities vary in different directions under the influence of an electric field. This phenomenon is called *birefringence*.

2-2-3   Index Ellipsoid

The wave velocity along a given direction in a dielectric crystal depends on the direction of its polarization. According to Snell's law, the wave velocity is inversely proportional to the refractive index of the medium. The refractive indices in the x, y, and z directions are related by

$$\frac{x^2}{n_x^2} + \frac{y^2}{n_y^2} + \frac{z^2}{n_z^2} = 1 \tag{2-2-12}$$

where $n_x = \sqrt{\epsilon_{r11}} = \sqrt{\epsilon_{11}/\epsilon_0}$ is the refractive index in the x direction

$n_y = \sqrt{\epsilon_{r22}} = \sqrt{\epsilon_{22}/\epsilon_0}$ is the refractive index in the y direction

$n_z = \sqrt{\epsilon_{r33}} = \sqrt{\epsilon_{33}/\epsilon_0}$ is the refractive index in the z direction

This is the equation of a generalized index ellipsoid. Its major axes are parallel to the x, y, and z directions and its axis lengths are $2n_x$, $2n_y$, and $2n_z$, respectively. If the medium is an uniaxial crystal with its optic axis in the z direction, Eq. (2-2-12) becomes

$$\frac{x^2}{n_o^2} + \frac{y^2}{n_o^2} + \frac{z^2}{n_e^2} = 1 \tag{2-2-13}$$

where n_o = refractive index of the ordinary wave in the x and y direction

n_e = refractive index of the extraordinary wave in the z direction

This is an ellipsoid of revolution with the circular symmetry axis parallel to z. The major axes in the x and y directions are of length $2n_o$ and that in the z direction is $2n_e$.

2-2-4   Refractive Indices

A uniaxial crystal is a crystal having a single axis of threefold, fourfold, or sixfold symmetry. The crystal KD*P (potassium dideuterium phosphate, KD_2PO_4) has a fourfold axis of symmetry. If the optic axis of the crystal KD*P is oriented in the z direction and an electric field is applied to the z direction, its electrooptic coefficients $r_{41} = r_{52} = r_{63}$ are the only nonvanished elements of the crystal electrooptic tensor, and the index ellipsoid is expressed [9] by

$$\frac{x^2}{n_o^2} + \frac{y^2}{n_o^2} + \frac{z^2}{n_e^2} + 2r_{63}xyE_z = 1 \tag{2-2-14}$$

In order to obtain the index ellipsoid equation in the form of Eq. (2-2-13) it is necessary to eliminate the mixed term of Eq. (2-2-14). If so, it is desirable to transform the old coordinate system of x, y, and z into a new set of coordinate x', y', and z' by

$$x = x'\cos 45° - y'\sin 45° \tag{2-2-15}$$

$$y = x'\sin 45° + y'\cos 45° \tag{2-2-16}$$

$$z = z' \tag{2-2-17}$$

That is,

$$\frac{x'^2}{n_{x'}^2} + \frac{y'^2}{n_{y'}^2} + \frac{z'^2}{n_{z'}^2} = 1 \tag{2-2-18}$$

Substitution of Eqs. (2-2-15) through (2-2-17) into Eq. (2-2-14) yields

$$\left(\frac{1}{n_o^2} + r_{63}E_z\right)x'^2 + \left(\frac{1}{n_o^2} - r_{63}E_z\right)y'^2 + \frac{z'^2}{n_e^2} = 1 \tag{2-2-19}$$

By comparing the coefficients of the first terms in Eqs. (2-2-18) and (2-2-19) and using the binomial expansion for $n_o^2 r_{63}E_z \ll 1$, we find that the new refractive index in the x' direction is

$$n_{x'} = \left(\frac{n_o^2}{1 + n_o^2 r_{63}E_z}\right)^{1/2} = n_o\left(1 - \frac{n_o^2}{2}r_{63}E_z\right) \tag{2-2-20}$$

$$= n_o - \tfrac{1}{2}n_o^3 r_{63}E_z$$

Similarly,

$$n_{y'} = n_o\left(1 + \frac{n_o^2}{2}r_{63}E_z\right) = n_o + \frac{1}{2}n_o^3 r_{63}E_z \tag{2-2-21}$$

$$n_z = n_e \tag{2-2-22}$$

where x', y', and z are the principal axes of the index ellipsoid when an electric field is applied along the z direction. The factor r_{63} is the electrooptic constant of the second-rank electrooptic tensor r_{ij} for a uniaxial crystal.

Example 2-2-1   Change of Dielectric Refractive Index by Electric Field

A dielectric crystal KD*P has the following parameters:

Refractive index with no field applied $n_o = 1.52$
Electrooptical constant $r_{63} = 26.4 \times 10^{-12}$ m/V
Applied voltage $V = 4000$ V
Crystal width $d = 1$ cm

(a) Calculate the changing refractive index after the electric field is applied.
(b) Determine the new refractive index in the x' and y' directions.

Solution. (a) Using Eq. (2-2-20) or Eq. (2-2-21), we obtain the changing refractive index as

$$|n_{x'} - n_o| = \frac{1}{2} n_o^3 r_{63} E_z$$

$$= \frac{1}{2}(1.52)^3(26.4 \times 10^{-12})\left(\frac{4000}{10^{-2}}\right)$$

$$= 1.854 \times 10^{-5}$$

(b) The new refractive indexes in the x' and y' directions are

$$n_{x'} = 1.52 - 1.854 \times 10^{-5}$$

$$n_{y'} = 1.52 + 1.854 \times 10^{-5}$$

2-3 *LIGHT-WAVE PROPAGATION IN MAGNETIC CRYSTALS (FERRITES)*

Ferrites are solids with a particular type of crystal structure. They are made up of atoms of oxygen, iron, and another element which might be lithium, magnesium, zinc, or any others. Ferrites have the important characteristics of low loss and strong magnetic effects at microwave frequencies.

The fundamental theory of microwave ferrite devices is based on the motion of the magnetization vector. In general, the total effective magnetic field consists of the dc magnetic field $\mathbf{H}_0$ and the ac magnetic field $\mathbf{H}$. That is,

$$\mathbf{H}_t = \mathbf{H}_0 + \mathbf{H} \tag{2-3-1}$$

The total magnetization consists of the dc magnetization $\mathbf{M}_0$ and the ac magnetization $\mathbf{M}$. That is,

$$\mathbf{M}_t = \mathbf{M}_0 + \mathbf{M} \tag{2-3-2}$$

Figure 2-3-1 shows the diagram of magnetic moment precession about a dc magnetic field $\mathbf{H}_0$.

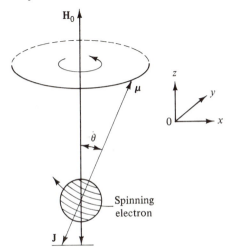

Spinning electron

Figure 2-3-1 Magnetic moment precession about a dc magnetic field $\overline{\mathbf{H}}_0$ (After J. Helszajn [12]; reprinted by permission of John Wiley & Sons, Ltd.)

The dc magnetic field $\mathbf{H}_0$ is assumed in the z direction. Under equilibrium conditions the magnetic dipole moment vector $\boldsymbol{\mu}$ lies in the direction $\mathbf{H}_0$. When a small external force is applied, however, the magnetic dipole $\boldsymbol{\mu}$ is tilted by an angle θ with $\mathbf{H}_0$. The magnetic torque $\mathbf{T}$ exerted on the magnetic dipole $\boldsymbol{\mu}$ is then

$$\mathbf{T} = \boldsymbol{\mu} \times \mathbf{H}_0 \tag{2-3-3}$$

The magnetic dipole $\boldsymbol{\mu}$ can be expressed in terms of the angular momentum $\mathbf{J}$ as

$$\boldsymbol{\mu} = \gamma \mathbf{J} \tag{2-3-4}$$

where $\gamma = (g\mu_0 e/(2m)) = 2.21 \times 10^5$ (rad/s)/(amperes/m) is the gyromagnetic ratio

g = the spectroscopic-splitting factor or the Landé g factor, approximately equal to 2 for a free electron

e = the value of the electron charge

m = the electron mass

$\mu_0 = 4\pi \times 10^{-7}$ henry/m is the permeability of free space.

Magnetic torque can be also written

$$\mathbf{T} = \frac{1}{\gamma} \frac{d\boldsymbol{\mu}}{dt} \tag{2-3-5}$$

Then the equation of motion for a single magnetic dipole is

$$\frac{d\boldsymbol{\mu}}{dt} = \gamma(\boldsymbol{\mu} \times \mathbf{H}_0) \tag{2-3-6}$$

Similarly, the equation of motion for a magnetization vector is

$$\frac{d\mathbf{M}_0}{dt} = \gamma(\mathbf{M}_0 \times \mathbf{H}_0) \tag{2-3-7}$$

where $\mathbf{M}_0 = N\boldsymbol{\mu}$ is the total dc magnetization in webers per square meter

N = the number of unbalanced spin per unit volume

$\mathbf{H}_0$ = the dc magnetic field in amperes per meter.

In matrix forms the magnetic fields and magnetization vectors for dc and ac cases can be expressed, respectively, as

$$\hat{H}_0 = \begin{bmatrix} 0 \\ 0 \\ H_0 \end{bmatrix} \qquad \hat{H} = \begin{bmatrix} H_x \\ H_y \\ H_z \end{bmatrix} \tag{2-3-8}$$

and

$$\hat{M}_0 = \begin{bmatrix} 0 \\ 0 \\ M_0 \end{bmatrix} \qquad \hat{M} = \begin{bmatrix} M_x \\ M_y \\ M_z \end{bmatrix} \tag{2-3-9}$$

The magnetic effects of ferrite crystals have many applications in microwave devices such as Faraday-rotation isolators, Faraday-rotation circulators, magnetic switches, and magnetic phaseshifters.

2-3-1 Tensor Permeability

The constitutive equation of a magnetic crystal in time-varying fields is given by Eq. (2-2-2) as

$$\mathbf{B} = \hat{\mu}\mathbf{H} = \mu_0(\mathbf{H} + \mathbf{M}) \tag{2-3-10}$$

where $\hat{\mu} = \mu_0(1 + \hat{\chi}_m)$ is the magnetic tensor permeability of the crystal

$\hat{\chi}_m =$ the magnetic tensor susceptibility

$\mathbf{M} = \hat{\chi}_m\mathbf{H}$ is the magnetic polarization in dipole moment per unit volume

The conventional notation for the tensor permeability is

$$\hat{\mu} = \mu_0 \begin{bmatrix} \mu_{xx} & -jk & 0 \\ jk & \mu_{yy} & 0 \\ 0 & 0 & 1 \end{bmatrix} \tag{2-3-11}$$

where

$$\mu_{xx} = 1 + \chi_{xx}$$

$$\mu_{yy} = 1 + \chi_{yy}$$

$$jk = -\chi_{xy} = j\mu_{xy}$$

Both χ and k are, in general, complex. This is the well-known Polder tensor permeability. Magnetic tensor susceptibility is given by

$$\hat{\chi}_m = \begin{bmatrix} \chi_{xx} & \chi_{xy} & 0 \\ \chi_{yx} & \chi_{yy} & 0 \\ 0 & 0 & 0 \end{bmatrix} \tag{2-3-12}$$

where $\chi_{xx} = \chi_{yy} = \dfrac{\omega_m \omega_0}{\omega_0^2 - \omega^2}$ for isotropic crystal $\qquad$ (2-3-13)

$\chi_{yx} = -\chi_{xy} = \dfrac{j\omega_m}{\omega_0^2 - \omega^2}$ for isotropic crystal $\qquad$ (2-3-14)

$\omega_m = \gamma M_0$ $\qquad$ is the resonant angular frequency in the dc magnetization field $\qquad$ (2-3-15)

$\omega_0 = \gamma H_0$ $\qquad$ is the resonant angular frequency, or the Larmor precessional frequency of the electron in the dc magnetic field $\qquad$ (2-3-16)

$\omega = 2\pi f$ $\qquad$ is the driving angular frequency

The scalar permeability can be written

$$[\mu]^{\pm} = \begin{bmatrix} \mu \pm k & 0 & 0 \\ 0 & \mu \pm k & 0 \\ 0 & 0 & 1 \end{bmatrix} \tag{2-3-17}$$

where

$$\mu \pm k = 1 + \frac{\omega_m}{\omega_0 \mp \omega} \tag{2-3-18}$$

2-3-2 Wave Propagation in Infinite Isotropic Ferrites

An isotropic magnetic crystal exhibits a symmetrical tensor permeability as shown in Eqs. (2-3-13) and (2-3-14), which follow

$$\chi_{xx} = \chi_{yy} \tag{2-3-19}$$

and

$$\chi_{yx} = -\chi_{xy} \tag{2-3-20}$$

If the dc magnetic field is assumed to be in the form $e^{j(\omega t - \beta z)}$, then $k_x = k_y = 0$ and $k_z = \beta$. The propagation constants in the z direction can be obtained by solving Maxwell's equation:

$$\beta_+ = \omega \sqrt{\epsilon \mu_0 (\mu + k)} \tag{2-3-21}$$

and

$$\beta_- = \omega \sqrt{\epsilon \mu_0 (\mu - k)} \tag{2-3-22}$$

where $\epsilon = \epsilon_0 \epsilon_r$ is the dielectric permittivity of the medium

$$k = \mp \left(1 + \frac{\omega_m}{\omega_0 \mp \omega} - \mu \right) \text{ is defined in Eq. (2-3-18)}$$

The Faraday rotation for a length z is

$$\theta = \frac{1}{2} (\beta_+ - \beta_-) z \tag{2-3-23}$$

If the light wave propagates in a direction perpendicular to the dc magnetic field (say in the x direction), then $k_y = k_z = 0$ and $k_x = \beta$. The propagation constants will be

$$\beta_\perp = \omega \sqrt{\epsilon \mu_0 \mu_{\text{eff}}} \tag{2-3-24}$$

and

$$\beta_\parallel = \omega \sqrt{\epsilon \mu_0} \tag{2-3-25}$$

where $\mu_{\text{eff}} = \dfrac{\mu^2 - k^2}{\mu}$ is the effective magnetic permeability.

Example 2-3-2 Faraday Rotation

A certain infinite isotropic ferrite is used for a Faraday rotator under the following parameters:

Operating frequency	$f = 10$ GHz
Gyromagnetic ratio	$\gamma = 2.21 \times 10^5$ rad/s/A/m
Relative permittivity	$\epsilon_r = 10$

dc magnetic field	$H_0 = 1000$ A/m
dc total magnetization	$M_0 = 10,000$ A/m
Length of rotator	$z = 6$ cm

(a) Compute the magnetic resonant angular frequency ω_m.
(b) Calculate the dc Larmor resonant angular frequency ω_0.
(c) Find the effective permeabilities $\mu + k$ and $\mu - k$.
(d) Determine the phase constants β_+ and β_-.
(e) Compute the rotated angle in radians and degrees.

Solution.   (a) From Eq. (2-3-15) the magnetic angular frequency is

$$\omega_m = \gamma M_0 = 2.21 \times 10^5 \times 10^4 = 2.21 \times 10^9 \text{ rad/s}$$

(b) From Eq. (2-3-16) the dc Larmor resonant angular frequency is

$$\omega_0 = \gamma H_0 = 2.21 \times 10^5 \times 10^3 = 0.221 \times 10^9 \text{ rad/s}$$

(c) From Eq. (2-3-18) the effective permeabilities are

$$\mu + k = 1 + \frac{\omega_m}{\omega_0 - \omega} = 1 + \frac{2.21 \times 10^9}{0.221 \times 10^9 - 62.832 \times 10^9}$$

$$= 0.96$$

$$\mu - k = 1 + \frac{\omega_m}{\omega_0 + \omega} = 1 + \frac{2.21 \times 10^9}{0.221 \times 10^9 + 62.832 \times 10^9}$$

$$= 1.04$$

(d) From Eqs. (2-3-21) and (2-3-22) the phase constants are

$$\beta_+ = \omega\sqrt{\epsilon\mu_0(\mu + k)} = \frac{6.2832 \times 10^{10}}{3 \times 10^8}\sqrt{10 \times 0.96}$$

$$= 649.26 \text{ rad/m}$$

$$\beta_- = \omega\sqrt{\epsilon\mu_0(\mu - k)} = \frac{6.2832 \times 10^{10}}{3 \times 10^8}\sqrt{10 \times 1.04}$$

$$= 675.42 \text{ rad/m}$$

(e) From Eq. (2-3-23) the rotated angle is

$$\theta = \frac{1}{2}(\beta_+ - \beta_-)z = \frac{1}{2}(649.26 - 675.42) \times 0.06$$

$$= -0.7848 \text{ rad} = -44.9 \text{ degrees}$$

2-3-3   *Wave Propagation in Infinite Anisotropic Ferrites*

When an electric field is applied to a magnetic crystal, the tensor permeability is not always symmetrical. Consequently,

$$\chi_{xx} \neq \chi_{yy} \tag{2-3-26}$$

The magnetic tensor permeability as given by Eq. (2-3-11) can be rewritten

$$\hat{\mu} = \mu_0 \begin{bmatrix} 1+\chi_{xx} & \chi_{xy} & 0 \\ -\chi_{yx} & 1+\chi_{yy} & 0 \\ 0 & 0 & 1 \end{bmatrix} \tag{2-3-27}$$

where

$$\mu_{xx} = 1 + \chi_{xx}$$
$$\mu_{yy} = 1 + \chi_{yy}$$

The propagation constants in the z direction can be found by solving Maxwell's equation.

$$\beta_{\pm}^2 = \omega^2 \epsilon \mu_0 \left[\left(\frac{\mu_{xx} - \mu_{yy}}{2} \right) \pm \sqrt{\left(\frac{\mu_{xx} - \mu_{yy}}{2} \right)^2 + \mu_{xy}^2} \right] \tag{2-3-28}$$

2-4 PLANE WAVE PROPAGATION IN THE IONOSPHERE

The ionosphere is that region of the earth's atmosphere in which the constituent gases are ionized by solar radiation. This region extends from about 50 km above the earth to several earth radii and has different layers designated C, D, E, and F layers in order of height. The electron density distribution of each layer varies with the time of day, season, year, and geographical location. During the day the electron density N is approximately 5×10^{12} electrons per cubic meter (or 5×10^6 electrons per cubic centimeter) at an altitude between 90 and 100 km. The E and F layers have a permanent existence, but the D layer is present only during the day. The electron density determines the reflection and refraction of radio frequency waves. Figure 2-4-1 shows a plot of the electron density versus height [10] and Table 2-4-1 tabulates the heights of the ionosphere regions and layers [11].

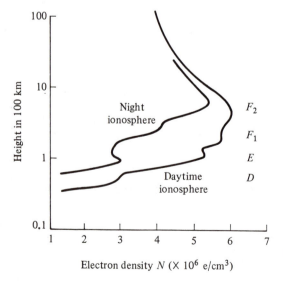

Figure 2-4-1 Electron density profiles of ionosphere

TABLE 2-4-1   HEIGHTS OF IONOSPHERE REGIONS AND LAYERS

Region	Layer	Height (km)
D	D	50–90
E	E_1, E_2, E_3	90 to 120–140
F	F_1, F_2	Above 120–140

2-4-1 Effective Dielectric Constant of the Ionosphere

The permittivity of the ionosphere is developed by assuming that the ionosphere is an electrically neutral plasma consisting of N electrons per unit volume and N protons per unit volume. It is further assumed that there are no collisions between the protons and electrons and that the effect of the earth's magnetic field is neglected. In the ionosphere electrons are free to move under the influence of an applied electric field and protons are considered to remain stationary. If an electric field $\mathbf{E}$ is incident on the ionosphere, a force exerted on each electron is given by

$$\mathbf{F} = -e\mathbf{E} = m\mathbf{a} = m\frac{d\mathbf{v}}{dt} \tag{2-4-1}$$

For a sinusoidal wave, d/dt is equal to $j\omega$. Then Eq. (2-4-1) can be expressed as

$$-e\mathbf{E} = j\omega m\mathbf{v} \tag{2-4-2}$$

where $\mathbf{v}$ equals the velocity of the electrons in meters per second.

For a surface 1 meter square, the convection-current density and the displacement-current density can be written

$$\mathbf{J}_c + \mathbf{J}_d = -Ne\mathbf{v} + j\omega\epsilon_0\mathbf{E} \tag{2-4-3}$$

Substitution of Eq. (2-4-2) into Eq. (2-4-3) yields the total current density as

$$\mathbf{J} = \mathbf{J}_c + \mathbf{J}_d = j\omega\epsilon_0\left(1 - \frac{Ne^2}{\omega^2 m\epsilon_0}\right)\mathbf{E} \tag{2-4-4}$$

The permittivity of the ionosphere is then given by

$$\epsilon = \epsilon_0\epsilon_r = \epsilon_0\left(1 - \frac{\omega_p^2}{\omega^2}\right) \tag{2-4-5}$$

where $\omega_p^2 = \dfrac{Ne^2}{m\epsilon_0}$ is called the plasma frequency.

If the values of e, m, and ϵ_0 are inserted into Eq. (2-4-5), the effective relative dielectric constant of the ionosphere becomes

$$\epsilon_r = 1 - \frac{81N}{f^2} \tag{2-4-6}$$

where N = number of electrons per cubic meter
f = frequency in hertz

The possible values of ϵ_r vary from unity through zero to a very large negative number, depending on the relationship between f and N.

2-4-2 Reflection from and Refraction in the Ionosphere

When a radio wave is incident on the ionosphere from free space, the wave will refract into the ionosphere and change direction according to Snell's law. For simplicity, Snell's law can be expressed as

$$n_0 \sin \theta_0 = n \sin \theta \tag{2-4-7}$$

where $\theta_0 =$ incident angle between the propagation direction of the incident wave and the line normal to the boundary

$\theta =$ refractive angle between the propagation direction of the refracted wave and the line normal to the boundary

$n_0 = 1$ is the index of refraction for free space

$n = \sqrt{\epsilon_r}$ is the index of refraction for the ionosphere

Figure 2-4-2 shows a diagram for wave propagation in the ionosphere.

At some point in the ionosphere the refractive index $\sqrt{\epsilon_r}$ may be such as to refract the wave traveling parallel to the earth's surface and then continue to return it toward the earth. The turning point occurs at the point where the refractive angle θ is $90°$. Substitution of Eq. (2-4-6) into Eq. (2-4-7) for $\theta = 90°$ yields

$$\sin^2 \theta_0 = 1 - \frac{81N}{f^2} \tag{2-4-8}$$

If the incident wave is vertical—that is, $\theta_0 = 0$—the frequency for a vertical wave to

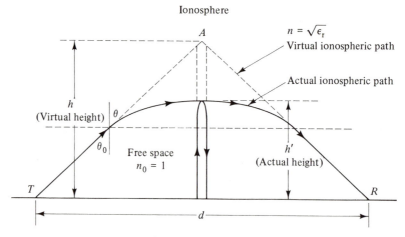

Figure 2-4-2 Wave propagation in ionosphere

return toward the earth at an electron density N is given by

$$f_v = 9N^{1/2} \qquad \text{Hz} \tag{2-4-9}$$

where N is the number of electrons per cubic meter.

If the electron density reaches its maximum, the critical frequency is expressed as

$$f_{cr} = 9\sqrt{N_{max}} \tag{2-4-10}$$

If the signal frequency is less than or equal to the critical frequency, $f \leq f_{cr}$, reflections occur for all incident angles θ_0 because $\sin\theta_0$ is imaginary. If the signal frequency is greater than the critical frequency, $f > f_{cr}$, reflections may occur for all incident angles, depending on the frequency that is lower than or equal to the maximum usable frequency (MUF). From Eq. (2-4-8) the maximum usable frequency is given by

$$f_{muf} = 9\sqrt{N_{max}} \sec\theta_0 = f_{cr}\sec\theta_0 \tag{2-4-11}$$

The maximum usable frequency is the highest frequency of a wave that is returned toward the earth.

The virtual height of a layer, as shown in Fig. 2-4-2, is the path for a wave to travel from transmitter T through reflection point A to receiver site R. The virtual height h is related to the frequencies f, f_{cr} and the incident angle θ_0 by

$$\sec\theta_0 = \frac{\left[h^2 + (d/2)^2\right]^{1/2}}{h} = \frac{f}{f_{cr}} \tag{2-4-12}$$

Assuming that the wave travels at the speed of light in vacuum, the virtual height is

$$h = \frac{1}{2}ct \qquad \text{meters} \tag{2-4-13}$$

where $c = 3 \times 10^8$ m/s

t = time in seconds required for the signal to travel to reflection point A in the ionosphere and back to the transmitter site

2-4-3 Characteristic Equations of the Ionosphere

The characteristic equations of the ionosphere for a uniform plane wave can be expressed as follows.
Constitutive Parameters:

$$\text{Relative dielectric constant} \qquad \epsilon_r = 1 - \frac{81N}{f^2} \tag{2-4-14}$$

$$\text{Relative permeability constant} \qquad \mu_r = 1$$

$$\text{Conductivity} \qquad \sigma = 0$$

The wave propagation constant is

$$\gamma = \alpha + j\beta = j\omega\sqrt{\mu_0\epsilon_0\epsilon_r} = j\omega\sqrt{\mu_0\epsilon_0}\sqrt{1 - \frac{81N}{f^2}}$$

$$= j\frac{\omega}{c}\sqrt{1 - \frac{81N}{f^2}} = j\beta_0\sqrt{1 - \frac{81N}{f^2}} \tag{2-4-15}$$

And the wave impedance is

$$\eta = \sqrt{\frac{\mu}{\epsilon}} = \eta_0\left(1 - \frac{81N}{f^2}\right)^{-1/2} \tag{2-4-16}$$

For the wave velocity in the ionosphere, we have

$$v_p = \frac{\omega}{\beta} = c\left(1 - \frac{81N}{f^2}\right)^{-1/2} \tag{2-4-17}$$

Then the group velocity v_g is given by

$$v_g = \frac{d\omega}{d\beta} = c\left(1 - \frac{81N}{f^2}\right)^{1/2} \tag{2-4-18}$$

This group velocity is the velocity of energy transmission in the ionosphere.

Example 2-4-3   Characteristics of Ionosphere

At a 150-km height in the ionosphere the electron density at night is about 2×10^{12} m^{-3} and the signal maximum usable frequency (MUF) is $f_{muf} = 1.5 f_{cr}$ for a transmission distance of 600 km. Determine the following items.

(a) Critical frequency
(b) Relative dielectric constant
(c) Phase constant
(d) Wave impedance
(e) Wave velocity
(f) Group velocity
(g) Incident angle

Solution.   (a) From Eq. (2-4-10) the critical frequency is

$$f_{cr} = 9(2 \times 10^{12})^{1/2} = 12.73 \text{ MHz}$$

(b) From Eq. (2-4-14) the relative dielectric constant is

$$\epsilon_r = 1 - \frac{f_{cr}^2}{f^2} = 1 - \left(\frac{1}{1.5}\right)^2 = 0.56$$

(c) From Eq. (2-4-15) the phase constant is

$$\beta = \beta_0\left(1 - \frac{f_{cr}^2}{f^2}\right)^{1/2} = \frac{2\pi f}{c}(0.56)^{1/2} = 0.30 \text{ rad/m}$$

(d) From Eq. (2-4-16) the wave impedance is

$$\eta = \eta_0 (0.56)^{-1/2} = \frac{377}{0.75} = 502.67 \text{ ohms}$$

(e) From Eq. (2-4-17) the wave velocity is

$$v_p = \frac{c}{(0.75)} = \frac{3 \times 10^8}{0.75} = 4 \times 10^8 \text{ m/s}$$

(f) From Eq. (2-4-18) the group velocity is

$$v_g = 3 \times 10^8 \times 0.75 = 2.25 \times 10^8 \text{ m/s}$$

(g) From Eq. (2-4-11) the incident angle is

$$\theta_0 = \text{arcsec}\left(\frac{f}{f_{cr}}\right) = \text{arcsec}(1.5) = 48.2°$$

2-5 PLANE WAVE PROPAGATION IN RADAR SYSTEMS

The word RADAR is an acronym for RAdio Detection And Ranging and it refers to the process of using radio waves to detect and locate some material object or target. A target is located by determining the distance and direction from the radar site to the target. The determination of a target location requires, in general, the measurement of three coordinates—usually the range, angle of azimuth, and angle of elevation. The radar detection depends on the reflection of microwaves from a target. The microwaves reflected from a target are generally called microwave echoes.

2-5-1 Radar Cross Section

The radar cross section σ of a target is defined as the area intercepting the amount of electromagnetic power that when reradiated isotropically by the target produces an echo at the source of radiation equal to that observed from the target (see Fig. 2-5-1).

Let the electric field incident on the target be E_i and the echo electric field at the radar reflected by the target be E_r. Then the radar cross section σ can be expressed as

$$\sigma = 4\pi R^2 \frac{|E_r|^2}{|E_i|^2} \tag{2-5-1}$$

where R is the range between the radar and the target in meters. For a matched antenna, the power received in a matched load is

$$P_i = \frac{|E_i|^2}{\eta_0} A_e \tag{2-5-2}$$

where $\eta_0 = 120\pi$ ohms is the free space impedance

$A_e = \dfrac{\lambda^2}{4\pi} g_a$ m^2 is the effective antenna aperture in square meters

g_a = numeric value is the antenna gain

When the target reradiates its received power isotropically, the power density at the radar is

$$p_{di} = \frac{P_i g_a}{4\pi R^2} \tag{2-5-3}$$

Then the electric field at the radar reflected from the target is

$$E_r = \sqrt{120\pi p_{di}} = \frac{1}{R}\sqrt{30 p_i g_a} \tag{2-5-4}$$

Substitution of Eqs. (2-5-2) and (2-5-3) into Eq. (2-5-1) yields the radar cross section σ as

$$\sigma = \frac{\lambda^2}{4\pi} g_a^2 \qquad \text{in m}^2 \tag{2-5-5}$$

To illustrate, the very short dipole antenna has a gain of 1.5 and its radar cross section is

$$\sigma = \frac{\lambda^2}{4\pi}(1.5)^2 = 0.18\lambda^2 \tag{2-5-6}$$

2-5-2   Radar Equation

A simple radar consists of a transmitter and a receiver. The transmitter generates electrical power for radiation and the receiver receives electrical power from radiation. Both the transmitter and the receiver use the same antenna as shown in Fig. 2-5-1.

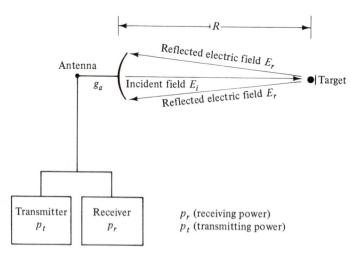

Figure 2-5-1   Power radiation for radar

Let P_t be the power of the radar transmitter. Then the power density p_d at a distance R from the transmitting antenna with a gain of g_t is

$$p_d = \frac{P_t g_t}{4\pi R^2} \qquad (2\text{-}5\text{-}7)$$

When the signal impinges the target, the power density reradiated by the target toward the receiving antenna—that is, the transmitting antenna in this case—is

$$p_{rd} = \frac{p_d \sigma}{4\pi R^2} = \frac{P_t g_t \sigma}{(4\pi R^2)^2} \qquad (2\text{-}5\text{-}8)$$

The power received by the receiving antenna with an effective aperture A_{er} is given by

$$P_r = \frac{P_t g_t \sigma}{(4\pi R^2)^2} A_{er} \qquad (2\text{-}5\text{-}9)$$

If the transmitting and receiving antennas are the same, then

$$g_t = \frac{4\pi}{\lambda^2} A_{et} = \frac{4\pi}{\lambda^2} A_{er} = g_r = g = \frac{4\pi}{\lambda^2} A_e \qquad (2\text{-}5\text{-}10)$$

and the received power can be written

$$P_r = \frac{P_t g^2 \lambda^2 \sigma}{(4\pi)^3 R^4} = \frac{P_t A_e^2 \sigma}{4\pi \lambda^2 R^4} \qquad (2\text{-}5\text{-}11)$$

The maximum radar range $R_{\max}$ is the distance beyond which the target can no longer be detected. It occurs when the received echo signal power P_r is just equal to a minimum detectable signal power $P_{\min}$ and therefore the maximum range is

$$R_{\max} = \left[\frac{P_t g^2 \lambda^2 \sigma}{(4\pi)^3 P_{\min}}\right]^{1/4} = \left[\frac{P_t A_e^2 \sigma}{4\pi \lambda^2 P_{\min}}\right]^{1/4} = \left[\frac{P_t g \sigma A_e}{(4\pi)^2 P_{\min}}\right]^{1/4} \qquad (2\text{-}5\text{-}12)$$

where P_t = power generated by the transmitter in watts
g = antenna gain (numeric value)
σ = radar cross section in square meters
λ = wavelength in meters
A_e = effective antenna aperture in square meters
$P_{\min}$ = minimum detectable signal power in watts

Maximum range.   For a radar set and an isolated object in space, the maximum range of detection depends upon the transmitting power, the minimum detectable echo-pulse signal power, the effective antenna aperture, the square of the wavelength, and the radar cross section of the object. The minimum detectable echo-pulse power depends mainly on the quality of the receiver input circuit, the pulse repetition frequency, and the number of pulses returned from the object as the antenna beam scans past it.

Minimum range. Objects at a very short range, as well as those at an extremely great range, cannot be detected by radar. One reason for the minimum range limitation is that most transmitters cannot end a pulse with a perfect suddenness. Another effect that limits minimum range is the recovery time of the duplexer. This factor is limiting because a transmitting-receiving device functions as a switch to connect the antenna to the transmitter during the time of the pulse width and to the receiver during the remainder of the repetition period.

Example 2-5-2 Radar Range

A certain radar has a transmitter power of 100 MW (peak) at a frequency of 3 GHz and a minimum detectable power of 1.00 μW. The radar antenna has a power gain of 20 dB. (The receiver power gain is not considered.)

(a) Calculate the radar cross section in square meters.
(b) Compute the effective antenna aperture in square meters.
(c) Determine the maximum radar range in meters.

Solution. (a) The radar cross section is

$$\sigma = \frac{\lambda^2}{4\pi}g_a^2 = \frac{\left[3\times10^8/(3\times10^9)\right]^2}{4\pi}(100)^2 = 7.96 \text{ m}^2$$

(b) The effective antenna aperture is

$$A_e = \frac{\lambda^2}{4\pi}g_a = \frac{\left[3\times10^8/(3\times10^9)\right]^2}{4\pi}(100) = 7.96\times10^{-2} \text{ m}^2$$

(c) The maximum radar range is

$$R_{max} = \left[\frac{P_t g \sigma A_e}{(4\pi)^2 P_{min}}\right]^{1/4}$$

$$= \left[\frac{100\times10^6\times100\times7.96\times7.96\times10^{-2}}{(4\pi)^2\times1\times10^{-6}}\right]^{1/4}$$

$$= 2.52 \text{ km} = 1.56 \text{ miles}$$

2-5-3 Duty Cycle and Signal-to-Noise Ratio

Duty cycle. Duty cycle is defined as the ratio of pulse duration over repetition period for a pulse train and it may be expressed

$$\text{Duty cycle} = \frac{\text{pulse duration}}{\text{pulse repetition period}} = \frac{\tau}{T} = \tau f$$

$$= \frac{\text{average power}}{\text{peak power}} = \frac{P_{av}}{P_{pk}} \qquad (2\text{-}5\text{-}13)$$

Figure 2-5-2 shows a pulse train for determining the duty cycle.

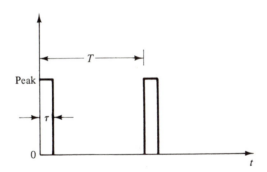

Figure 2-5-2   Pulse train

Example 2-5-3   Duty Cycle and Peak Power

The pulse duration of a radar signal is 1 μs and its pulse period is 2 ms.
(a) Calculate the duty cycle.
(b) Compute the peak power if its average power is 50 W.

Solution.   (a) From Eq. (2-5-13) the duty cycle is

$$\text{Duty cycle} = \frac{10^{-6}}{2 \times 10^{-3}} = 0.0005$$

(b) The peak power is

$$P_{\text{pk}} = \frac{50}{0.0005} = 100 \text{ kW}$$

Signal-to-noise ratio.   The signal-to-noise ratio at the output within a narrow frequency band can be expressed as

$$\text{SN} = \frac{S_o}{N_o} \tag{2-5-14}$$

The noise N_o at the output of the network within the same frequency band arises from amplification of the input noise and from noise N_n generated within the network. It may be written

$$N_o = N_n + GkTB \tag{2-5-15}$$

where kTB is the available noise power of the standard source in bandwidth B at room temperature T of 300°K and G is the available power gain of the network at the frequency of the band considered. The signal-to-noise ratio at the output is

$$\text{SN} = \frac{S_i G}{N_n + GkTB} \tag{2-5-16}$$

where S_i is input signal power and $S_i G$ is the power of the amplified input signal.

Because noise does limit receiver sensitivity, the input circuits of the receiver are the most critical parts of the entire radar system. To permit the detection of a given object at maximum range, the input circuits must use the smallest echo signal as effectively as possible and combine with the signal the least possible amount of noise. That is, the highest obtainable signal-to-noise ratio is desired in order that the signal at the indicator can be detected. When the signal falls below the level required by the signal-to-noise ratio, the signals are obscured, and any increase of amplification is then useless because both signal and noise are amplified together. Radar receivers are always designed with enough stages so that large-amplitude noise voltages appear in the output with full gain; therefore the limit of sensitivity is set by the noise.

2-5-4 Doppler Frequency

When the radar transmitter and receiver are colocated, a moving target along the direction of wave propagation causes each frequency component of the transmitted wave that strikes the target to be shifted by an amount

$$f_{d1} = \frac{v_r}{c} f_t \qquad (2\text{-}5\text{-}17)$$

where v_r = relative or radial velocity of the target with respect to radar
f_t = frequency of transmitted signal
$c = 3 \times 10^8$ m/s is the velocity of light in vacuum

When this signal is reflected or reradiated from the moving target back to the radar, the total Doppler frequency shift of each component is

$$f_d = \frac{2v_r}{c} f_t \qquad (2\text{-}5\text{-}18)$$

This situation is shown in Fig. 2-5-3. Thus when the relative velocity of the target is 300 m/s and the transmitted frequency is 10 GHz, the Doppler frequency is 20 kHz.

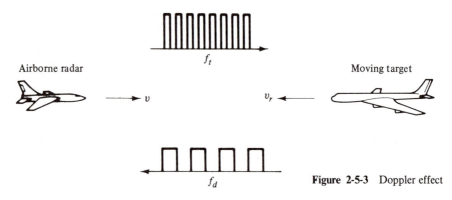

Figure 2-5-3 Doppler effect

Alternatively, when a moving target is approaching a radar with a relative velocity of v_r with respect to the radar, the transmitted wave from the radar will travel a speed of $c - v_r$ toward the target and the reflected wave from the target will travel a speed of $c + v_r$ back to the radar. Then the Doppler frequency is given by

$$f_d = f_t \frac{c + v_r}{c - v_r} - f_t \doteq \frac{2v_r}{c} f_t \qquad (2\text{-}5\text{-}19)$$

Conversely,

$$f_d = f_t - f_t \frac{c - v_r}{c + v_r} \doteq \frac{2v_r}{c} f_t \tag{2-5-20}$$

In an airborne radar the radar transmits electromagnetic energy toward the ground and uses the Doppler shift of the received energy to determine two or three of the velocity components of the aircraft. This situation is illustrated schematically in Fig. 2-5-4. The basic Doppler equation is given by

$$f_d = \frac{2vf_t}{c} \cos \theta \tag{2-5-21}$$

where v = the velocity of the aircraft
        θ = the angle between the velocity vector and the direction of propagation

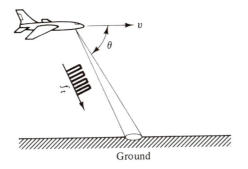

Ground

Figure 2-5-4   Doppler beam toward ground

Because the antenna beam has a finite width and scattering from the ground is randomlike, the information from the ground is not a single frequency but instead is in the form of a noiselike frequency spectrum. A certain amount of smoothing time of about 1 sec is required to determine the quasi-instantaneous velocity for a given frequency.

Example 2-5-4   Doppler Frequency

An airborne radar has a velocity of 700 m/s and a transmitted signal of 9 GHz. A moving-target airplane has a relative velocity of 200 m/s with respect to the radar and a stationary target on the ground with an angle of 60° between the radar velocity vector and the direction of radar-beam propagation.
(a) Determine the Doppler frequency when the radar sees the moving target in the air.
(b) Find the Doppler frequency when the radar sees the stationary target on the ground.

Solution.   (a) From Eq. (2-5-18) the Doppler frequency for the moving target in the air is

$$f_d = \frac{2v_r}{c} f_t = \frac{2 \times 200}{3 \times 10^8} (9 \times 10^9) = 12 \text{ kHz}$$

(b) From Eq. (2-5-21) the Doppler frequency for the stationary target on the ground is

$$f_d = \frac{2vf_t}{c} \cos\theta = \frac{2 \times 700 \times 9 \times 10^9}{3 \times 10^8} \cos 60°$$

$$= 21 \text{ kHz}$$

2-5-5 Radar Systems

A radar system contains two major generators. One is the transmitting oscillator, which provides the high-frequency powerful pulses needed for echo detection. The other is the local oscillator of the superheterodyne receiver. This oscillator produces a low-power continuous oscillation of a frequency nearly equal to the transmitting frequency. In a microwave radar system the transmitting oscillator is usually a magnetron and the receiving oscillator is a reflex klystron. The traveling-wave tube (TWT) is used as an amplifier in the receiver set. The transmitting antenna is also used as the receiving antenna. A radar system is shown in Fig. 2-5-5.

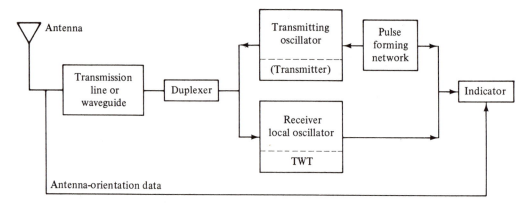

Figure 2-5-5   Radar system

A radar system should have the following characteristics.

1. Reliability of Detection. Reliability of detection includes both the maximum detection range plus the probability or percentage of time that the desired target can be detected at any range. Because detection is inherently a statistical problem, this measure of performance must also include the probability of mistaking an unwanted target or noise for a true target.

2. Accuracy. Accuracy is measured with respect to the target parameter estimates. These parameters include the target range and the angular coordinates.

3. Certainty. The third quality of a radar system is the extent to which the accuracy parameters can be measured without ambiguity or, alternatively, the difficulty encountered in resolving any ambiguities that may be present.

4. Resolution. Resolution is the degree to which two or more targets can be separated in one or more spatial coordinates, in radial velocity, or in acceleration. In a simple sense, resolution measures the ability of a radar system to distinguish radar echoes from similar aircraft in a formation, or to distinguish a missile from a possible decoy. In a more sophisticated sense, resolution in a ground-mapping radar includes the separation of a multitude of targets with widely divergent radar-echoing areas without cross talk between various reflectors.

5. Discrimination. Discrimination is the ability to detect or track (or acquire) a target echo in the presence of environmental echoes. It is convenient to include here the discrimination of a missile or an aircraft from manmade dipoles or decoys, and the ability to separate the echoes of a reentry body from its wake.

6. Countermeasure Immunity. Countermeasure immunity is the selection of a transmitted signal to give an enemy the least possible information. It includes those processing techniques that make the least use of the identifying characteristics of the desired signals. In some cases, the information received by two or more receivers, or derived by two or more complete systems at different locations, or using different principles, parameters, or operation may be compared (correlated) to provide useful discrimination between desired and undesired signals.

7. Immunity to Radio Frequency Interference (RFI). Immunity to radio frequency (RF) interference means the ability of a radar system to perform its mission in close proximity to other radar systems. It includes both the ability to inhibit detection or display the transmitted signals from another radar and the ability to detect desired targets in the presence of another radar signal. The RF interference is also called electromagnetic compatibility (EMC).

REFERENCES

[1] Fuchs, K. The Conductivity of Thin Metallic Films According to the Electron Theory of Metals. *Proc. Camb. Phil. Soc.*, **30** (1938), 100.

[2] Sondheimer, E. H. The Mean-Free-Path of Electrons in Metals. *Advances in Physics*, **1** (1952), 1.

[3] *American Institute of Physics Handbook*, pp. 6-12, 6-119 to 6-121, and 6-138. New York: McGraw-Hill Book Company, 1972.

[4] Schulz, Richard B., et al. Shielding Theory and Practice. *Proc. 9th Tri-Service Conference on Electromagnetic Compatibility*, October 1963.

[5] Vasaka, C. S. Problems in Shielding Electrical and Electronic Equipments. Johnsville, Pa.: U.S. Naval Air Development Center, Rept. No. NACD-EL-N5507, June 1955.

[6] Liao, Samuel Y.
   1. Design of a Gold Film on a Glass Substrate for Maximum Light Transmittance and RF Shielding Effectiveness. *IEEE Electromagnetic Compatibility Symposium Record* 75CH1002-5 EMC. San Antonio, Texas, October 1975.
   2. Light Transmittance and Microwave Attenuation of a Gold-Film Coating on a Plastic Substrate. *IEEE Trans. on Microwave Theory and Techniques*, vol. MTT-23, no. 10 (October 1975).
   3. Light Transmittance and RF Shielding Effectiveness of a Gold Film on a Glass Substrate. *IEEE Trans. on Electromagnetic Compatibility*, vol. EMC-17, no. 4 (November 1975).

[7] Hawthorne, E. I. Electromagnetic Shielding with Transparent Coated Glass. *Proc. IRE*, **42** (March 1954) 548–553.

[8] Liao, Samuel, Y. RF Shielding Effectiveness and Light Transmittance of Copper or Silver Coating on Plastic Substrate. *IEEE Trans. on Electromagnetic Compatibility*, vol. EMC-18, no. 4 (November 1976).

[9] Yariv, Ammon. *Introduction to Optical Electronics*, p. 230. New York: Holt, Rinehart & Winston, 1971.

[10] Davis, K. *Ionospheric Radio Propagation*, Chapter 1, National Bureau of Standards Monograph 80. Washington, D.C., 1965.

[11] Kelson, John M. *Radio Ray Propagation in the Ionosphere*, Chapter 3. New York: McGraw-Hill Book Company, 1964.

[12] Helszajn, J. *Principles of Microwave Ferrite Engineering*, p. 2, New York: John Wiley & Sons, 1969.

SUGGESTED READINGS

1. Budden, K. G. *Radio Waves in the Ionosphere*. Cambridge: Cambridge University Press, 1961.

2. Garriott, Owen K. *Introduction to Ionospheric Physics*. New York: Academic Press, 1969.

3. Helszajn, J. *Principles of Microwave Ferrite Engineering*. New York: John Wiley & Sons, 1969.

4. Kraus, John D. *Radio Astronomy*. New York: McGraw-Hill Book Company, 1966.

5. Lax, B., and K. J. Button. *Microwave Ferrites and Ferrimagnetics*. New York: McGraw-Hill Book Company, 1962.

6. Soohoo, R. F. *Theory and Applications of Ferrites*. Englewood Cliffs, N.J.: Prentice-Hall, Inc., 1960.

7. Von Aulock, W. H., and C. E. Fay. *Linear Ferrite Devices for Microwave Applications*. New York: Academic Press, 1968.

8. Yeh, K. C., and C. H. Liu. *Theory of Ionospheric Waves*. New York: Academic Press, 1972.

PROBLEMS

2-1   Wave propagation in metallic-film coating on glass

2-1-1. Bulk gold has a conductivity of 4.1×10^7 mhos/m, a resistivity of 2.44×10^{-8} ohm-m, and an electron mean-free-path of 570 Å. Calculate the surface conductivity, surface resistivity, and surface resistance of gold film for thicknesses of 10 to 100 Å with an increment of 10 Å for each step.

2-1-2. Bulk silver has a conductivity of 6.170×10^7 mhos/m, a resistivity of 1.620×10^{-8} ohm-m, and an electron mean-free-path of 570 Å. Calculate the surface conductivity, surface resistivity, and surface resistance of silver film for thicknesses of 10 to 100 Å with an increment of 10 Å for each step.

2-1-3. Bulk copper has a conductivity of 5.800×10^7 mhos/m, a resistivity of 1.724×10^{-8} ohm-m, and an electron mean-free-path of 420 Å Calculate the surface conductivity, surface resistivity, and surface resistance of copper film for thicknesses of 10 to 100 Å with an increment of 10 Å for each step.

2-1-4. Write a FORTRAN program to compute the light transmittance and light reflection of a gold-film coating on a nonabsorbing plastic glass for thicknesses of 10 to 100 Å with an increment of 10 Å for each step. Wavelengths vary from 2000 to 10,000 Å with an increment of 500 Å each step. The values of the refractive index n and the extinction index k of a gold film are listed in Table 2-1-2. (The values for 2200 to 4000 Å may be projected.) The refractive index n of the nonabsorbing plastic glass is 1.50. Use F10.5 format for numerical outputs, Hollerith format for character outputs, and data statements to read the input values.

2-1-5. Write a FORTRAN program to compute the light transmittance and light reflection of a silver-film coating on a nonabsorbing plastic glass for wavelengths from 2000 to 3700 Å with an increment of 100 Å each step. The other requirements are the same as specified in Problem 2-1-4.

2-1-6. Write a FORTRAN program to compute the light transmittance and light reflection of a copper-film coating on a nonabsorbing plastic glass for wavelengths from 4500 to 10,000 Å with an increment of 500 Å each step. The other requirements are the same as specified in Problem 2-1-4.

2-2   Wave Propagation in Dielectric Crystals

2-2-1. The transformed index ellipsoid equation is shown in Eq. (2-2-19). Derive the equation.

2-2-2. The refractive indices of a dielectric crystal in the x, y, and z directions are related by a generalized index ellipsoid equation as shown in Eq. (2-2-12). Derive Eq. (2-2-14) from Eq. (2-2-12) by assuming that the electric field is in the x, y, and z directions.

2-2-3. When the optic axis of a dielectric crystal is oriented to the z direction and an electric field is applied in the same direction, the induced indices of the crystal are given by Eqs. (2-2-20), (2-2-21), and (2-2-22). Verify these three equations.

2-3   Wave propagation in magnetic crystals

2-3-1. The gyromagnetic ratio is given by the formula $g\mu_0 e/(2m)$ as shown in Eq. (2-3-4). Calculate its numerical value.

2-3-2. The resonant angular frequency (or the Larmor precessional frequency) ω_0 of the electron in the dc magnetic field H_0 is defined as $\omega_0 = \gamma H_0$. Calculate the resonant frequency in GHz for $H_0 = 10^5$ A/m.

2-3-3. A Faraday-rotation device has the following parameters:

$$M_0 = 7.95 \times 10^4 \text{ A/m} \qquad \omega = 2 \times 10^{10} \text{ rad/s}$$

$$\epsilon_r = 10 \qquad H_0 = 2 \times 10^5 \text{ A/m}$$

(a) Calculate the magnetization angular frequency ω_m.
(b) Compute the resonant angular frequency ω_0.
(c) Find the factor $(\mu \pm k)$.
(d) Calculate the positive and negative phase constants (β_+ and β_-).
(e) Determine the Faraday rotation angle in degrees for a length of 1 cm.

2-4 Wave propagation in ionosphere

2-4-1. At a 100-km height in the ionosphere the electron density at night is about 3×10^{12} per cubic meter.
(a) Calculate the critical frequency in gigahertz.
(b) Explain the significant impact of the critical frequency.

2-4-2. At a 100-km height in the ionosphere the electron density during daytime is about 5×10^{12} per cubic meter.
(a) Compute the critical frequency in gigahertz.
(b) Explain the significant meaning of the critical frequency.

2-4-3. Derive Eq. (2-4-11) from Eq. (2-4-8).

2-4-4. At a 300-km height in the ionosphere the electron density at night is about 3×10^{12} m^{-3} and the signal maximum usable frequency (MUF) is $f = 2f_{cr}$ for a transmission distance of 700 km. Determine the following items.
(a) Critical frequency
(b) Relative dielectric constant
(c) Phase constant
(d) Wave impedance
(e) Wave velocity
(f) Group velocity
(g) Incident angle

2-5 Wave propagation in radar systems

2-5-1. The distance between a target and a transmitter is 3 km. If the target requires a minimum peak power density of 10 mW/m² for detection, determine the minimum radiated peak power from the transmitter for operation.

2-5-2. If the duty cycle of a signal is 0.001, what is the average power for a peak power of 2 MW?

2-5-3. If the transmitted frequency is 9 GHz and the relative velocity of a target is 400 m/s, determine the Doppler frequency.

2-5-4. A radar antenna has a power gain of 10 dB and is operating at a frequency of 10 GHz. Calculate the radar cross section.

2-5-5. The minimum power p_{min} for a radar is 0.001 mW and the transmitting power is 100 MW (peak). The radar antenna has a power gain of 20 dB and is operating at a frequency of 10 GHz. Determine the maximum radar range.

Chapter 3

Microwave Solid-State Junction-Effect Devices

3-0 INTRODUCTION

Ever since Shockley and his coworkers invented the transistor (contraction for transfer resistor) in 1948 [1], electronic technology has had a revolutionary impact on solid-state devices. Since then transistors and related semiconductor devices have replaced vacuum tubes for lower power sources. Moreover, microwave power transistor technology has advanced significantly during the past decade and today has many applications, such as in L-band transmitters for telemetry systems, phased array radar systems, and L- and S-band transmitters for communications systems. The tunnel diode is a *p-n* junction device that operates in tunneling processes and is useful in high-speed switching and logic circuits. The metal-semiconductor (MES) diode (Schottky-barrier diode) has many applications in integrated circuits as modulators or detectors at microwave frequencies. The metal-insulator-semiconductor (MIS) diode is the basis of the MOSFETs and has many uses in microprocessors and charge-coupled devices (CCDs). In this chapter these four types of solid-state devices are described.

In studying microwave solid-state devices, the electrical behavior of solids is the first item to be investigated. In the following chapters it will be seen that the transport of charge through a semiconductor depends on both the properties of the electron and on the arrangement of atoms in the solids. Semiconductors are a group of substances having electrical conductivities that are intermediate between metals and insulators. Because the conductivity of the semiconductors can be varied over wide ranges by changes in their temperature, optical excitation, and impurity content, they are the natural choice for electronic devices. The properties of important semiconductors are tabulated in Table 3-0-1.

TABLE 3-0-1 PROPERTIES OF IMPORTANT SEMICONDUCTORS

Semiconductor	Bandgap energy (eV)		Mobility at 300°K (cm²/volt-sec)		Relative Dielectric Constant
	0°K	300°K	Holes	Electrons	
C	5.51	5.47	1600	1800	5.5
Ge	0.89	0.80	1900	3900	16
Si	1.16	1.12	600	1350	11.8
AlSb	1.75	1.63	420	200	11
GaSb	0.80	0.67	1400	4000	15
GaAs	1.52	1.43	400	8500	13.1
GaP	2.40	2.26	75	110	10
InSb	0.26	0.18	750	78,000	17
InAs	0.46	0.33	460	33,000	14.5
InP	1.34	1.29	150	4600	14
CdS	2.56	2.43	50	300	10
CdSe	1.85	1.70		800	10
ZnO		3.20		200	9
ZnS	3.70	3.60		165	8

The energy bands of a semiconductor play a major role in their electrical behavior. Any semiconductor has a forbidden energy region in which no allowable states can exist. The energy band above the forbidden region is called the *conduction band* and the bottom of the conduction band is designated by E_c. The energy band below the forbidden region is called the *valence band* and the top of the valence band is designated by E_v. The separation between the energy of the lowest conduction band and that of the highest valence band is called the *energy bandgap E_g*, which is the most important parameter in semiconductors.

Electron energy is conventionally defined as positive when measured upwards; and the hole energy is positive when measured downward. A simplified band diagram is shown in Fig. 3-0-1.

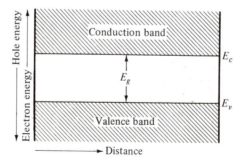

Figure 3-0-1   Energy-band diagram

3-1 *MICROWAVE TRANSISTORS*

The microwave transistor is a nonlinear device and its principle of operation is similar to that of the low-frequency transistor; requirements for dimensions, process control, heat sinking, and packaging, however, are more severe. The power output of a microwave transistor decreases rapidly while its operating frequencies are above 3 GHz. At about 12 GHz its power level is limited to a few milliwatts. The primary use of these devices is in stripline-type applications for local oscillator operation. Efficiencies are generally high, ranging from 15% and up.

3-1-1 *Physical Structures*

All microwave transistors are now planar in form and almost all are of the silicon *n-p-n* type. The geometry can be characterized as follows: (1) interdigitated, (2) overlay, and (3) matrix, which is also called mesh or emitter grid (see Fig. 3-1-1). The interdigitated type is for a small signal and power; both the overlay type and the matrix type are for small power only.

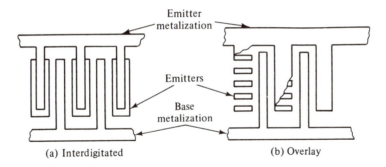

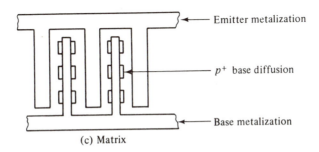

Figure 3-1-1 Surface geometries of microwave power transistors (After H. Sobol and F. Sterzer [3]; reprinted by permission of the IEEE, Inc.)

The figure of merit for the three surface geometries shown in Fig. 3-1-1 is listed in Table 3-1-1 [3]. The state of the art of microwave transistor fabrication today limits emitter width w to about 1 μm and the base thickness t to about 0.2 μm. Emitter length ℓ is of the order of 25 μm for the overly and matrix. Figure 3-1-2 shows a double-diffused silicon epitaxial n-p-n microwave transistor with its impurity profile [5].

TABLE 3-1-1　FIGURE OF MERIT (M) OF VARIOUS SURFACE GEOMETRIES

Surface Geometry and Unit Cell	$M = \dfrac{EP}{BA}$
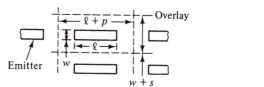	$\dfrac{2(\ell + w)}{(w + s)(\ell + p)}$
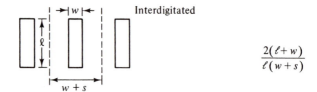	$\dfrac{2(\ell + w)}{\ell(w + s)}$
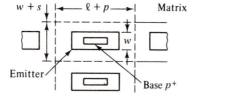	$\dfrac{2(\ell + w)}{(w + s)(\ell + p)}$

(After Sobol and Sterzer [3]; reprinted by permission of the IEEE, Inc.)

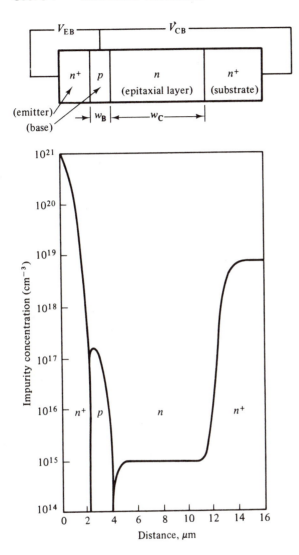

Figure 3-1-2 Double-diffused silicon epitaxial *n-p-n* microwave transistor (From S. M. Sze [5]; reprinted by permission of John Wiley & Sons, Ltd.)

3-1-2   Principles of Operation

The operation of a microwave power transistor is similar to that of the low-frequency transistor, but class C operation is much better than operation in class A or AB mode. Figure 3-1-3 shows a simple model of a microwave transistor.

When the transistor is biased for class C operation, both the emitter-base and collector-base junctions are reverse-biased, and no current flows in the absence of an

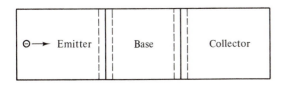

Figure 3-1-3 Transit-time model of a microwave transistor

applied signal. The depletion layer at the emitter-base junction is much smaller than that at the collector-base junction. When an RF voltage of sufficient magnitude is applied to the emitter-base junction, the junction is forward-biased for a fraction of an RF cycle. Thus the electrons are injected into the base. The injected carriers transit the base by combined diffusion and drift flow and are then accelerated in the collector-base depletion region. The electric field in the collector-base depletion region, even at the minimum swing of the RF signal, is sufficiently large to accelerate electrons to their saturation velocity. The flow of energetic electrons injected during the time of forward emitter-base bias represents a pulse of current in the collector circuit. The peak of the current pulse occurs when the electrons traverse the collector-base depletion region with a saturation velocity. If the collector circuit is a real load at the driving-signal frequency, an output power can be taken from the load at the fundamental frequency.

3-1-3 Microwave Characteristics

The characteristics of microwave transistors can be analyzed in several ways. The most common approach is a combination of internal-parameter and two-port analyses. Figure 3-1-4 shows the current-equivalent circuits of a microwave transistor.

It can be seen from the equivalent circuits in the figure that

$r_e =$ the emitter resistance

$r_b = \dfrac{w}{\ell} r_0$ is the base resistance, where w is the emitter strip width and ℓ is the emitter strip length

$r_0 = \dfrac{\rho_b}{w}$ and ρ_b is the average resistivity of the base layer

$C_e =$ the emitter depletion-layer capacitance

$C_c = C_0 w \ell$ is the collector depletion-layer capacitance, where C_0 is the collector capacitance per unit area

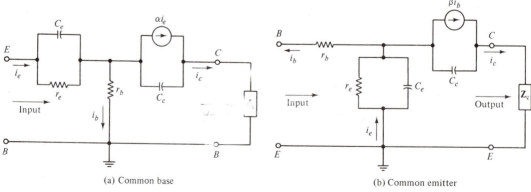

(a) Common base   (b) Common emitter

Figure 3-1-4   Current-equivalent circuits of microwave transistors

(A) Current gain.   The dc common-base current gain α_0, also referred to as h_{FB} from the four-terminal hybrid parameters (where subscripts F and B refer to forward and common-base, respectively), is defined by

$$\alpha_0 = h_{FB} = \frac{\Delta I_C}{\Delta I_E} \tag{3-1-1}$$

Similarly, the dc common-emitter current gain β_0, also referred to as h_{FE}, is defined by

$$\beta_0 = h_{FE} = \frac{\Delta I_C}{\Delta I_B} \tag{3-1-2}$$

Because $I_E = I_B + I_C$, it follows that α_0 and β_0 are related to each other by

$$\beta_0 = \frac{\alpha_0}{1 - \alpha_0} \tag{3-1-3}$$

For a small signal, the common-base current gain α is defined as

$$\alpha = h_{fb} = \left. \frac{\partial I_C}{\partial I_E} \right|_{\Delta V_{CB} = 0} \tag{3-1-4}$$

Similarly, the small-signal common-emitter current gain β is defined as

$$\beta = h_{fe} = \left. \frac{\partial I_C}{\partial I_B} \right|_{\Delta V_{CE} = 0} \tag{3-1-5}$$

Substitution of Eqs. (3-1-1) and (3-1-2) into Eqs. (3-1-4) and (3-1-5) yields

$$\alpha = \alpha_0 + I_E \frac{\partial \alpha_0}{\partial I_E} \tag{3-1-6}$$

$$\beta = \beta_0 + I_B \frac{\partial \beta_0}{\partial I_B} \tag{3-1-7}$$

and

$$\beta = \frac{\alpha}{1 - \alpha} \tag{3-1-8}$$

(B) Cutoff frequency.   The charge-carrier transit-time cutoff frequency f_T of a charge-control type of microwave transistor is defined [6] by the relation

$$f_T = \frac{1}{2 \pi \tau} \tag{3-1-9}$$

where τ is the average time for a charge carrier moving at an average velocity v to traverse the emitter-collector distance L. This delay time represents the sum of four delays encountered sequentially by the minority charge carrier. That is,

$$\tau = \tau_E + \tau_B + \tau_X + \tau_C \tag{3-1-10}$$

where τ_E = emitter-base junction charging time
τ_B = base transit time
τ_X = base-collector depletion layer transit time
τ_C = collector-base junction charging time

The base and base-collector depletion-layer transit times for typical microwave transistors are much greater than the junction charging times. Therefore the average time for a carrier is approximately equal to

$$\tau = \tau_B + \tau_X \qquad (3\text{-}1\text{-}11)$$

The base-collector depletion-layer transit time is dependent of the collector-base bias and the epitaxial doping level. In general, a CW class C microwave power transistor would have a dc breakdown voltage about two times higher than the bias voltage. The typical 28-V microwave power transistor requires an epitaxial layer with an impurity level of $N_d = 5 \times 10^{15}/\text{cm}^3$ or 1-ohm·cm resistivity. The depletion-layer transit time corresponding to 28-V operation and 1-ohm·cm resistivity is

$$\tau_X = 1.7 \times 10^{-11} \text{ sec} \qquad (3\text{-}1\text{-}12)$$

The base transit time is a function of the base width δ and is given by

$$\tau_B = 8.3 \times 10^{-10} \delta^2 \text{ sec} \qquad (3\text{-}1\text{-}13)$$

Therefore the charge-carrier transit-time cutoff frequency can be simplified [3] and expressed as

$$f_T = \frac{1}{2\pi(\tau_X + \tau_B)} \qquad (3\text{-}1\text{-}14)$$

3-1-4 Design Example

(A) Power gain G_p. When frequencies are in the microwave region, it is very difficult to achieve short and open circuits for measuring the Z, Y, and H parameters. An alternative method is to use the S parameters in traveling waves for solving the power gain at microwave frequencies. The two-port network of a microwave transistor amplifier is shown in Fig. 3-1-5, and the two S-parameter equations are expressed as

$$\begin{aligned} \mathbf{b}_1 &= \mathbf{S}_{11}\mathbf{a}_1 + \mathbf{S}_{12}\mathbf{a}_2 \\ \mathbf{b}_2 &= \mathbf{S}_{21}\mathbf{a}_1 + \mathbf{S}_{22}\mathbf{a}_2 \end{aligned} \qquad (3\text{-}1\text{-}15)$$

Power gain G_p of a microwave transistor amplifier is defined as the ratio of the output power P_ℓ delivered to the load $\mathbf{Z}_\ell$ over the input power P_{avs} available from the source to the network. That is,

$$G_p = \frac{P_\ell}{P_{\text{avs}}} \qquad (3\text{-}1\text{-}16)$$

The power delivered to the load is the result of the power incident on the load minus the power reflected from the load:

$$P_\ell = \tfrac{1}{2}|\mathbf{b}_2|^2 - \tfrac{1}{2}|\mathbf{a}_2|^2 = \tfrac{1}{2}|\mathbf{b}_2|^2\left(1 - |\Gamma_\ell|^2\right) \qquad (3\text{-}1\text{-}17)$$

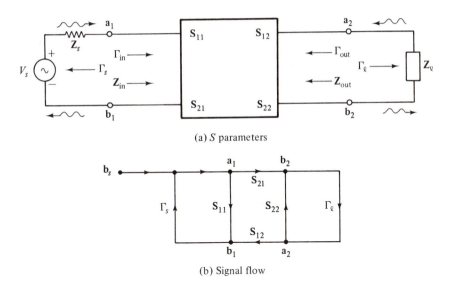

(a) S parameters

(b) Signal flow

Figure 3-1-5   Two-port network of a microwave transistor amplifier

where

$$\Gamma_{\ell} = \frac{Z_{\ell} - Z_0}{Z_{\ell} + Z_0}$$

is the reflection coefficient of the load. The power available from the source is given by

$$P_{\text{avs}} = \frac{\frac{1}{2}|b_s|^2}{\left(1 - |\Gamma_s|^2\right)} \tag{3-1-18}$$

where

$$\Gamma_s = \frac{Z_s - Z_0}{Z_s + Z_0}$$

is the reflection cofficient of the source, and b_s is a function of b_2 which is to be determined. Then the power gain is expressed by

$$G_p = \frac{|b_2|^2}{|b_s|^2}\left(1 - |\Gamma_s|^2\right)\left(1 - |\Gamma_{\ell}|^2\right) \tag{3-1-19}$$

The transfer function from b_s to b_2 can be derived by using the nontouching loop rules of signal-flow theory. The nontouching loop rules, often called Mason's rules [17], include the following terms:

1. *Path.* A path is a series of directed lines followed in sequence and in the same direction in such a way that no node is touched more than once. The value of the path is the product of all coefficients encountered enroute. In Fig. 3-1-5(b) there is only one path from b_s to b_2, and the value of the path is S_{21}. There are two paths from b_s to b_1, and the values are S_{11} and $S_{21}\Gamma_{\ell}S_{12}$.

2. *First-order loop.* A first-order loop is defined as the product of all coefficients along the paths that start from a node and move in the direction of the arrows back to that original node without passing the same node twice. In Fig. 3-1-5(b) there are three first-order loops, and the values are $S_{11}\Gamma_s$, $S_{22}\Gamma_\ell$, and $S_{21}\Gamma_s S_{12}\Gamma_\ell$.

3. *Second-order loop.* A second-order loop is defined as the product of any two nontouching first-order loops. In Fig. 3-1-5(b) there is only one second-order loop, and its value is $S_{11}\Gamma_s S_{22}\Gamma_\ell$.

4. *Third-order loop.* A third-order loop is the product of any three nontouching first-order loops. Figure 3-1-5(b) has no third-order loop.

Then the transfer function for the ratio of a dependent variable in question to an independent variable of the source is expressed by

$$T = \frac{\begin{aligned}P_1\left[1 - \Sigma L(1)^1 + \Sigma L(2)^1 - \Sigma L(3)^1 + \cdots\right] \\ + P_2\left[1 - \Sigma L(1)^2 + \Sigma L(2)^2 \cdots\right] + P_3\left[1 \cdots\right]\end{aligned}}{1 - \Sigma L(1) + \Sigma L(2) - \Sigma L(3) + \cdots} \tag{3-1-20}$$

where $P_1, P_2, P_3 \ldots$ are the various paths connecting these variables
$\Sigma L(1), \Sigma L(2), \Sigma L(3) \ldots$ are the sums of all first-order, second-order, third-order loops..., respectively
$\Sigma L(1)^1, \Sigma L(2)^1, \Sigma L(3)^1 \ldots$ are the sums of all first-order, second-order, third-order loops that do not touch the first path between the variables, and
$\Sigma L(1)^2, \Sigma L(2)^2, \Sigma L(3)^2 \ldots$ are the sums of all first-order, second-order, third-order loops that do not touch the second path

From Fig. 3-1-5(b) the transfer function of $\mathbf{b}_2$ over $\mathbf{b}_s$ is given by

$$\frac{\mathbf{b}_2}{\mathbf{b}_s} = \frac{S_{21}}{1 - S_{11}\Gamma_s - S_{22}\Gamma_\ell - S_{21}\Gamma_s S_{12}\Gamma_\ell + S_{11}\Gamma_s S_{22}\Gamma_\ell} \tag{3-1-21}$$

Substitution of Eq. (3-1-21) into Eq. (3-1-19) yields the power gain

$$G_p = \frac{|S_{21}|^2\left(1 - |\Gamma_s|^2\right)\left(1 - |\Gamma_\ell|^2\right)}{|(1 - S_{11}\Gamma_s)(1 - S_{22}\Gamma_\ell) - S_{21}S_{12}\Gamma_s\Gamma_\ell|^2} \tag{3-1-22}$$

(B) Maximum available power gain G_{max}. In order to maximize the forward power gain G_{max} of a microwave transistor amplifier, the input and output networks must be conjugately matched. Before discussing this subject, let us describe the stability of a transistor amplifier. Stability or resistance to oscillation is most important in amplifier design and is determined by the S parameters, the synthesized source, and the load impedances. Oscillations are only possible if either the input or the output port, or both, have negative resistance. This situation occurs if either $|\mathbf{S}_{11}|$ or $|\mathbf{S}_{22}|$ is greater than unity. Even with negative resistance, however, the amplifier might still be stable.

There are two traditional expressions for stability: conditional stability and unconditional stability. A network is conditionally stable if the real part of input impedance $\mathbf{Z}_{in}$ and output impedance $\mathbf{Z}_{out}$ is greater than zero for some positive real source and load impedances at a specific frequency. A network is unconditionally stable if the real part of input impedance $\mathbf{Z}_{in}$ and output impedance $\mathbf{Z}_{out}$ is greater than zero for all positive real sources and load impedances at a specific frequency.

The maximum power gain G_{max} that can be realized for a microwave transistor amplifier without external feedback produces the forward power gain when both input and output are simultaneously and conjugately matched. Conjugately matched conditions mean that the reflection coefficient Γ_s of the source is equal to the conjugate of the input reflection coefficient Γ_{in}, and the reflection coefficient Γ_ℓ of the load is equal to the conjugate of the output reflection coefficient Γ_{out}. These are

$$\Gamma_s = \Gamma_{in}^* \quad \text{and} \quad \Gamma_\ell = \Gamma_{out}^*$$

If a transistor amplifier is to be unconditionally stable, then the magnitudes of $\mathbf{S}_{11}$, $\mathbf{S}_{22}$, Γ_{in}, and Γ_{out} must be smaller than unity and the transistor's inherent stability factor K must be greater than unity and positive. K is computed from

$$K = \frac{1 + |\Delta|^2 - |\mathbf{S}_{11}|^2 - |\mathbf{S}_{22}|^2}{2|\mathbf{S}_{12}\mathbf{S}_{21}|} > 1 \tag{3-1-23}$$

where

$$\Delta = \mathbf{S}_{11}\mathbf{S}_{22} - \mathbf{S}_{12}\mathbf{S}_{21}$$

The input and output reflection coefficients are given by

$$\Gamma_{in} = \mathbf{S}_{11} + \frac{\mathbf{S}_{21}\mathbf{S}_{12}\Gamma_\ell}{1 - \mathbf{S}_{22}\Gamma_\ell} \tag{3-1-24}$$

and

$$\Gamma_{out} = \mathbf{S}_{22} + \frac{\mathbf{S}_{21}\mathbf{S}_{12}\Gamma_s}{1 - \mathbf{S}_{11}\Gamma_s} \tag{3-1-25}$$

Also, the boundary conditions for stability are given by

$$|\Gamma_{in}| = 1 = \left| \mathbf{S}_{11} + \frac{\mathbf{S}_{21}\mathbf{S}_{12}\Gamma_\ell}{1 - \mathbf{S}_{22}\Gamma_\ell} \right| \tag{3-1-26}$$

and

$$|\Gamma_{out}| = 1 = \left| \mathbf{S}_{22} + \frac{\mathbf{S}_{21}\mathbf{S}_{12}\Gamma_s}{1 - \mathbf{S}_{11}\Gamma_s} \right| \tag{3-1-27}$$

Substitution of real and imaginary values for the S parameters in Eqs. (3-1-26) and (3-1-27) yields the solutions of Γ_s and Γ_ℓ as

$$R_s \ (\text{radius of } \Gamma_s \text{ circle}) = \frac{|S_{12}S_{21}|}{|S_{11}|^2 - |\Delta|^2} \tag{3-1-28}$$

$$C_s \ (\text{center of } \Gamma_s \text{ circle}) = \frac{C_s^*}{|S_{11}|^2 - |\Delta|^2} \tag{3-1-29}$$

$$R_\ell \ (\text{radius of } \Gamma_\ell \text{ circle}) = \frac{|S_{12}S_{21}|}{|S_{22}|^2 - |\Delta|^2} \tag{3-1-30}$$

$$C_\ell \ (\text{center of } \Gamma_\ell \text{ circle}) = \frac{C_\ell^*}{|S_{22}|^2 - |\Delta|^2} \tag{3-1-31}$$

where
$$\Delta = S_{11}S_{22} - S_{12}S_{21}$$
$$C_s = S_{11} - \Delta S_{22}^*$$
$$C_\ell = S_{22} - \Delta S_{11}^*$$

The reflection coefficient of the source impedance required to match the input of the transistor conjugately for maximum power gain is

$$\Gamma_{sm} = C_s^* \left(\frac{B_s \pm \sqrt{B_s^2 - 4|C_s|^2}}{2|C_s|^2} \right) \tag{3-1-32}$$

where
$$B_s = 1 + |S_{11}|^2 - |S_{22}| - |\Delta|^2 \tag{3-1-33}$$

The reflection coefficient of the load impedance required to match the output of the transistor conjugately for maximum power gain is

$$\Gamma_{\ell m} = C_\ell^* \left(\frac{B_\ell \pm \sqrt{B_\ell^2 - 4|C_\ell|^2}}{2|C_\ell|^2} \right) \tag{3-1-34}$$

where
$$B_\ell = 1 + |S_{22}|^2 - |S_{11}|^2 - |\Delta|^2 \tag{3-1-35}$$

If the computed values of B_s and B_ℓ are negative, then the plus sign should be used in front of the radical in Eqs. (3-1-32) and (3-1-34). Conversely, if B_s and B_ℓ are positive, then the negative sign should be used.

Stability circles can be plotted directly on a Smith chart. These circles separate the output or input planes into stable and potentially unstable regions. A stability circle plotted on the output plane indicates the values of all loads that provide negative real input impedance, thereby causing the circuit to oscillate. A similar circle can be plotted on the input plane which again indicates the values of all loads that provide negative real output impedance and causes oscillation. A negative real impedance produces a reflection coefficient that has a magnitude greater than unity. The regions of instability occur within the circles whose centers and radii are expressed by Eqs. (3-1-28) through (3-1-31).

By using an appropriate sign, only one answer is possible in either Eq. (3-1-32) or (3-1-34) and a value of less than unity is obtained. The maximum available power gain possible is expressed as

$$G_{max} = \frac{|\mathbf{S}_{21}|}{|\mathbf{S}_{12}|}|K \pm (K^2 - 1)^{1/2}| \tag{3-1-36}$$

Hence, maximum available power gain is obtained only if the microwave transistor amplifier is loaded with Γ_{sm} and $\Gamma_{\ell m}$ as reflection coefficients. The maximum frequency of oscillation is determined after the maximum available power gain is achieved.

(C) Unilateral power gain G_u.   The unilateral power gain G_u refers to the forward power gain in a feedback amplifier that has its reverse power gain set to zero ($|\mathbf{S}_{12}|^2 = 0$) by adjusting a lossless reciprocal feedback network connected around the microwave transistor amplifier. That is,

$$G_u = \frac{|\mathbf{S}_{21}|^2 (1 - |\Gamma_s|^2)(1 - |\Gamma_\ell|^2)}{|1 - \mathbf{S}_{11}\Gamma_s|^2 |1 - \mathbf{S}_{22}\Gamma_\ell|^2} \tag{3-1-37}$$

The maximum unilateral power gain is obtained when $\Gamma_s = \mathbf{S}_{11}^*$ and $\Gamma_\ell = \mathbf{S}_{22}^*$. Then

$$G_{u\,max} = \frac{|\mathbf{S}_{21}|^2}{(1 - |\mathbf{S}_{11}|^2)(1 - |\mathbf{S}_{22}|^2)} \tag{3-1-38}$$

(D) Amplifier design with S parameters.   Figure 3-1-6 shows a microwave transistor-amplifier circuit with input and output matching networks. When $|\mathbf{S}_{12}|$ is very small or zero, maximum gain can be achieved if the input and output of the transistor are terminated in impedances that are the complex conjugates of $\mathbf{S}_{11}$ and $\mathbf{S}_{22}$, respectively.

The unilateral power gain G_u as indicated by Eq. (3-1-37) can be written

$$\begin{aligned} G_u &= \frac{1 - |\Gamma_s|^2}{|1 - \mathbf{S}_{11}\Gamma_s|^2} \cdot |\mathbf{S}_{21}|^2 \cdot \frac{1 - |\Gamma_\ell|^2}{|1 - \mathbf{S}_{22}\Gamma_\ell|^2} \\ &= g_s \cdot g_f \cdot g_\ell \quad \text{(numerical values)} \\ &= G_s + G_f + G_\ell \quad \text{dB} \end{aligned} \tag{3-1-39}$$

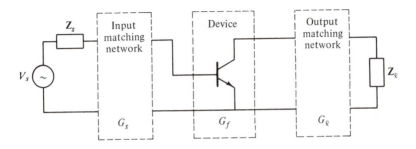

Figure 3-1-6   Microwave transistor-amplifier circuit

where $g_s = \dfrac{1 - |\Gamma_s|^2}{|1 - \mathbf{S}_{11}\Gamma_s|^2}$ is the additional power gain or loss resulting from the input impedance matching network between the device and source,

$g_f = |\mathbf{S}_{21}|^2$ is the forward power gain of the device with the input and output terminated in matching loads, and

$g_\ell = \dfrac{1 - |\Gamma_\ell|^2}{|1 - \mathbf{S}_{22}\Gamma_\ell|^2}$ is the additional power gain or loss due to the impedance matching network between the output of the device and the load.

Maximum unilateral transducer gain can be achieved by choosing impedance matching networks such that $\Gamma_s = \mathbf{S}_{11}^*$, $\mathbf{Z}_s = Z_0$, $\Gamma_\ell = \mathbf{S}_{22}^*$, and $\mathbf{Z}_\ell = \mathbf{Z}_0$ (see Fig. 3-1-7). Then Eq. (3-1-38) can be expressed as

$$G_{u\,\text{max}} = \frac{1}{1 - |\mathbf{S}_{11}|^2} \cdot |\mathbf{S}_{21}|^2 \cdot \frac{1}{1 - |\mathbf{S}_{22}|^2}$$

$$= g_{s\,\text{max}} \cdot g_f \cdot g_{\ell\,\text{max}} \qquad \text{(numerical values)} \qquad (3\text{-}1\text{-}40)$$

$$= G_{s\,\text{max}} + G_f + G_{\ell\,\text{max}} \qquad \text{dB}$$

where

$$g_{s\,\text{max}} = \frac{1}{1 - |\mathbf{S}_{11}|^2}$$

$$g_f = |\mathbf{S}_{21}|^2$$

$$g_{\ell\,\text{max}} = \frac{1}{1 - |\mathbf{S}_{22}|^2}$$

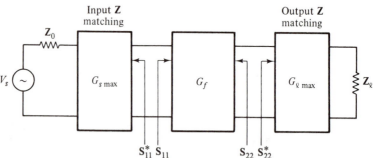

Figure 3-1-7 Amplifier circuit for maximum gain

Constant Gain Circles.   It is obvious that power gain G_s or G_ℓ is equal to maximum, respectively, for $\Gamma_s = \mathbf{S}_{11}^*$ or $\Gamma_\ell = \mathbf{S}_{22}^*$. It is also clear that power gain G_s or G_ℓ has a value of zero for $|\Gamma_s| = 1$ or $|\Gamma_\ell| = 1$. For any arbitrary value of G_s or G_ℓ between these extremes of zero and $G_{s\,\text{max}}$ or $G_{\ell\,\text{max}}$, solutions for Γ_s or Γ_ℓ lie on a circle.

Then for $0 < G_s < G_{s\,max}$ dB, we have

$$g_s = \frac{1 - |\Gamma_s|^2}{|1 - S_{11}\Gamma_s|^2} \qquad \text{(numerical value)} \qquad (3\text{-}1\text{-}41)$$

It is convenient to plot these circles on a Smith chart. The circles have their centers located on the vector drawn from the center of the Smith chart to point S_{11}^* or S_{22}^*. For example, the distance from the center of the Smith chart to the center of the constant gain circle along vector S_{11}^* is given by

$$d_s = \frac{g_{ns}|S_{11}|}{1 - |S_{11}|^2(1 - g_{ns})} \qquad (3\text{-}1\text{-}42)$$

The radius of the constant gain circle is expressed by

$$r_s = \frac{\sqrt{1 - g_{ns}}\,(1 - |S_{11}|^2)}{1 - |S_{11}|^2(1 - g_{ns})} \qquad (3\text{-}1\text{-}43)$$

where g_{ns} is the normalized gain value for the gain circle g_s or g_ℓ, respectively. That is,

$$g_{ns} = \frac{g_s}{g_{s\,max}} = g_s(1 - |S_{11}|^2) \qquad (3\text{-}1\text{-}44)$$

To illustrate, let $\Gamma_s = S_{11}^* = 0.606\,\underline{/-155°}$ as shown on the Smith chart. The constant gain circles are drawn on a Smith chart as shown in Fig. 3-1-8. Any value of Γ_s along a 1-dB circle would result in a power gain G_s of 1 dB and so on. The maximum power $G_{s\,max}$ is 3 dB.

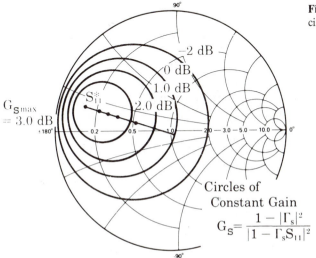

Figure 3-1-8 Constant power gain circles

Design Example for $G_{u\,max}$.   The S parameters of a microwave transistor can be measured by using the HP-8410S network analyzer system. For a frequency at 1 GHz, a certain transistor measured with a 50-ohm resistance matching the input and output has the following parameters:

$$\mathbf{S}_{11} = 0.606 \underline{/-155°}$$
$$\mathbf{S}_{22} = 0.48 \underline{/-20°}$$
$$\mathbf{S}_{21} = 6 \underline{/180°}$$

The desired amplifier circuit is shown in Fig. 3-1-9.

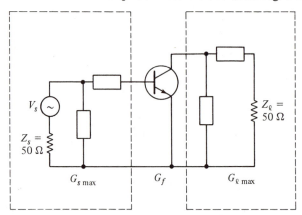

Figure 3-1-9 Desired amplifier circuit (courtesy Hewlett-Packard Co.)

Step 1. *Calculation and Plot of Input and Output Constant Power Gain Circles.* From Eq. (3-1-40) we have

$$G_{u\,max} = 10\log\left(\frac{1}{1-|\mathbf{S}_{11}|^2}\right) + 10\log|\mathbf{S}_{21}|^2 + 10\log\left(\frac{1}{1-|\mathbf{S}_{22}|^2}\right)$$
$$= 10\log\left(\frac{1}{1-|0.606|^2}\right) + 10\log|6|^2 + 10\log\left(\frac{1}{1-|0.48|^2}\right)$$
$$= 1.99 \text{ dB} + 15.60 \text{ dB} + 1.10 \text{ dB}$$
$$= 18.69 \text{ dB}$$

Only the two circles representing maximum gain are needed for this example. These circles have zero radius and are located at $\mathbf{S}_{11}^*$ and $\mathbf{S}_{22}^*$ (see Fig. 3-1-10).

Step 2. *Overlap of Two Smith Charts.* To facilitate the design of the matching networks, overlap two Smith charts 180° out of phase with each other. The original chart can be used to read impedances and the overlaid chart to read admittances (see Fig. 3-1-10).

Step 3. *Determination of the Output Matching Network.* To determine the matching network for the output, start the load impedance of 50 ohms at the center of the Smith chart and proceed along the constant unit resistance circle counterclockwise until reaching the constant conduc-

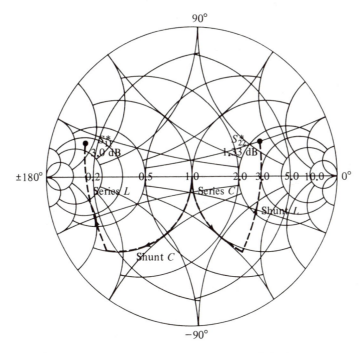

Figure 3-1-10   Constant gain circles

tance circle that intersects the point representing S_{22}^*.

$$jX = -j1.35 \times 50 = -j67.5 \ \Omega$$
$$C_\ell = 2.36 \text{ pF}$$

Then add an inductive susceptance along the constant conductance circle so that the impedance looking into the matching network will be equal to S_{22}^*.

$$jB = -j0.63\left(\frac{1}{50}\right) = -j12.6 \text{ millimhos}$$
$$L_\ell = 12.63 \text{ nH}$$

Step 4. Determination of the Input Matching Network. The same procedures can be applied to the input side; the results are a shunt capacitor and a series inductor.

$$jX = j0.62 \times 50 = j31 \ \Omega$$
$$L_s = 4.92 \text{ nH}$$
$$jB = \frac{j2.4}{50} = j48 \text{ millimhos}$$
$$C_s = 7.64 \text{ pF}$$

Step 5. Final Amplifier Network. Figure 3-1-11 shows the final microwave transistor-amplifier circuit.

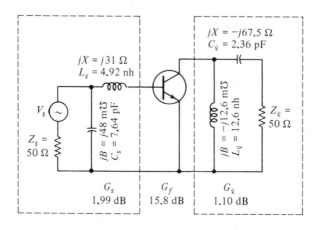

Figure 3-1-11 Transistor amplifier circuit (courtesy Hewlett-Packard Co.)

3-1-5 Power-Frequency Limitations

Do microwave transistors have any limitations on their frequency and output power? The answer is yes. Several authors have discussed this subject. Early [6] first introduced the frequency-power limitations inherent in (1) the limiting velocity of carriers in semiconductors and (2) the maximum fields attainable in semiconductors without the onset of avalanche multiplication. These basic ideas were later developed and discussed in detail by Johnson [7], who made three assumptions:

1. There is a maximum possible velocity of carriers in a semiconductor. This is the "saturated drift velocity v_s," and is on the order of 6×10^6 cm/sec for electrons and holes in silicon and germanium.
2. There is a maximum electric field E_m that can be sustained in a semiconductor without having dielectric breakdown. This field is about 10^5 V/cm in germanium and 2×10^5 V/cm in silicon.
3. The maximum current that a microwave power transistor can carry is limited by the base width.

Using these three postulates, Johnson derived four basic equations for the power-frequency limitations on microwave power transistors.

First equation. Voltage-frequency limitation

$$V_m f_T = \frac{E_m v_s}{2\pi} = \begin{cases} 2 \times 10^{11} & \text{V/sec for silicon} \\ 1 \times 10^{11} & \text{V/sec for germanium} \end{cases} \qquad (3\text{-}1\text{-}45)$$

where $f_T = \dfrac{1}{2\pi\tau}$ is the charge-carrier transit-time cutoff frequency

$\tau = \dfrac{L}{v}$ is the average time for a charge carrier moving at an average velocity v to traverse the emitter-collector distance L

$V_m = E_m L_{min}$ is the maximum allowable applied voltage

$v_s =$ the maximum possible saturated drift velocity

$E_m =$ the maximum electric field

With the carriers moving at a velocity v_s of 6×10^6 cm/sec, transit time can be reduced even further by decreasing the distance L. The lower limit on L can be reached when the electric field becomes equal to the dielectric breakdown field. The present state-of-the-art of microwave transistor fabrication, however, limits emitter-collector length L to about 25 μm for overlay and matrix devices and nearly 250 μm for interdigitated devices [3]. Consequently, there is an upper limit on cutoff frequency. In practice, the attainable cutoff frequency is considerably less than the maximum possible frequency indicated by Eq. (3-1-45) because the saturated velocity v_s and electric field intensity are not uniform.

Second equation. Current-frequency limitation

$$(I_m X_c) f_T = \frac{E_m v_s}{2\pi} \tag{3-1-46}$$

where $I_m =$ the maximum current of the device

$X_c = \dfrac{1}{\omega_T C_0} = \dfrac{1}{2\pi f_T C_0}$ is the reactive impedance, and

$C_0 =$ the collector-base capacitance

Note that the relationship $2\pi f_T \tau_0 \simeq 2\pi f_T \tau = 1$ was used in deriving Eq. (3-1-46) from Eq. (3-1-45). In practice, no maximum current exists because the area of the device cannot be increased without bound. If the impedance level is zero, maximum current through a velocity-saturated sample might be infinite. However, the limited impedance does limit maximum current for a maximum attainable power.

Third equation. Power-frequency limitation

$$(P_m X_c)^{1/2} f_T = \frac{E_m v_s}{2\pi} \tag{3-1-47}$$

This equation is obtained by multiplying Eq. (3-1-45) by Eq. (3-1-46) and replacing $V_m I_m$ by P_m. It is very significant that, for a given device impedance, the power capacity of a device must be decreased as the device cutoff frequency is increased. For a given product of $E_m v_s$—that is, a given material—the maximum power that can be delivered to the carriers traversing the transistor is infinite if the cross section of the transistor can be made as large as possible. In other words, the value of the reactance X_c must approach zero. Thus Eq. (3-1-47) allows the results to be predicted. Figure 3-1-12 shows a graph of Eq. (3-1-47) and of the experimental results reported from the manufacturers [8].

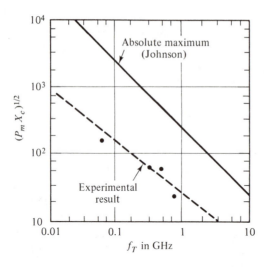

Figure 3-1-12 $(P_m X_c)^{1/2}$ versus f_T (After B. C. De Loach, Jr. [8]; reprinted by permission of the author and Academic Press.)

Fourth equation. Power gain-frequency limitation

$$(G_m V_{th} V_m)^{1/2} f = \frac{E_m v_s}{2\pi} \qquad (3\text{-}1\text{-}48)$$

where $G_m =$ the maximum available power gain,

$V_{th} = \dfrac{kT}{e}$ is the thermal equivalent voltage,

$k = 1.38 \times 10^{-23}$ J/°K   is the Boltzmann constant,

$T =$ the absolute temperature in degrees Kelvin,

$e = 1.60 \times 10^{-19}$ coulomb   and is the charge of electron

The maximum available power gain of a transistor was derived by Johnson [7] as

$$G_m = \left(\frac{f_T}{f}\right)^2 \frac{\mathbf{Z}_{out}}{\mathbf{Z}_{in}} \qquad (3\text{-}1\text{-}49)$$

where $\mathbf{Z}_{out}$ and $\mathbf{Z}_{in}$ are the output and input impedances, respectively.

If the electrode series resistances are assumed to be zero, then the ratio of the output impedance to the input impedance can be expressed as

$$\frac{\mathbf{Z}_o}{\mathbf{Z}_{in}} = \frac{C_{in}}{C_{out}} \qquad (3\text{-}1\text{-}50)$$

where C_{in} is input capacitance and C_{out} is output (base-collector) capacitance. When the maximum total carrier charges Q_m move to the collector in a carrier base transit time τ_b and with thermal voltage V_{th}, the input capacitance C_{in} and emitter diffusion capacitance C_d are related by

$$C_{in} = C_d \cong \frac{Q_m}{V_{th}} = \frac{I_m \tau_b}{V_{th}} \qquad (3\text{-}1\text{-}51)$$

Output capacitance is given by

$$C_{\text{out}} = \frac{I_m \tau_0}{V_m} \tag{3-1-52}$$

Substitution of Eqs. (3-1-45), (3-1-51), and (3-1-52) into Eq. (3-1-49) yields Eq. (3-1-48). The actual performance of a microwave transistor falls far short of that predicted by Eq. (3-1-48). At the present state-of-the-art the high-frequency limit of a 28-V silicon n-p-n transistor operating at the 1-W level is approximately 10 GHz. Typical power gains of microwave transistors lie in the 6 to 10 dB range.

Example 3-1-5   Power-Frequency Limitation

A certain Si microwave transistor has the following parameters:

$$X_c = 1 \ \Omega \qquad E_m = 2 \times 10^5 \ \text{V/cm}$$
$$f_T = 5 \ \text{GHz} \qquad v_s = 4 \times 10^5 \ \text{cm/s}$$

Determine the maximum allowable power that the transistor can carry.

Solution.   From Eq. (3-1-47) we have

$$P_m = \frac{1}{X_c f_T^2} \left(\frac{E_m v_s}{2\pi} \right)^2$$

$$= \frac{1}{1 \times (5 \times 10^9)^2} \left(\frac{2 \times 10^5 \times 4 \times 10^5}{2\pi} \right)^2$$

$$= 6.48 \ \text{W}$$

3-1-6 Bias Circuit Design

Microwave transistor-amplifier design requires biasing the transistor into the active region of performance and holding this bias (or quiescent point) constant over variations in temperature. However, the reverse leakage currents (I_{CBO} and I_{CEO}) will double for every 10°C temperature rise in a transistor. Also, at microwave frequencies the bypass capacitor becomes a problem because of RF instability. Most microwave circuit designs for best gain and low noise require that the emitter lead be dc grounded as close to the package as possible so that emitter series feedback is kept at an absolute minimum. There are two basic bias circuits for microwave transistor-amplifier design: voltage feedback and voltage feedback with constant base current (see Fig. 3-1-13).

The voltage feedback circuit shown in part (a) of Figure 3-1-13 uses fewer components and is almost temperature independent. The bias circuit (b) has a constant base current so that reverse leakage current I_{CEO} is maintained constant. Two inductors are used to impede current flow over a specified frequency range.

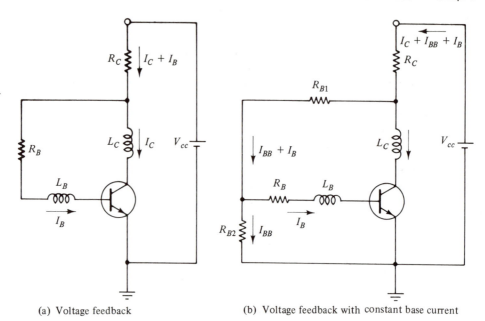

(a) Voltage feedback   (b) Voltage feedback with constant base current

Figure 3-1-13   Bias circuits for microwave transistor amplifier

Example 3-1-6   A certain microwave transistor has the following parameters:
$V_{CC} = 25$ V; $V_{BB} = 2.5$ V; $V_{BE} = 0.8$ V; $I_{BB} = 1$ mA; $I_{CBO} = 0$; $h_{FE} = 50$
Quiescent point (or operating bias point): $V_{CE} = 10$ V; $I_C = 10$ mA
To obtain thermal stabilization, it is necessary to design a bias circuit by using the voltage feedback or the voltage feedback with constant base current in Fig. 3-1-13.

(a) Determine the values of R_B and R_C for a voltage feedback bias circuit.
(b) Find the values of R_B, R_{B1}, R_{B2}, and R_C for a bias circuit of the voltage feedback with constant base current.

Solution.   (a)
$$R_B = \frac{V_{CB} - V_{BE}}{I_B} = \frac{V_{CE} - V_{BE}}{I_C/h_{FE}} = \frac{10 - 0.8}{0.2 \times 10^{-3}} = 46 \text{ k}\Omega$$

$$R_C = \frac{V_{CC} - V_{CE}}{I_C + I_B} = \frac{25 - 10}{(10 + 0.2) \times 10^{-3}} = 1.47 \text{ k}\Omega$$

(b)
$$R_B = \frac{V_{BB} - V_{BE}}{I_B} = \frac{2.5 - 0.8}{0.2 \times 10^{-3}} = 8.5 \text{ k}\Omega$$

$$R_{B1} = \frac{V_{CE} - V_{BB}}{I_{BB} + I_B} = \frac{10 - 2.5}{(1 + 0.2) \times 10^{-3}} = 6.25 \text{ k}\Omega$$

$$R_{B2} = \frac{V_{BB}}{I_{BB}} = \frac{2.5}{1 \times 10^{-3}} = 2.5 \text{ k}\Omega$$

$$R_C = \frac{V_{CC} - V_{CE}}{I_C + I_{BB} + I_B} = \frac{25 - 10}{(10 + 1 + 0.2) \times 10^{-3}} = 1.34 \text{ k}\Omega$$

3-2 *MICROWAVE TUNNEL DIODES*

The publication of Esaki's classic paper on tunnel diodes in January 1958 [9] quickly established the potential of tunnel diodes for microwave applications. Prior to 1958 the anomalous characteristic of some *p-n* junctions was reportedly observed by many solid-state scientists, but the irregularities were rejected immediately because they did not follow the "classic" diode equation. Esaki, however, described this anomalous phenomenon by using a quantum tunneling theory. The tunneling phenomenon is a majority carrier effect. The tunneling time of carriers through the potential energy barrier is not governed by the classic transit time concept—that the transit time is equal to the barrier width divided by the carrier velocity—but rather by the quantum transition probability per unit time. Tunnel diodes are very useful in many circuit applications in microwave amplification, microwave oscillation, and binary memory because of their low cost, light weight, high speed, low noise, high efficiency, frequency stability, low-power operation, and high-peak current-to-valley-current ratio. Yet because the dynamic negative resistance of a tunnel diode is exhibited at a very low voltage range of 0.1 to 0.3 V, its power output is limited to the low milliwatt range, with a frequency of operation up to about 8 GHz.

3-2-1 *Principles of Operation*

The tunnel diode is a negative-resistance semiconductor *p-n* junction diode. The negative resistance is created by the tunnel effect of electrons in the *p-n* junction. The doping of both the *p* and *n* regions of the tunnel diode is very high—impurity concentrations of 10^{19} to 10^{20} atoms/cm^3 being used—and the depletion-layer barrier at the junction is very thin, on the order of 100 Å or 10^{-6} cm. Classically, it is only possible for those particles to pass over the barrier if and only if they have an energy equal to or greater than the height of the potential barrier. Quantum mechanically, however, if the barrier is less than 3 Å, there is an appreciable probability that particles will tunnel through the potential barrier even though they do not have enough kinetic energy to pass over the same barrier. In addition to barrier thinness, filled energy states are also needed on the side from which particles will tunnel, plus allowed empty states on the other side into which particles penetrate at the same energy level. To understand tunnel effects fully, let us analyze the energy band pictures of a heavily doped *p-n* diode. Figure 3-2-1 shows the energy band diagrams of a tunnel diode.

Under open-circuit conditions or at zero-bias equilibrium the upper levels of electron energy of both *p* type and *n* type are lined up at the same Fermi level as shown in Fig. 3-2-1(a). Because no filled states on one side of the junction are at the same energy level as empty allowed states on the other side, there is no flow of charge in either direction across the junction and the current is zero, as shown at point (a) of the volt-ampere characteristic curve of a tunnel diode in Fig. 3-2-2.

In ordinary diodes the Fermi level exists in the forbidden band. Because the tunnel diode is heavily doped, the Fermi level exists in the valence band in *p*-type

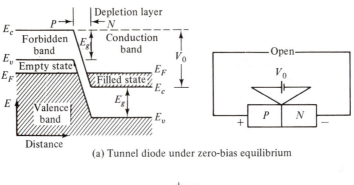

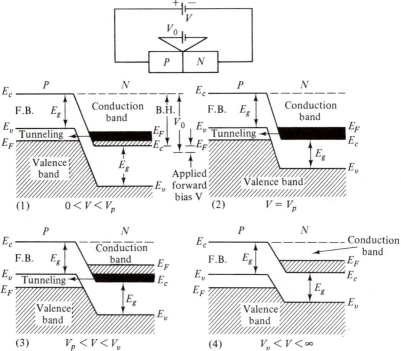

(a) Tunnel diode under zero-bias equilibrium

(b) Tunnel diode with applied forward bias

E_F is the Fermi level representing the energy state with 50%
   probability of being filled if no forbidden band exists
V_0 is the potential barrier of the junction
E_g is the energy required to break a covalent bond, which is
   0.72 eV for germanium and 1.10 eV for silicon
E_c is the lowest energy in the conduction band
E_v is the maximum energy in the valence band
V is the applied forward bias
F.B. stands for the forbidden band
B.H. represents the barrier height

Figure 3-2-1 Energy-band diagrams of tunnel diode

and in the conduction band in *n*-type semiconductors. When the tunnel diode is forward-biased by a voltage between zero and the value that would produce peak tunneling current I_p ($0 < V < V_p$), the energy diagram is as shown in sector (1) of Fig. 3-2-1(b). Accordingly, the potential barrier is decreased by the magnitude of the applied forward-bias voltage. A difference of Fermi levels in both sides is created. Because there are filled states in the conduction band of the *n* type at the same energy level as allowed empty states in the valence band of the *p* type, the electrons tunnel through the barrier from the *n* type to the *p* type, giving rise to a forward tunneling current from the *p* type to the *n* type, as shown in sector (1) of Fig. 3-2-2(a). As the forward bias is increased to V_p, the picture of the energy band is as shown in sector (2) of Fig. 3-2-1(b). A maximum number of electrons can tunnel through the barrier from the filled states in the *n* type to the empty states in the *p* type, producing the peak current I_p in Fig. 3-2-2(a). If bias voltage is increased further, the condition shown in sector (3) of Fig. 3-2-1(b) is reached. The tunneling current decreases as shown in sector (3) of Fig. 3-2-2(a). Finally, at very large bias voltage the band structure of sector (4) in Fig. 3-2-1(b) is obtained. Because there

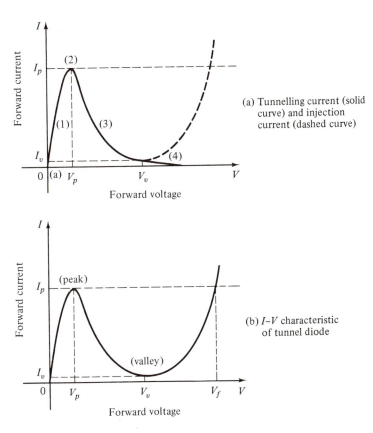

(a) Tunnelling current (solid curve) and injection current (dashed curve)

(b) *I–V* characteristic of tunnel diode

Figure 3-2-2   Voltage-current characteristics of a tunnel diode

are now no allowed empty states in the *p* type at the same energy level as filled states in the *n* type, no electrons can tunnel through the barrier so the tunneling current drops to zero as shown at point (4) of Fig. 3-2-2(a).

When forward-bias voltage V is increased above the valley voltage V_v, the ordinary injection current I at the *p-n* junction starts to flow. This injection current is increased exponentially with the forward voltage, as indicated by the dashed curve of Fig. 3-2-2(a). The total current, given by the sum of the tunneling current and the injection current, results in the volt-ampere characteristic of the tunnel diode shown in Fig. 3-2-2(b). It can be seen from the figure that total current reaches a minimum value I_v (or valley current), somewhere in the region where the tunnel diode characteristic meets the ordinary *p-n* diode characteristic. The ratio of peak current to valley current (I_p/I_v) can reach 50 to 100 (theoretically). In practice, this ratio is about 15.

3-2-2   Microwave Characteristics

The tunnel diode is very useful in microwave oscillators and amplifiers because this diode exhibits a negative-resistance characteristic in the region between peak current I_p and valley current I_v. The *V-I* characteristic of a tunnel diode with the load line is shown in Fig. 3-2-3.

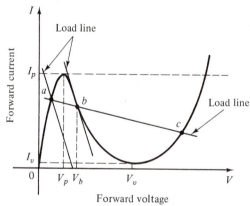

Figure 3-2-3   *V-I* characteristic of a tunnel diode with load lines

The *abc* load line intersects the characteristic curve in three points. Points *a* and *c* are stable points; point *b* is unstable. If the voltage and current vary about *b*, the final value of I and V would be given by point *a* or *c* but not by *b*. Because the tunnel diode has two stable states for this load line, the circuit is called *bistable* and can be used as a binary device in switching circuits. However, microwave oscillation or amplification generated by the tunnel diode is the major concern in this section. The second load line intersects the *V-I* curve at point *b* only. This point is stable and shows a dynamic negative conductance that enables the tunnel diode to function as a microwave amplifier or oscillator. The circuit with a load line crossing point *b* in the negative-resistance region is called *astable*. Another load line crossing point *a* in the positive-resistance region indicates a *monostable* circuit. The negative conductance in Fig. 3-2-3 is given by

$$-g = \frac{\partial i}{\partial v}\bigg|_{V_b} = \frac{1}{-R_n} \tag{3-2-1}$$

where R_n is the magnitude of negative resistance.

The negative resistance is constant and the diode circuit behavior is stable for a small variation of the forward voltage about V_b. A small-signal equivalent circuit for the tunnel diode operated in the negative-resistance region is shown in Fig. 3-2-4.

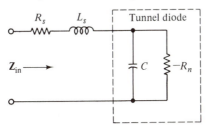

Figure 3-2-4   Equivalent circuit of a tunnel diode

Here R_s and L_s denote the inductance and resistance of the packaging circuit of a tunnel diode. Junction capacitance C of this diode is usually measured at the valley point. R_n is the negative resistance of the diode. Typical values of these parameters for a tunnel diode with a peak current I_p of 10 mA are

$$-R_n = -30\ \Omega, \quad R_s = 1\ \Omega, \quad L_s = 5\ \text{nH}, \quad \text{and} \quad C = 20\ \text{pF}$$

The input impedance $\mathbf{Z}_{\text{in}}$ of the equivalent circuit as shown in Fig. 3-2-4 is given by

$$\mathbf{Z}_{\text{in}} = R_s + j\omega L_s + \frac{R_n[j/(\omega C)]}{-R_n - j[1/(\omega C)]}$$

$$= R_s - \frac{R_n}{1+(\omega R_n C)^2} + j\left[\omega L_s - \frac{\omega R_n^2 C}{1+(\omega R_n C)^2}\right] \tag{3-2-2}$$

The real part of the input impedance $\mathbf{Z}_{\text{in}}$ must be zero for the resistive cutoff frequency. Consequently, the resistive cutoff frequency is given from Eq. (3-2-2) by

$$f_c = \frac{1}{2\pi R_n C}\sqrt{\frac{R_n}{R_s} - 1} \tag{3-2-3}$$

The imaginary part of the input impedance must be zero for the self-resonance frequency. Thus

$$f_r = \frac{1}{2\pi R_n C}\sqrt{\frac{R_n^2 C}{L_s} - 1} \tag{3-2-4}$$

The tunnel diode can be connected either in parallel or in series with a resistive load as an amplifier; and its equivalent circuits are shown in Fig. 3-2-5.

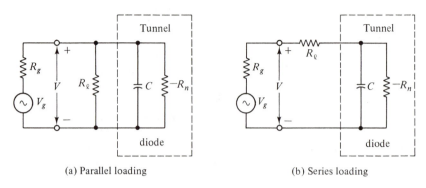

(a) Parallel loading   (b) Series loading

Figure 3-2-5   Equivalent circuit of tunnel diode with a parallel or series load

(a) Parallel loading.   From Fig. 3-2-5(a) it can be seen that the output power in the load resistance is given by

$$P_{\text{out}} = \frac{V^2}{R_\ell} \qquad (3\text{-}2\text{-}5)$$

One part of this output power is generated by the small input power through the tunnel-diode amplifier with a gain of A and this part can be expressed as

$$P_{\text{in}} = \frac{V^2}{AR_\ell} \qquad (3\text{-}2\text{-}6)$$

Another part of the output power is generated by the negative resistance and is expressed as

$$P_n = \frac{V^2}{R_n} \qquad (3\text{-}2\text{-}7)$$

Therefore

$$\frac{V^2}{AR_\ell} + \frac{V^2}{R_n} = \frac{V^2}{R_\ell} \qquad (3\text{-}2\text{-}8)$$

and the gain equation of a tunnel diode amplifier is given by

$$G = \frac{R_n}{R_n - R_\ell} \qquad (3\text{-}2\text{-}9)$$

When negative resistance R_n of the tunnel diode approaches load resistance R_ℓ, gain G approaches infinity, and the system goes into oscillation.

(b) Series loading.   In the series circuit shown in Fig. 3-2-5(b) power gain G is given by

$$G = \frac{R_\ell}{R_\ell - R_n} = \frac{1}{1 - R_n/R_\ell} \qquad (3\text{-}2\text{-}10)$$

The device remains stable in the negative-resistance region without switching if $R_\ell < R_n$.

A tunnel diode can be connected to a microwave circulator to make a negative-resistance amplifier (see Fig. 3-2-6). A microwave circulator is a multiport junction in which power can flow only from port 1 to port 2, port 2 to port 3, and so on in the direction shown. Although the number of ports is restricted, microwave circulators with four ports are most common. If the circulator is perfect and has a positive real characteristic impedance R_0, an amplifier with infinite gain can be built by selecting a negative-resistance tunnel diode whose input impedance has a real part equal to $-R_0$ and an imaginary part equal to zero. The reflection coefficient from Fig. 3-2-5 is infinite. In general, the reflection coefficient is given by

$$\Gamma = \frac{-R_n - R_0}{-R_n + R_0} \qquad (3\text{-}2\text{-}11)$$

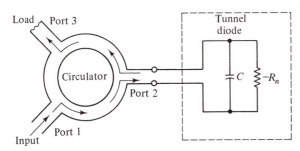

Figure 3-2-6 Tunnel diode connected to a circulator

Example 3-2-2   Amplification of a Circulator-Tunnel-Diode Amplifier

A certain tunnel diode has a negative resistance R_n of 25 ohms and is connected to port 2 of a perfect three-port circulator. The circulator has a positive real characteristic resistance R_0 of 20 ohms. Input signal power of 1 mW is fed into port 1, and the output is taken from port 3.

(a) Determine the output power in milliwatts (mW) and decibels referred to one milliwatt (dBm).

(b) Find the power gain in decibels.

Solution.   (a) From Eq. (3-2-11) the reflection coefficient is

$$\Gamma = \frac{-25 - 20}{-25 + 20} = 9$$

The output power is

$$P = P_{\text{in}}|\Gamma|^2 = 10^{-3} \times 81 = 81 \text{ mW} = 19 \text{ dBm}$$

(b) The power gain is

$$\text{Gain} = 19 \text{ dB}$$

3-2-3   *Power-Frequency Limitations*

Because the operational principle of a tunnel diode is quantum mechanism which allows electrons to tunnel through potential barriers in order to exhibit a negative resistance, there are severe limitations on both the power output and the oscillation frequency. Negative resistance occurs only in the forward-bias direction at a voltage intermediate between zero volt and the diode's cut-in voltage. It is apparent that the maximum voltage that can be applied to a tunnel diode in a negative-resistance region is less than the bandgap voltage. For Si, Ge, GaAs, and GaSb, bandgap voltage is 1.12, 0.80, 1.43, and 0.67 V at room temperature, respectively. Consequently, tunnel diodes have low-power output.

3-3 MICROWAVE METAL-SEMICONDUCTOR (MES) DIODES

A simple metal-semiconductor (MES) junction exhibits a Schottky-barrier effect that is often useful in diodes operating at microwave frequencies. In 1938 Schottky proposed that this potential barrier could arise from stable space charges in the semiconductor alone without the presence of a chemical layer [11]. The model derived from his theory is known as the *Schottky barrier*. Thus metal-semiconductor (MES) diodes are often referred to as *Schottky-barrier diodes*, or hot-carrier diodes, or surface-carrier diodes.

Early types of metal-semiconductor devices were point-contact diodes which were used as radar mixers during World War II. A point-contact diode is usually fabricated by pressing a finely pointed tungsten wire onto a piece of *p*-type silicon. These devices were used in most microwave mixer and detector applications in military and space microwave systems until the late 1960s. In addition, metal semiconductor diodes are particularly useful as detector, modulator, high-speed microwave rectifier, and high-speed claming and switching devices.

The metal-semiconductor diode is a fundamental component of the metal-semiconductor field-effect transistor (MESFET). In this section microwave metal semiconductor diodes are described.

3-3-1 Physical Structures

Any metal or semiconductor has a specific work function for its electron movement. The work function Φ_m (in eV) of a metal or Φ_s (in eV) of a semiconductor is the minimum energy required for an electron to escape from the metal or semiconductor at the Fermi level into a vacuum at rest. Figure 3-3-1 shows the metal work functions for clean metal surfaces in vacuum versus their atomic numbers [12].

Einstein equation.   When a light beam is incident on metal, each photon in the incident beam is assumed to give up all its energy $h\nu$ to an electron. In order for this electron to get out of the metal, it must expend some energy against the potential barrier of the metal. This energy is called the *work function*. The Einstein photoelectric equation can be written

$$h\nu = \Phi_m + \tfrac{1}{2}mv_m^2 \tag{3-3-0}$$

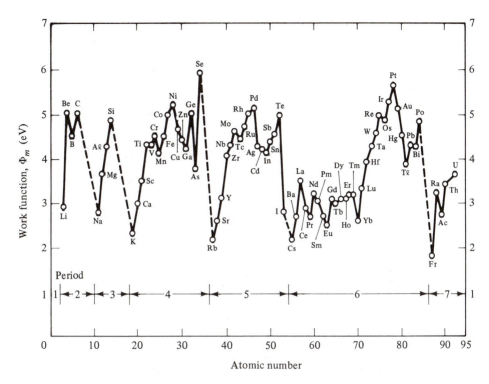

Figure 3-3-1 Work functions of metals. (After S. M. Sze [12]; reprinted by permission of John Wiley & Sons, Ltd.)

where $h\nu$ = photon energy in eV (electron volts)
$h = 6.626 \times 10^{-34}$ J-s is Planck's constant
ν = frequency in Hz
Φ_m = work function of the metal in eV
$\frac{1}{2}mv_m^2$ = kinetic energy of electron in eV
$m = 9.11 \times 10^{-31}$ kg is the mass of electron
v_m = maximum velocity of electron in m/sec

Most photoelectrons give up more energy than Φ_m in escaping from the metal surface by undergoing collisions within the metal. The value of work function Φ_m is least in the electropositive metals, such as sodium, potassium, rubidium, and cesium, as shown in Fig. 3-3-1. Consequently, these metals are generally used in photoelectric cells. From the Einstein equation it is seen that if Φ_m is larger than $h\nu$, then no photoelectrons would be emitted. This is the case with zinc and visible light. The photoelectric effect is an important means by which X rays and γ rays lose energy in passing matter.

Example 3-3-1   Einstein Equation

A violet light of 0.35 μm is incident on a piece of rubidium with a work function of 2.2 eV.

(a) Determine the kinetic energy in eV of the ejected electron.

(b) Compute the maximum velocity of the ejected electron.

Solution.   (a) From Eq. (3-3-0) the electron kinetic energy is

$$\frac{1}{2}mv_m^2 = \frac{6.626 \times 10^{-34}}{1.6 \times 10^{-19}} \times \frac{3 \times 10^8}{0.35 \times 10^{-6}} - 2.2 = 1.35 \quad \text{eV}$$

(b) The maximum velocity of the ejected electron is

$$v_m = \left(\frac{2 \times 2.16 \times 10^{-19}}{9.11 \times 10^{-31}}\right)^{1/2} = 6.89 \times 10^5 \quad \text{m/s}$$

In general, metal-semiconductor junction diodes consist of four types, depending on the differences of their work functions:

1. Metal and n-type semiconductor junction with a positive work function difference between the metal and semiconductor—that is, $\Phi_m > \Phi_{sn}$.

2. Metal and p-type semiconductor junction with a negative work function difference between the metal and semiconductor—that is, $\Phi_m < \Phi_{sp}$.

3. Metal and n-type semiconductor contact with a negative work function difference between the metal and semiconductor—that is, $\Phi_m < \Phi_{sn}$.

4. Metal and p-type semiconductor contact with a positive work function difference between the metal and semiconductor—that is, $\Phi_m > \Phi_{sp}$.

Figure 3-3-2 shows the four types of metal-semiconductor junction structures. Various metal-semiconductor junction diodes as used in practice are shown in Fig. 3-3-3 [12].

3-3-2 Principles of Operation

The principles of operation for the four types of metal-semiconductor junction diodes are similar, depending on the work function differences between them and the bias-voltage polarities across them.

Schottky-barrier height.   When a metal with a greater work function Φ_m comes in contact with an n-type semiconductor with a smaller work function Φ_{sn} (that is, $\Phi_m > \Phi_{sn}$), as shown in Type I of Fig. 3-3-2, the electron energy level in the semiconductor must be lowered in order to align the two Fermi levels at equilibrium. In other words, the electrons must diffuse from the semiconductor conduction band into the metal. A depletion region W is then formed near the semiconductor side of

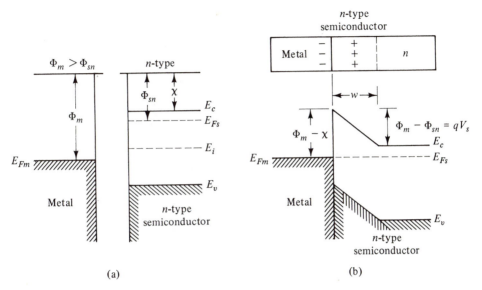

Type I: $\Phi_m > \Phi_{sn}$

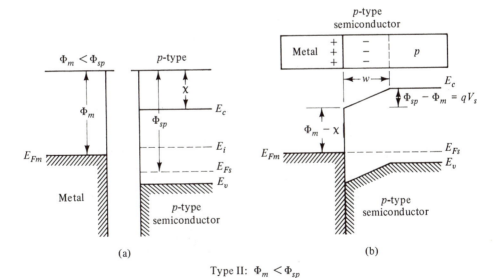

Type II: $\Phi_m < \Phi_{sp}$

Figure 3-3-2 Schottky-barrier diodes: physical structures with energy band diagrams

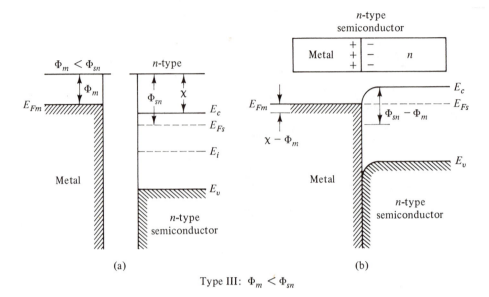

Type III: $\Phi_m < \Phi_{sn}$

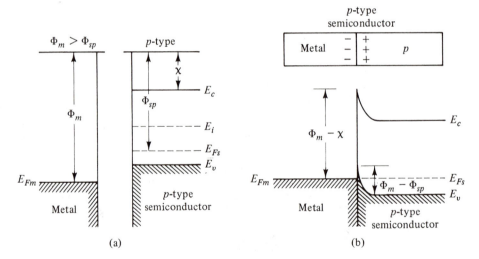

Type IV: $\Phi_m > \Phi_{sp}$

Figure 3-3-2 (continued) Schottky-barrier diodes

this junction. The positive charges due to uncompensated donor ions within the depletion region match the negative charges in the metal. The *electron affinity* χ (in eV) of a semiconductor is an energy measure from its conduction energy band edge to the vacuum level. The difference between the work function Φ_m of the metal and the electron affinity χ of the semiconductor determines the Schottky-barrier height. So the Schottky-barrier height for a metal-n-type-semiconductor junction diode is

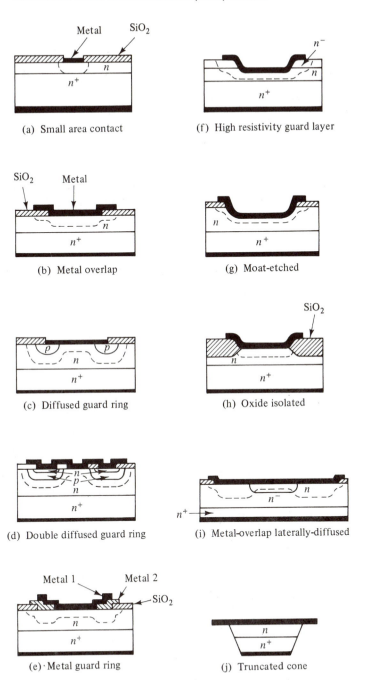

Figure 3-3-3 Various structures of Schottky-barrier diodes. (After S. M. Sze [12]; reprinted by permission of John Wiley & Sons, Ltd.)

given by

$$\Phi_{Bn} = \Phi_m - \chi \quad \text{eV} \tag{3-3-1}$$

where $\chi = \Phi_s + E_F - E_g$ is the semiconductor electron affinity

Φ_s = semiconductor work function

E_F = Fermi energy level

E_g = bandgap energy

Table 3-3-1 tabulates Schottky barrier heights for metals on semiconductors [12]. The Schottky-barrier height in the semiconductor itself is expressed as

$$qV_B = \Phi_m - \Phi_{sn} \quad \text{eV} \tag{3-3-2}$$

The potential difference V_B prevents further electron diffusion from the semiconduc-

TABLE 3-3-1   SCHOTTKY-BARRIER HEIGHTS (VOLTS AT 300 K).

Semiconductor	Type	E_g (eV)	Ag	Al	Au	Cr	Cu	Hf	In	Mg	Mo	Ni	Pb	Pd	Pt	Ta	Ti	W
Diamond	p	5.47			1.71													
Ge	n	0.80	0.54	0.48	0.59		0.52		0.64			0.49	0.38					0.48
Ge	p		0.50		0.30				0.55									
Si	n	1.12	0.69		0.81	0.61	0.69	0.58		0.40	0.68	0.66	0.60	0.81	0.90		0.50	0.67
Si	p		0.53		0.57	0.34	0.50	0.46	0.63			0.42	0.50	0.54			0.61	0.45
SiC	n	2.90		2.00	1.95													
AlAs	n	2.16			1.20										1.00			
AlSb	p	1.63			0.55													
BN	p	7.50			3.10													
BP	p	6.00			0.87													
GaSb	n	0.67			0.60													
GaAs	n	1.43	0.88	0.80	0.90		0.82	0.72							0.84	0.85		0.80
GaAs	p		0.63		0.42			0.68										
GaP	n	2.26	1.20	1.07	1.30	1.06	1.20	1.84		1.04	1.13	1.27			1.45		1.12	
GaP	p				0.72													
InSb	n	0.18	0.18		0.17													
InAs	p	0.33			0.47													
InP	n	1.29	0.54		0.52													
InP	p				0.76													
CdS	n	2.43	0.56	Ohmic	0.78		0.50					0.45	0.59	0.62	1.10			0.84
CdSe	n	1.70	0.43		0.49		0.33								0.37			
CdTe	n		0.81	0.76	0.71										0.76			
ZnO	n	3.20		0.68	0.65		0.45		0.30						0.68	0.75	0.30	
ZnS	n	3.60	1.65	0.80	2.00		1.75		1.50	0.82					1.87	1.84	1.10	
ZnSe	n		1.21	0.76	1.36		1.10		0.91					1.16	1.40			
PbO	n		0.95						0.93					0.96	0.95			

(From S. M. Sze [12]; reprinted by permission of John Wiley & Sons, Ltd.)

tor into the metal. The potential difference V_B can be increased or decreased, however, by applying either a reverse-bias or a forward-bias voltage.

For ideal contact between a metal and a p-type semiconductor, the Schottky-barrier height is given by

$$\Phi_{Bp} = E_g - (\Phi_m - \chi) = E_g - \Phi_{Bn} \quad eV \qquad (3\text{-}3\text{-}3)$$

where E_g is the bandgap energy (in eV) of the semiconductor.

For a given semiconductor and for any metal, the sum of the Schottky-barrier heights on n-type and p-type substrates is thus expected to be equal to the gap energy:

$$\Phi_{Bn} + \Phi_{Bp} = E_g \quad eV \qquad (3\text{-}3\text{-}4)$$

Table 3-3-2 lists the Schottky-barrier heights for seven metals on n-type or p-type silicon that has an electron affinity energy of 4.01 eV.

TABLE 3-3-2   SCHOTTKY-BARRIER HEIGHTS ON n OR p SILICON.

Metal	Work function Φ_m (eV)	Barrier height in eV		Bandgap energy E_g (eV)
		Φ_{Bn}	Φ_{Bp}	
Ag	4.70	0.69	0.53	1.22
Al	4.69	0.68	0.57	1.25
Au	4.82	0.81	0.34	1.15
Cr	4.62	0.61	0.50	1.11
Cu	4.70	0.69	0.46	1.15
Ni	4.67	0.66	0.50	1.16
Pb	4.61	0.60	0.54	1.14

Example 3-3-2   Schottky-barrier Heights of a GaAs MES Diode

The Schottky-barrier height for a metal-semiconductor junction diode is a function of the metal work function and the electron affinity of the semiconductor.

(a) Determine the Schottky-barrier heights for Au(n-GaAs) and Au(p-GaAs) junction diodes.

(b) Find the gap energy of GaAs.

Solution.   (a) From Appendix 6 the electron affinity of GaAs is 4.07, and from Table 3-3-2 the work function of Au is 4.82. Then from Eq. (3-3-1) Schottky-barrier heights are

$$\Phi_{Bn} = \Phi_m - \chi_n = 4.82 - 4.07 = 0.75 \quad eV$$

and

$$\Phi_{Bp} = \Phi_m - \chi_p = 4.82 - 4.07 = 0.75 \quad eV$$

(b) From Eq. (3-3-4) the gap energy of GaAs is

$$E_g = 0.75 + 0.75 = 1.50 \quad \text{eV}$$

It should be carefully noted that the values of the work functions and the electron affinities sometimes vary for different sources because of the temperatures and the doping materials.

Schottky-barrier effect.   The Schottky-barrier effect results from the induced-image-potential lowering when an electric field is applied across the metal-semiconductor junction. Consider a simple metal-vacuum interface. When an electron is at distance X from the metal surface, an image positive charge is induced inside the metal at a distance $-X$ from the metal surface. This is shown in Fig. 3-3-4(a). According to Coulomb's law of electric charge force, the attractive image force between the two charges is given by

$$F = -\frac{q^2}{4\pi\epsilon_0(2X)^2} = -\frac{q^2}{16\pi\epsilon_0 X^2} \quad \text{newtons} \tag{3-3-5}$$

where $\epsilon_0 = 8.854 \times 10^{-12}$ F/m is the permittivity of the vacuum or free space
$\qquad q = 1.6 \times 10^{-19}$ coulomb is the magnitude of charge

The electric field created by these two charges is

$$E = \frac{F}{q} = -\frac{q}{16\pi\epsilon_0 X^2} \quad \text{V/m} \tag{3-3-6}$$

Then the voltage due to the two charges may be found by integrating the field equation Eq. (3-3-6) from infinity to X.

$$V_B = -\int_\infty^x E \, dx = \frac{q}{16\pi\epsilon_0 X} \quad \text{volts} \tag{3-3-7}$$

The potential work done by the electron for distance X is

$$W(x) = qV_B = \frac{q^2}{16\pi\epsilon_0 X} \quad \text{eV} \tag{3-3-8}$$

This image potential energy is shown in Fig. 3-3-4(b).

When an external electric field E is applied across the interface, the total potential energy is given by

$$W(x)_{\text{total}} = \frac{q^2}{16\pi\epsilon_0 X} + qEX \quad \text{eV} \tag{3-3-9}$$

Total potential energy is measured downward from the X axis as shown in Fig. 3-3-4(c).

The Schottky-barrier effect ΔV_B (also called Schottky-barrier lowering) can be determined by differentiating Eq. (3-3-9) and equating the result to zero. Then the location of the lowering X_m is

$$X_m = \left(\frac{q}{16\pi\epsilon_0 E}\right)^{1/2} \quad \text{meters} \tag{3-3-10}$$

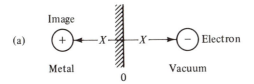

(a)

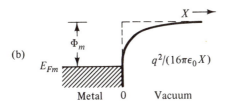

(b)

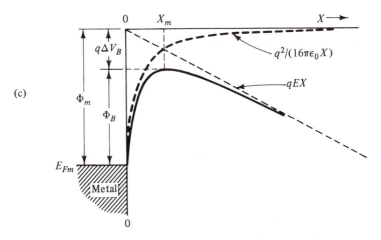

(c)

Figure 3-3-4 Schottky-barrier effect (After S. M. Sze [5]; reprinted by permission of John Wiley & Sons, Ltd.)

and the potential lowering ΔV_B is

$$\Delta V_B = \left(\frac{q\mathrm{E}}{4\pi\epsilon_0} \right)^{1/2} = 2\mathrm{E}X_m \quad \text{volts} \qquad (3\text{-}3\text{-}11)$$

Both X_m and ΔV_B are shown in Fig. 3-3-4(c).

In conclusion, the Schottky-barrier height ($\Phi_B = \Phi_m - q\Delta V_B$) is considerably lowered at high electric fields, and the effective metal work function for thermionic emission is then reduced greatly as shown in Fig. 3-3-4(c). This result can apply to the metal-semiconductor (MES) diodes.

Example 3-3-2A   Schottky-barrier Potential Lowering

If the applied electric field across a gold-silicon (n-type) MES diode is 10^4 V/m,
(a) determine the Schottky-barrier lowering potential ΔV_B, and
(b) find the location of the lowering X_m.

Solution.   (a) From Eq. (3-3-11) the potential lowering is

$$\Delta V_B = \left(\frac{q\mathrm{E}}{4\pi\epsilon_s} \right)^{1/2}$$

$$= \left(\frac{1.6 \times 10^{-19} \times 10^4}{4\pi \times 8.854 \times 10^{-12} \times 11.8} \right)^{1/2} = 1.10 \quad \text{mV}$$

(b) Then the location of the lowering is

$$X_m = \frac{\Delta V_B}{2E} = \frac{1.10 \times 10^{-3}}{2 \times 10^4} = 550 \text{ Å}$$

3-3-3 Current-Voltage Characteristics

The current-voltage equation for a MES diode is similar in form to the one for a p-n junction diode and it can be written

$$I = I_r \left[\exp\left(\frac{qV}{\eta kT} \right) - 1 \right] \tag{3-3-12}$$

where I_r = reverse saturation current
V = applied voltage
$k = 1.38 \times 10^{-23}$ W-s/°K is the Boltzmann constant
T = absolute temperature in degrees Kelvin
η = ideality factor between 1 and 2 for a specific material. $\eta = 1.02$ for the W-Si diode and $\eta = 1.04$ for the W-GaAs diode.

The reverse saturation current under the reverse bias condition for a metal-semiconductor junction diode is determined by thermionic emission over the Schottky-barrier height Φ_B and it is expressed as

$$I_r = AA^*T^2 \exp\left(\frac{-\Phi_B}{kT} \right) \tag{3-3-13}$$

where A =   junction area in centimeters squared
A^* =   120A/ cm^2/°K^2 is the effective Richardson constant for n-type Si and 30A/cm^2/°K^2 for p-type Si
$\Phi_B = \Phi_m - \chi$ is the Schottky-barrier height as defined by Eq. (3-3-1)

When Schottky-barrier lowering is considered, Eq. (3-3-13) is expressed as

$$I_r = AA^*T^2 \exp\left(\frac{-\Phi_B}{kT}\right) \exp\left(\frac{q\,\Delta V_B}{kT}\right) \qquad (3\text{-}3\text{-}14)$$

In conclusion, the current-voltage equation for a metal-semiconductor junction diode is given by

$$I = AA^*T^2 \exp\left(\frac{-\Phi_B}{kT}\right) \exp\left(\frac{q\,\Delta V_B}{kT}\right)\left[\exp\left(\frac{qV}{\eta kT}\right) - 1\right] \qquad (3\text{-}3\text{-}15)$$

Example 3-3-3   Current of a Metal-Semiconductor Diode

A metal-semiconductor diode is formed by silver and an n-type silicon semiconductor. The diode cross section is 3×10^{-4} cm^2 and operates at room temperature.

(a) Determine the reverse saturation current in nanoamperes without considering the potential lowering effect.

(b) Calculate the diode current for a forward voltage of 0.45 V.

Solution.   (a) From Eq. (3-3-13) the reverse saturation current is

$$I_r = AA^*T^2 \exp\left(\frac{-\Phi_B}{kT}\right)$$

$$= 3 \times 10^{-4} \times 120 \times (300)^2 \exp\left(\frac{-0.69}{26 \times 10^{-3}}\right)$$

$$= 9.40 \text{ nA}$$

(b) From Eq. (3-3-12) the diode current is

$$I = I_r\left[\exp\left(\frac{qV}{\eta kT}\right) - 1\right]$$

$$\approx 9.4 \times 10^{-9} \exp\left(\frac{0.45}{1 \times 26 \times 10^{-3}}\right)$$

$$= 9.4 \times 10^{-9} \times 3.26 \times 10^7$$

$$= 306 \quad \text{mA}$$

3-3-4 Microwave Applications

Metal-semiconductor diodes have many applications in microwave circuits at microwave frequencies. When these diodes are forward-biased, the current consists entirely of majority carriers and minority-carrier effects are virtually eliminated. When they are reverse-biased, the current is extremely low. Because these diodes have essentially no minority-carrier storage and a minimum capacitance, they are particularly useful for high-speed logic gates, such as microwave rectifiers or down-converters.

The saturation time constant of the MES diode is very small. In microwave circuit design a MES diode is often clamped between the base and collector terminals of a transistor. (See Fig. 3-3-5.) The saturation time of the circuit is

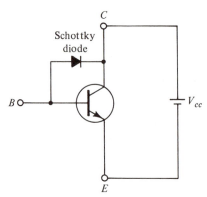

Figure 3-3-5 A Schottky diode incorporated with a bipolar transistor

reduced markedly compared with that of the original transistor because excessive base current flows through the MES diode to the collector terminal and eliminates storage of the minority carriers. As a result, the response time of a high-speed switching circuit incorporated with a MES diode is extremely fast at microwave frequencies.

Because Schottky-barrier height Φ_B is much less than the energy gap of the semiconductor, the voltage that will draw a given current is reduced accordingly. The knee voltage of its current-voltage curve is much lower than the ordinary *p-n* junction diode, and the turn-on voltage is very low. When MES diodes are in ohmic contact form, their contact resistance is low, so their ohmic losses and noise figures are low also.

Rectifying contacts.   The early version of metal-semiconductor diodes was the point-contact form, which used a small, sharp-pointed metal wire in contact with a semiconductor. Its current-voltage characteristics were nonlinear and unpredictable. Modern MES diodes are structured in planar form by depositing an appropriate metal film on a semiconductor surface. When a forward-bias voltage V is applied across the MES diode, contact potential is reduced from V_B to $V_B - V$. Thus the forward current flows through the diode. Conversely, a reverse-bias voltage V_r increases the contact potential from V_B to $V_B + V_r$ and eliminates the current flow, so the MES diodes are often used as rectifiers in microwave circuits.

Ohmic contacts.   The ohmic contact between a metal and semiconductor is defined as the contact that has a negligible contact resistance compared with the bulk resistance of a semiconductor. This type of contact can be formed by doping the semiconductor heavily in the contact region. As a result, the semiconductor has no depletion region of the rectifying contact type. The physical structures and energy band diagrams of the ohmic-contact MES diodes are shown in Types III and IV of Fig. 3-3-2. The ohmic contact has minimal resistance, which means that it is not useful for signal rectification. It is often used to interconnect *p* and *n* regions in microwave integrated circuits, however.

3-4 MICROWAVE METAL-INSULATOR-SEMICONDUCTOR (MIS) DIODES

The metal-insulator-semiconductor (MIS) diode is a fundamental component of many microwave solid-state devices, such as MOSFETs, VMOSFETs, microprocessors, and charge-coupled devices (CCDs). The MIS diode structure was first proposed in 1959 by Moll [14] as a voltage-variable capacitor. In 1962 Terman [15] constructed the first MIS diode for investigation of the surface charge effect. MIS diode technology advanced rapidly once microwave integrated circuits were developed. Of all MIS diodes the metal-SiO₂-Si (MOS) diode is by far the most important one.

3-4-1 Physical Structure

The performance of a metal-insulator-semiconductor (MIS) diode generally depends on the surface effects of the three materials and the surface effects are based on the applied voltage across them. Figure 3-4-1 is a schematic diagram of a MIS diode.

The metal material may be Al, the insulator may be SiO_2 or Al_2O_3 with a thickness d of about 0.1 μm, and the semiconductor may be either the n type or p type of Si or GaAs. The three materials form a typical capacitor structure.

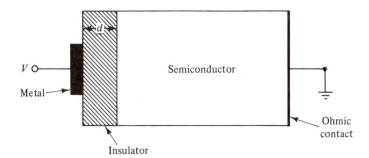

Figure 3-4-1 Schematic diagram of a MIS diode

3-4-2 Electronic Mechanism

The electronic mechanism of a MIS diode is perhaps best explained by means of energy-band theory. Depending on the applied voltage (the gate voltage for the MOSFET), the surface effects of an ideal MIS diode can be classified as four kinds for p-type semiconductors: equilibrium, accumulation, depletion, and inversion.

1. Equilibrium Case at Zero Applied Voltage. For an ideal MIS diode, the metal work function Φ_m and the semiconductor work function Φ_s are equal

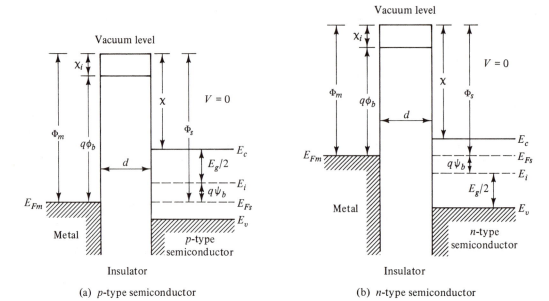

Figure 3-4-2   Energy-band diagrams of ideal MIS diodes with zero applied voltage (flat band)

(see Fig. 3-4-2). The relationships of the two work functions can be written

$$\Phi_{ms} = \Phi_s - \left(\chi + \frac{E_g}{2} + q\psi_b\right) = 0 \quad \text{for } p \text{ type} \tag{3-4-1}$$

$$\Phi_{ms} = \Phi_s - \left(\chi + \frac{E_g}{2} - q\psi_b\right) = 0 \quad \text{for } n \text{ type} \tag{3-4-2}$$

where $\Phi_m =$ metal work function in eV

$\Phi_s =$ semiconductor work function in eV

$\chi =$ semiconductor electron affinity in eV

$E_g =$ semiconductor bandgap energy in eV

$\psi_b = \dfrac{E_i - E_F}{q}$ is the potential difference between the Fermi level E_F and the intrinsic Fermi level E_i

When the applied voltage is zero, the band is flat and it is called the *flatband condition*. Under this condition the resistivity of the insulator is infinity and the dc current flowing through the diode is zero.

2. Accumulation Case at Negative Applied Voltage. When a negative voltage is applied across the *p*-type ideal MIS diode as shown in Fig. 3-4-3(a), a number of negative charges (electrons) is immediately deposited on the metal and an equal number of induced positive charges (holes) is accumulated at the surface of the *p*-type semiconductor. Because the negative applied voltage depresses the electrostatic potential of the metal and raises the electron energy in the

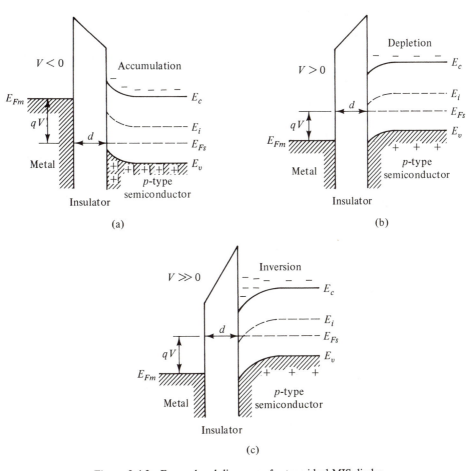

Figure 3-4-3 Energy-band diagrams of p-type ideal MIS diodes

metal with respect to the semiconductor, the top of the valence band at the insulator-semiconductor interface is bent upward to the Fermi level. This band bending causes an accumulation of majority carriers (holes) near the semiconductor surface and is referred to as the *accumulation* case. The carrier holes can be written

$$p = p_0 \exp\left(\frac{E_i - E_{Fs}}{\eta k T} \right) \tag{3-4-3}$$

where p_0 = holes concentration
E_i = intrinsic Fermi level
E_{Fs} = Fermi level of the semiconductor
$k = 1.38 \times 10^{-23}$ J/°K is the Boltzmann constant
T = absolute temperature in degrees Kelvin
η = factor. 1 is for low bias voltage and 2 is for high bias voltage

3. Depletion Case at Positive Applied Voltage. When a small positive voltage is applied across the *p*-type ideal MIS diode as shown in Fig. 3-4-3(b), positive charges are deposited on the metal and negative charges are induced at the semiconductor surface. The potential of the metal is raised up and the potential of the semiconductor is lowered down. As a result, the bands bend downward and the majority carriers are depleted. This is called the *depletion* case.

4. Inversion Case at Very Large Positive Applied Voltage. When a very large positive voltage is applied across the *p*-type ideal MIS diode as shown in Fig. 3-4-3(c), the bands bend even further downward and the intrinsic Fermi level E_i crosses over the Fermi level E_{Fs} at the surface. In effect, the amount of electrons (minority carriers) at the surface is larger than that of holes (majority carriers) and this situation is called the *inversion* case.

The polarities of applied voltages for an *n*-type ideal MIS diode are interchanged and the signs of the corresponding quantities are reversed. (See Fig. 3-4-4.)

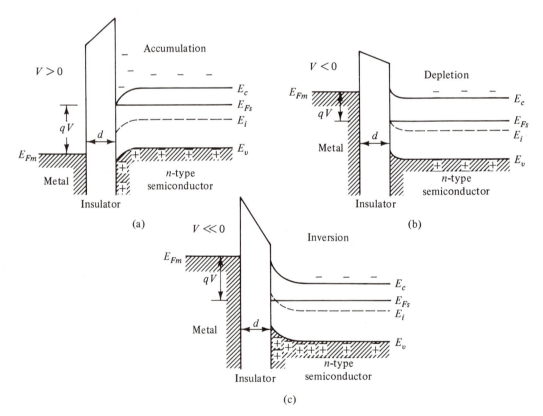

Figure 3-4-4 Energy-band diagrams of *n*-type ideal MIS diodes

Example 3-4-2   Carrier Concentration and Fermi Energy Level

A p-type GaAs has the following parameters:

Intrinsic density	$n_i = 10^6 \text{ cm}^{-3}$
Hole concentration	$p_0 = 10^{10} \text{ cm}^{-3}$
Bandgap energy	$E_g = 1.43 \text{ eV}$
Temperature	$T = 300 \text{ °K}$

(a) Calculate the electron concentration
(b) Compute the Fermi energy level with respect to the intrinsic Fermi level
(c) Determine the Fermi level relative to the maximum valence-band energy

Solution.   (a) The electron concentration is

$$n_0 = \frac{n_i^2}{p_0} = \frac{10^{12}}{10^{10}} = 100 \text{ cm}^{-3}$$

(b) From Eq. (3-4-3) the Fermi level is

$$E_i - E_F = kT\ell n\left(\frac{p_0}{n_i}\right) = 26 \times 10^{-3} \ell n\left(\frac{10^{10}}{10^6}\right)$$

$$= 26 \times 10^{-3} \times 9.21$$

$$= 0.24 \quad \text{eV}$$

That means the Fermi level is 0.24 eV below the intrinsic Fermi level
(c) The Fermi level is above the maximum valence-band energy by

$$E_F = E_g/2 - 0.24 = 0.715 - 0.24 = 0.475 \quad \text{eV}$$

3-4-3   Threshold Voltage for Strong Inversion

The threshold voltage for strong inversion is that voltage in which the semiconductor surface is as strongly n-type as the p-type substrate. In other words, the intrinsic Fermi level E_i should bend down below the Fermi level E_F by ψ_b, where ψ_b is the potential difference between the two levels E_i and E_F. (See Fig. 3-4-5.) When the two levels are equal, the condition is called *intrinsic*. Therefore the surface potential for strong inversion is given by

$$\psi_s(\text{inv}) = 2\psi_b = 2\frac{kT}{q}\ell n \frac{N_a}{n_i} \tag{3-4-4}$$

where n_i is the intrinsic hole or electron concentration.

The depletion width is dependent of surface potential and it may be written

$$W = \left(\frac{2\epsilon_s \psi_s}{qN_a}\right)^{1/2} \tag{3-4-5}$$

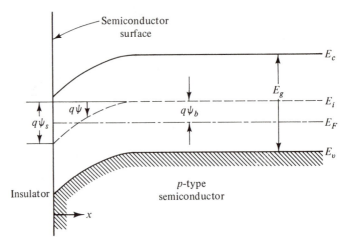

Figure 3-4-5 Energy-band diagram at the surface of a p-type MIS diode (After S. M. Sze [5]; reprinted by permission of John Wiley & Sons, Ltd.)

where ϵ_s = semiconductor permittivity
$\quad$ N_a = acceptor concentration per cm^3
$\quad$ ψ_s = surface potential

Then the surface potential may be expressed as

$$\psi_s = \frac{qN_aW^2}{2\epsilon_s} \tag{3-4-6}$$

Next, the maximum depletion width is given by

$$W_m = \left[\frac{2\epsilon_s\psi_s(\text{inv})}{qN_a}\right]^{1/2} = 2\left[\frac{\epsilon_s kT\ell n\,(N_a/n_i)}{q^2N_a}\right]^{1/2} \tag{3-4-7}$$

The surface minority space-charge density (electrons) in the depletion region at strong inversion for a p-type MIS diode is

$$Q_{sd} = -qN_aW_m = -2(\epsilon_s qN_a\psi_b)^{1/2} \tag{3-4-8}$$

The applied voltage will appear partially across the insulator and partially across the semiconductor. Thus

$$V = V_i + \psi_s \tag{3-4-9}$$

where $V_i = |Q_s|/C_i$ $\qquad$ is the potential across the insulator
$\quad$ $|Q_s| = Q_m = Q_{in} + qN_aW$ $\quad$ is the surface charge density
$\quad$ $Q_{in} =$ $\qquad\qquad\qquad$ surface charge density in the inversion region
$\quad$ $C_i = \dfrac{\epsilon_i}{d}$ $\qquad\qquad\quad$ is the capacitance of the insulator per unit area

The total capacitance of the MIS diode is

$$C \doteq \frac{C_i C_d}{C_i + C_d} \qquad (3\text{-}4\text{-}10)$$

where $C_d = \frac{\epsilon_s}{W}$ is the depletion-layer capacitance per unit area

$\epsilon_s =$ semiconductor permittivity

The threshold voltage required for strong inversion in a p-type ideal MIS diode becomes

$$V_{\text{th}} = -\frac{Q_{sd}}{C_i} + 2\psi_b = -\frac{2}{C_i}(\epsilon_s q N_a \psi_b)^{1/2} + 2\psi_b \qquad (3\text{-}4\text{-}11)$$

In practice, the work functions of the metal and semiconductor are not equal. Then the threshold voltage is given by

$$V_{\text{th}} = \frac{\Phi_{\text{ms}}}{q} - \left(\frac{Q_{\text{in}}}{C_i} + \frac{Q_{sd}}{C_i}\right) + 2\psi_b$$

$$= \frac{\Phi_{\text{ms}}}{q} - \frac{Q_{\text{in}}}{C_i} - \frac{2}{C_i}(\epsilon_s q N_a \psi_b)^{1/2} + 2\psi_b \qquad (3\text{-}4\text{-}12)$$

where $V_{\text{fb}} = \dfrac{\Phi_{\text{ms}}}{q} - \dfrac{Q_{\text{in}}}{C_i}$ is the flatband potential

$\Phi_{\text{ms}} =$ work function difference between the metal and semiconductor

Example 3-4-3   Calculations of a Silicon MIS Diode Characteristics

A certain MIS diode is formed by a metal-oxide-silicon (p-type) material. The carrier concentration N_a is 10^{16} cm^{-3} and the thickness of the insulator SiO$_2$ is 1000 Å. The intrinsic concentration of silicon at room temperature is 1.5×10^{10} cm^{-3}.
(a) Find the surface potential for strong inversion in volts.
(b) Determine the maximum depletion width in micrometers.
(c) Calculate the insulator capacitance per unit area ($\epsilon_{ir} = 3.9$).
(d) Compute the surface space-charge density per square meter.
(e) Determine the threshold voltage required for strong inversion in volts.

Solution. (a) From Eq. (3-4-4) the surface potential for strong inversion is

$$\psi_s(\text{inv}) = \frac{2kT}{q} \ell\text{n} \frac{N_a}{N_i}$$

$$= 2 \times 26 \times 10^{-3} \ell\text{n}\left(\frac{10^{16}}{1.5 \times 10^{10}}\right)$$

$$= 0.697 \text{ V}$$

(b) From Eq. (3-4-7) the maximum depletion width is

$$W_m = \left[\frac{2\epsilon_s \psi_s (\text{inv})}{qN_a} \right]^{1/2}$$

$$= \left[\frac{2 \times 11.8 \times 8.854 \times 10^{-12} \times 0.697}{1.6 \times 10^{-19} \times 10^{22}} \right]^{1/2}$$

$$= 0.30 \ \mu\text{m}$$

(c) The insulator capacitance is

$$C_i = \frac{\epsilon_{ir}}{d} = \frac{3.9 \times 8.854 \times 10^{-12}}{1000 \times 10^{-10}} = 3.45 \ 10^{-4} \ \text{F/m}^2$$

(d) The surface space-charge density is

$$Q_{sd} = -qN_a W_m$$

$$= -1.6 \times 10^{-19} \times 10^{22} \times 0.30 \times 10^{-6}$$

$$= -4.80 \times 10^{-4} \ \text{m}^{-2}$$

(e) From Eq. (3-4-11) the threshold voltage required for strong inversion is

$$V_{th} = -\frac{Q_{sd}}{C_i} + \psi_s (\text{inv})$$

$$= \frac{4.80 \times 10^{-4}}{3.45 \times 10^{-4}} + 0.697$$

$$= 2.087 \ \ \text{V}$$

3-4-4 Surface Space-Charge Density

The surface space-charge density in a semiconductor for an ideal MIS diode is a function of the surface potential ψ_s [16]. Figure 3-4-6 shows the variation of the surface space-charge density as a function of the surface potential ψ_s for a *p*-type MIS diode with $N_a = 4 \times 10^{15}$ cm^{-3} at room temperature.

1. The surface space-charge density in the accumulation region is approximately equal to

$$|Q_{sa}| \doteq (2\epsilon_s kTp_o)^{1/2} \exp\left(\frac{q|\psi_s|}{2kT} \right) \quad \text{for } \psi_s < 0 \qquad (3\text{-}4\text{-}13)$$

where ψ_s = surface potential as defined by Eq. (3-4-4)
        p_o = hole density at equilibrium for the *p*-type semiconductor

2. The surface space-charge density at the flatband point is

$$Q_{sf} = 0 \quad \text{for } \psi_s = 0 \qquad (3\text{-}4\text{-}14)$$

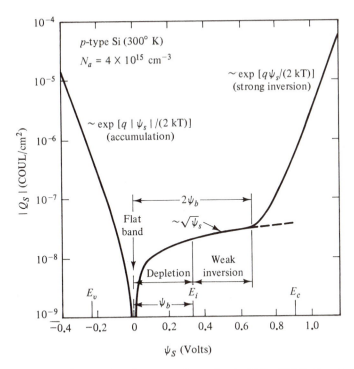

Figure 3-4-6 Variation of surface space-charge density [After C.G.B. Garrett and W. H. Brattain. Physical theory of semiconductor surface. *Phys. Rev.*, **99** (1955)]

3. The surface space-charge density in the depletion region is approximately given by

$$|Q_{sd}| \doteq (2\epsilon_s kTp_0)^{1/2}\sqrt{\psi_s} \quad \text{for } 2\psi_b > \psi_s > 0 \qquad (3\text{-}4\text{-}15)$$

where ψ_b is the potential difference between the Fermi level and the intrinsic Fermi level.

4. The surface space-charge density in the inversion region is approximately expressed as

$$|Q_{si}| \simeq (2\epsilon_s kTp_0)^{1/2}\exp\left(\frac{q\psi_s}{2kT}\right) \quad \text{for } \psi_s > 2\psi_b \qquad (3\text{-}4\text{-}16)$$

REFERENCES

[1] Shockley, W. The theory of *p-n* junction in semiconductors and *p-n* junction transistors. *Bell Syst. Tech. J.*, **28** (1949), 435.

[2] Cooke, Harry F. Microwave transistors: Theory and Design. *Proc. IEEE*, **59**, no. 8 (August 1971), 1163–1181.

[3] Sobol, H., and F. Sterzer. Solid-state microwave power sources. *IEEE Spectrum*, **9** (April 1972), 32.

[5] Sze, S. M. *Physics of Semiconductor Devices*, 2nd ed., p. 146, 250, and 366. New York: John Wiley & Sons, 1981.

[6] Early, J. M. Maximum rapidly switchable power density in junction triodes. *IRE Trans. on Electron Devices*, **ED-6** (1959), 322–325.

[7] Johnson, E. O. Physical limitation on frequency and power parameters of transistors. *RCA Review*, **26**, no. 6 (June 1965), 163–177.

[8] DeLoach, B. C., Jr. Recent advances in solid state microwave generators. *Advances in Microwaves*, vol. 2. New York: Academic Press, 1967.

[9] Esaki, L. New phenomenon in narrow germanium *p-n* junction. *Physical Review*, **109** (1959), 603–604.

[11] Schottky, W. Halbleitertheorie der sperrchicht, *Naturwissen-schaften*, **26** (1938), 843.

[12] Sze, S. M. *Physics of Semiconductor Devices*, 2nd ed., pp. 251, 299, and 291. New York: John Wiley & Sons, 1981.

[13] Milnes, A. G. *Semiconductor Devices and Integrated Electronics*, p. 100. New York: Van Nostrand Reinhold Company, 1980.

[14] Moll, J. L. Variable capacitance with large capacity charge. *Wescon. Conv. Rec.*, pt. 3 (1959), 32.

[15] Terman, L. M. An investigation of surface states at silicon/silicon dioxide interface employing metal-oxide-silicon diode. *Solid-state Electronics*, **5** (1962), 285.

[16] Garrett, C. G. B., and W. H. Brattain. Physical theory of semiconductor surface. *Phys. Rev.*, **99** (1955), 376.

[17] Mason, Samuel J., and Henry Zimmerman. *Electric Circuits, Signals and Systems*. Chap. 4. New York: John Wiley & Sons, Inc., 1964.

SUGGESTED READINGS

1. Howes, M. J., and D. V. Morgan. *Microwave Devices*, Chapter 4. New York: John Wiley & Sons, 1976.

2. Liao, Samuel Y. *Microwave Devices and Circuits*, Chapter 6. Englewood Cliffs, NJ: Prentice-Hall, Inc., 1980.

3. Milnes, A. G. *Semiconductor Devices and Integrated Electronics*, Chapters 2 and 3. New York: Van Nostrand Reinhold Company, 1980.

4. Seymour, J. *Electronic Devices and Components*, Chapters 4 and 8. New York: John Wiley & Sons, 1981.

5. Streetman, B. G. *Solid State Electronic Devices*, 2nd ed, Chapter 7. Englewood Cliffs, NJ: Prentice-Hall, Inc., 1980.

6. Sze, S. M. *Physics of Semiconductor Devices*, 2nd ed., Chapters 5 and 7. New York: John Wiley & Sons, 1981.

7. Vendelin, George D. *Design of Amplifiers and Oscillators by the S-Parameter Method*. New York: John Wiley & Sons, 1982.

PROBLEMS

3-1 Microwave Transistors

3-1-1. The S parameters of a certain microwave transistor measured at 3 GHz with a 50-Ω resistance matching the input and output are $S_{11} = 0.505\underline{/-150°}$, $S_{22} = 0.45\underline{/-20°}$, and $S_{21} = 7\underline{/180°}$. The transistor is to be used as an amplifier device in an amplifier circuit as shown in Fig. 3-1-9.

   (a) Calculate and plot the input and output constant power gain circles on a Smith chart.

   (b) Overlap two Smith charts to facilitate the design procedures. Use the original chart to read the impedance and the overlaid chart to read the admittance.

   (c) Determine the capacitance and inductance of the output matching network.

   (d) Find the capacitance and inductance of the input matching network.

3-1-2. A certain silicon microwave transistor has a maximum electric field intensity E_m of 3×10^5 V/cm and its carrier has a saturated drift velocity v_s of 4×10^6 cm/sec. The emitter-collector length L is 4 μm.

   (a) Calculate the maximum allowable applied voltage.

   (b) Compute the transit time for a charge to transverse the emitter-collector length L

   (c) Determine the maximum possible transit frequency.

3-1-3. A microwave transistor has voltage-frequency and current-frequency limitations as shown in Eqs. (3-1-45) and (3-1-46). Derive the power-frequency limitation as shown in Eq. (3-1-47).

3-1-4. A microwave transistor has a voltage-current-power limitation. Derive the power gain-frequency relationship as shown in Eq. (3-1-48).

3-2 Microwave Tunnel Diodes

3-2-1. A negative-resistance device is connected through a 1-kΩ resistor in series and a 0.01-μF capacitor in parallel to a combination of a supply voltage V of 10 V and a signal source V_s as shown in Fig. P3-2-1(a). Its V-I characteristic curve is shown in Fig. P3-2-1(b).

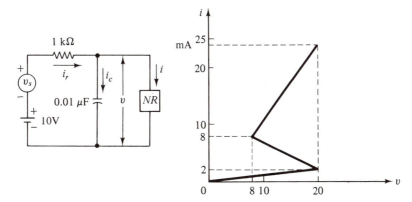

Figure P3-2-1

(a) Find the negative resistance and forward resistance of the device.

(b) Draw the load line on the *V-I* curve.

(c) Determine the quiescent operating point of the circuit by the values of voltage and current.

(d) Determine the new operating point by the values of voltage and current when a signal voltage of 14 V is applied to the circuit.

(e) Draw the new load line on the *V-I* curve.

(f) Find the time constant of the circuit.

(g) Compute v, i, i_c, and i_r as a function of time after the triggering signal is applied and before the transition takes place.

(h) Find transition time T in microseconds.

(i) Calculate v, i, i_c, and i_r immediately after transition.

3-2-2. A microwave tunnel diode has a negative resistance R_n and the resonant circuit has a circuit resistance R_ℓ. Derive an equation for the gain of a microwave tunnel diode amplifier.

3-2-3. A certain microwave tunnel diode has a negative resistance of $69 + j9.7$ Ω. Determine the resonant-circuit impedance so that the microwave tunnel diode amplifier will produce a power gain of 15 dB.

3-3 Microwave Metal-Semiconductor (MES) Diodes

3-3-1. A metal-semiconductor diode is formed of gold and *n*-type silicon. The work function Φ_m of gold is 4.82 eV and the electron affinity χ of silicon is 4.01 eV.

(a) Calculate the Schottky-barrier height in eV.

(b) Determine the Schottky-barrier height in eV for a MES diode formed by gold and *p*-type silicon.

3-3-2. The Schottky-barrier height of a gold-silicon diode can be lowered by increasing the electric field. Gold has a work function Φ_m of 4.82 eV.

(a) Calculate the Schottky-barrier height for an electric field of 4×10^7 V/m.

(b) Compute the Schottky-barrier height for an electric field of 4×10^9 V/m.

(c) Determine the Schottky-barrier lowering effect in percentage.

3-3-3. A certain MES diode has a low turn-on voltage, low reverse current, and very fast switching speed. The diode is formed of gold and *n*-type silicon.

(a) Calculate the reverse current without considering the Schottky barrier effect at room temperature. (The cross section of the diode is assumed to be 0.0254 cm^2.)

(b) Compute the forward current at 0.4 V.

(c) Estimate the time constant if the diode has a capacitance of 1 pF and a resistance of 40 Ω.

3-3-4. Describe the microwave application of MES diodes.

3-4 Metal-Insulator-Semiconductor (MIS) Diodes

3-4-1. The surface potential for strong inversion at the insulator-semiconductor interface is shown in Eq. (3-4-4). Derive Eq. (3-4-4).

3-4-2. If the *p*-type silicon semiconductor of a MIS diode has $N_a = 1.5 \times 10^{15}$ cm^{-3} and $n_i = 1.5 \times 10^{10}$ cm^{-3},

(a) compute the surface potential for strong inversion at room temperature, and

(b) determine the maximum depletion width W_m.

3-4-3. A certain p-type Si MIS diode has the following parameters:

$$N_a = 1.5 \times 10^{15} \text{ cm}^{-3} \qquad \text{Insulator SiO}_2: \quad \epsilon_{ir} = 4.0$$
$$\epsilon_s = 11.8 \qquad\qquad\qquad\qquad\qquad d = 100 \text{ Å}$$

(a) Determine surface charge Q_{Sd} in the depletion region.
(b) Compute the insulator capacitance.
(c) Calculate the threshold voltage required for strong inversion.

3-4-4. A metal-insulator-semiconductor (MIS) diode has a depletion width of 0.30 μm. The acceptor concentration N_a is 4×10^{15} cm^{-3} and the semiconductor relative dielectric constant ϵ_s is 11.8. Find the surface potential.

3-4-5. A n-type GaAs has the following parameters:

Intrinsic density	$n_i = 10^6$ cm^{-3}
Electron concentration	$n_o = 10^9$ cm^{-3}
Bandgap energy	$E_g = 1.43$ eV
Temperature	$T = 300 \, °$K

(a) Compute the hole density
(b) Calculate the Fermi energy level relative to the intrinsic Fermi level
(c) Determine the Fermi level with respect to the maximum valence-band energy level

Chapter 4

Microwave Solid-State Field-Effect Devices

4-0 INTRODUCTION

After the invention of the transistor in 1948, Shockley proposed a new type of field-effect transistor (FET) in 1952 in which the conductivity of a layer of a semiconductor is modulated by a transverse electric field [1]. In a conventional transistor both the majority and the minority carriers are involved; so this type of transistor is customarily referred to as a *bipolar* transistor. In a field-effect transistor the current flow is carried by majority carriers only; so this type is referred to as a *unipolar* transistor. In addition, field-effect transistors are controlled by a voltage at the third terminal rather than by a current as in a bipolar transistor. The purpose of this chapter is to describe the physical structures, operating principles, microwave characteristics, and power-frequency limitations of unipolar field-effect transistors. The unipolar field-effect transistor has several advantages over the bipolar junction transistor.

1. It may have voltage gain in addition to current gain.
2. Its efficiency is higher than that of a bipolar transistor.
3. Its noise figure is very low.
4. Its operating frequency is up to the X band.
5. Its input resistance is very high, up to several megohms.

Unipolar field-effect transistors can be classified into three types: *p-n* junction gate, MES gate (metal-semiconductor structure or Schottky-barrier gate), and MIS gate (metal-insulator-semiconductor structure). If the field-effect transistor is constructed of a metal-semiconductor, the device is called a metal-semiconductor field-effect transistor (MESFET). If an insulator is sandwiched between a metal and semiconductor, the structure is a form of MIS diode that was discussed in Section 3-4. This type is called the metal-insulator-semiconductor field-effect transistor (MISFET). Because the metal gate electrode is insulated from the source and drain, this device is also referred to as the insulated-gate field-effect transistor (IGFET).

If the insulator is oxide, for instance, the device is commonly referred to as a metal-oxide-semiconductor field-effect transistor (MOSFET). If the channel of a MOSFET is of vertical or U-shaped form, the device is called the VMOSFET or UMOSFET, respectively.

MESFETs are widely used as amplifiers in radar systems, MOSFETs are very useful in very large scale integration (VLSI) circuits, such as microprocessors and computer memories, and VMOSFETs and UMOSFETs are being used as high-speed and high-power devices in microwave systems. All these devices are discussed in detail in this chapter. A family tree of microwave solid-state field-effect transistors is shown in Fig. 4-0-1, and their performance characteristics are tabulated in Table 4-0-1 [2]. The figure and table appear on pages 106 and 107.

In addition, the charge-coupled device (CCD) is a metal-oxide-semiconductor (MOS) diode structure and the motion of its charge packets is transversely controlled by applied gate voltages. So the CCD has a fundamental similarity with the MOSFET. The CCD is also described in this chapter.

4-1 METAL-SEMICONDUCTOR FIELD-EFFECT TRANSISTORS (MESFETs)

In 1938 Schottky suggested that the potential barrier could arise from stable space charges in the semiconductor alone without the presence of a chemical layer [3]. The model derived from his theory is known as the Schottky barrier. The Schottky-barrier diode was described in Section 3-3. If the field-effect transistor is constructed by a metal-semiconductor Schottky-barrier diode, the device is called a metal-semiconductor field-effect transistor (MESFET). The material may be either silicon or gallium arsenide (GaAs) and the channel type may be either n channel or p channel. GaAs MESFET has higher electron mobility as compared to silicon, so it has higher power gain. Because its electric field and electron saturation drift velocity are higher, so its output power is also higher. Another special feature is its lower noise figure. This is due to the higher mobility of its electron carriers. Therefore the GaAs MESFETs are more commonly used as amplifiers in microwave integrated circuits for high-power, low-noise and broadband amplifier applications.

Because GaAs MESFETs are capable of amplifying small signals up to the frequency range of the X band with a low-noise figure, they have lately replaced

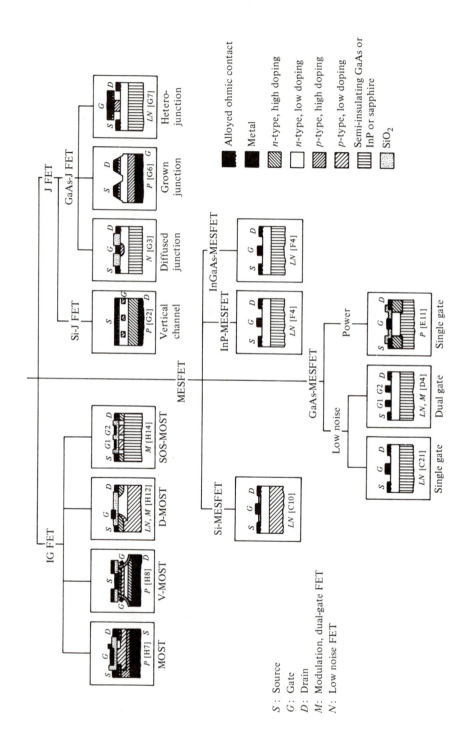

Figure 4-0-1 Family tree of microwave field-effect transistors. (After C. A. Liechti [2]; reprinted by permission of the IEEE, Inc.)

S : Source
G : Gate
D : Drain
M : Modulation, dual-gate FET
N : Low noise FET

Alloyed ohmic contact
Metal
n-type, high doping
n-type, low doping
p-type, high doping
p-type, low doping
Semi-insulating GaAs or InP or sapphire
SiO₂

TABLE 4-0-1 PERFORMANCE CHARACTERISTICS OF MICROWAVE FIELD-EFFECT TRANSISTORS.

Type	Semiconductor	Single/Dual Gate	Channel			Application	Frequency (GHz)	Output Power CW (W)	Associated Power Gain (dB)	Small-Signal Gain (dB)	Power Added Efficiency (%)	Noise Figure (dB)	Associated Gain (dB)	Cascaded Noise Figure (dB)	Maximum Available Gain (dB)
			Type	Length (μm)	Width (mm)										
MESFET															
Silicon	Si	SG	n	0.5		LN	10					5.8			5.9
LN SG GaAs	GaAs	SG	n	0.5		LN	10					2.7	10.5	3.1	13
LN SG GaAs	GaAs	SG	n	1		LN	10					3.2	8.0	3.6	10
LN DG GaAs	GaAs	DG	n	1		LN, M	10					4.0	12	4.2	18
P SG GaAs	GaAs	SG	n	1.5	5.2	P	8	2.2	3.2	4.2	22				
P SG GaAs	GaAs	SG	n	1.2	0.6	P	22	0.14	4.8	5.6	9				
LN SG InP	InP	SG	n	1		LN	10					4.7	6.6	5.4	7.8
LN SG InGaAs	InGaAs	SG	n	1		LN	7					5.7	5.0	7.0	18
JFET															
Silicon	Si	SG	n	1	1.8	P	2.7	0.2	6.0		19				
Diff. Junction	GaAs	SG	n	2		LN	4					2.5	10	2.7	10
Grown Junction	GaAs	SG	n	1.5	6.1	P	6	1.0	6.0	7.0	26				
Heterojunction	GaAs	SG	n	2		LN	6								
IGFET															
MOST	Si	SG	n	5	20	P	0.7	16	6	10	26				
V-MOST	Si	SG	n	1	18	P	2	4.0	5	6	32				
SG D-MOST	Si	SG	n	1		LN	1					3.0	9.0	3.3	
DG D-MOST	Si	DG	n	1		LN, M	1					4.5	14	4.6	15
DG SOS-MOST	Si	DG	n	4		M	0.5								25

LN—low noise amplification; P—power amplification; M—modulation, switching or amplification with controlled gain; SG—single-gate FET; DG—dual-gate FET.

(After C. A. Liechti [2]; reprinted by permission of the IEEE, Inc.)

parametric amplifiers in airborne radar systems because the latter are complicated to fabricate and expensive to produce.

Junction Field-Effect Transistors (JFETs): Unipolar field-effect transistors may be either *p-n* junction gate or Schottky-barrier gate. Before considering MESFETs, let us describe the JFETs for comparison. The JFET is the one originally proposed by Shockley [1]. Figure 4-1-1 shows the schematic diagram and circuit symbol for an *n*-channel junction field-effect transistor.

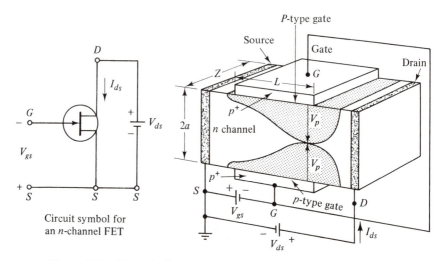

Figure 4-1-1   Schematic diagram and circuit symbol for an *n*-channel JFET

The *n*-type material is sandwiched between two highly doped layers of *p*-type material that is designated P^+. This type of device is called the *n*-channel JFET. If the middle part is a *p*-type semiconductor, the device is called the *p*-channel JFET. The two *p*-type regions in the *n*-channel JFET (see Fig. 4-1-1) are referred to as *gates*. Each end of the *n*-channel is joined by a metallic contact. Based on the directions of the biasing voltages as shown in Fig. 4-1-1, the left-hand contact, which supplies the source of the flowing electrons, is referred to as the *source* whereas the right-hand contact, which drains the electrons out of the material, is called the *drain*. The circuit symbol for an *n*-channel JFET is also shown in Fig. 4-1-1. The direction of the drain current I_{ds} is flowing from the drain to the device. For a *p*-channel JFET, the polarities of the two biasing voltages V_{gs} and V_{ds} are interchanged, the head of the arrow points away from the device, and the drain current I_{ds} flows away from the device.

4-1-1 Physical Structure

The metal-semiconductor field-effect transistor (MESFET) was developed by many scientists and engineers, such as Mead [4] and Hooper [5]; sometimes it is also called the Schottky-barrier field-effect transistor. Figure 4-1-2 shows the schematic diagram circuit of a metal-semiconductor field-effect transistor of the GaAs type.

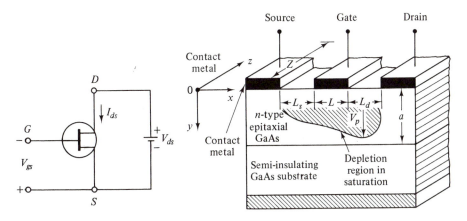

Figure 4-1-2   Schematic diagram and circuit symbol of a GaAs MESFET

The device is an interdigitated structure that was fabricated by using an n-type GaAs epitaxial film about 0.15 to 0.35 μm thick on a semi-insulating substrate about 100 μm. The n-channel layer is doped with either sulfur or tin in a doping concentration N between 8×10^{16} and 2×10^{17} per cubic centimeter. The electron mobility in the layer is in the range of 3000 to 4500 cm^2/V-s. The Schottky-barrier gate is evaporated aluminum. The source and drain contacts are Au-Ge, Au-Te, or Au-Te-Ge alloys. A contact metallization pattern of gold is used to bring the source, drain, and gate contacts out to bonding pads over the semi-insulating substrate. Between the active n-type epitaxial layer and the semi-insulating substrate a buffer layer of about 3 μm with a doping concentration of 10^{15} to 10^{16} cm^{-3} is often fabricated.

4-1-2 Principles of Operation

In the FETs of Figs. 4-1-1 and 4-1-2 a voltage is applied in the direction to reverse-bias the p-n or metal n-type junction between the source and the gate whereas the source and the drain electrodes are forward biased. Under this bias condition the majority carriers (electrons) flow in the n-type epitaxial layer from the source electrode, through the channel beneath the gate, to the drain electrode. The current in the channel causes a voltage drop along its length so that the Schottky barrier-gate electrode becomes progressively more reverse biased toward the drain electrode. As a result, a charge-depletion region is set up in the channel and gradually pinches off the channel against the semi-insulating substrate toward the drain end. As the reverse bias between the source and gate increases, so does the height of the charge-depletion region. The decrease of the channel height in the nonpinched-off region will increase the channel resistance. Consequently, the drain current I_{ds} will be modulated by the gate voltage V_{gs}. This phenomenon is analogous to the characteristics of the collector current I_c versus the collector voltage V_c with the base current I_b as parameter in a bipolar transistor. In other words, a family of curves of drain current I_{ds} versus voltage V_{ds} between the source and drain with the gate voltage V_{gs} as parameter will be generated in an unipolar gallium arsenide metal-semiconductor field-effect transistor (GaAs MESFET), as shown in Fig. 4-1-3.

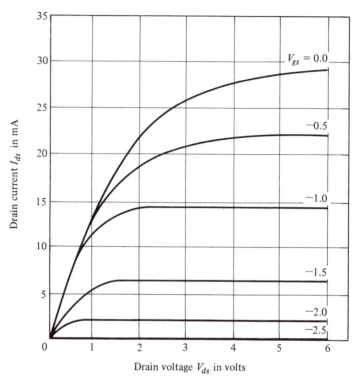

Figure 4-1-3   Voltage-current characteristics of a typical n-channel GaAs MESFET

The transconductance of a FET is expressed as

$$g_m = \left. \frac{dI_{ds}}{dV_{gs}} \right|_{V_{ds}=\text{constant}} \qquad \text{mhos} \qquad (4\text{-}1\text{-}1)$$

The drain current I_{ds} for a fixed drain-to-source voltage V_{ds} is a function of the reverse-biasing gate voltage V_{gs}. Because the drain current I_{ds} is controlled by the field effect of the gate voltage V_{gs}, this device is referred to as a *field-effect transistor* (FET). When the drain current I_{ds} is continuously increasing, the ohmic voltage drop between the source and the channel reverse-biases the p-n junction further. As a result, the channel is pinched off eventually. When the channel is pinched off, the drain current I_{ds} will remain almost constant even though the drain-to-source voltage V_{ds} is continuously increased.

Pinch-off voltage V_p.   The pinch-off voltage is the gate reverse voltage that removes all the free charge from the channel. From Fig. 4-1-2 Poisson's equation for voltage in the n-channel, in terms of the volume charge density, is given by

$$\frac{d^2V}{dy^2} = -\frac{\rho}{\epsilon} = -\frac{qN}{\epsilon} = -\frac{qN}{\epsilon_r \epsilon_0} \qquad (4\text{-}1\text{-}2)$$

where ρ = the volume charge density in coulombs per cubic meter
$\quad\;\; q$ = the charge in coulombs
$\quad\;\; N$ = the electron concentration in electrons per cubic meter
$\quad\;\; \epsilon$ = the permittivity of the material in farads per meter
$\quad\;\; \epsilon = \epsilon_r \epsilon_0, \epsilon_r$ is the relative dielectric constant
$\quad\;\; \epsilon_0 = 8.854 \times 10^{-12}$ F/m is the permittivity of free space

Integration of Eq. (4-1-2) once and application of the boundary condition of the electric field $E = -(dV/dy) = 0$ at $y = a$ yield

$$\frac{dV}{dy} = -\frac{qN}{\epsilon}(y - a) \qquad \text{volts/meter} \qquad (4\text{-}1\text{-}3)$$

Integration of Eq. (4-1-3) once and application of the boundary condition $V = 0$ at $y = 0$ result in

$$V = -\frac{qN}{2\epsilon}(y^2 - 2ay) \qquad \text{volts} \qquad (4\text{-}1\text{-}4)$$

Then the pinch-off voltage V_p at $y = a$ is expressed as

$$V_p = \frac{qNa^2}{2\epsilon} \qquad \text{volts} \qquad (4\text{-}1\text{-}5)$$

where a is the height of the channel in meters.

Equation (4-1-5) indicates that the pinch-off voltage is a function of the doping concentration N and the channel height a. Doping may be increased to the limit set by the gate breakdown voltage and the pinch-off voltage may be made large enough so that drift saturation effects just become dominant.

Example 4-1-2   Pinch-Off Voltage of a GaAs MESFET

A certain GaAs MESFET has the following parameters:

Channel height	$a = 0.1 \; \mu\text{m}$
Electron concentration	$N = 8 \times 10^{17} \text{cm}^{-3}$
Relative dielectric constant	$\epsilon_r = 13.10$

Calculate the pinch-off voltage.

Solution.   From Eq. (4-1-5) the pinch-off voltage is

$$V_p = \frac{qNa^2}{2\epsilon} = -\frac{1.6 \times 10^{-19} \times 8 \times 10^{23} \times (0.1 \times 10^{-6})^2}{2 \times 8.854 \times 10^{-12} \times 13.10}$$

$$= -5.52 \text{ volts}$$

4-1-3 Small-Signal Equivalent Circuit

For microwave frequencies, the metal-semiconductor field-effect transistor has a very short channel length and its velocity saturation occurs in the channel before reaching the pinched path. Microwave characteristics of a metal semiconductor field-effect transistor depend not only on intrinsic parameters, such as g_m, G_d, R_i, C_{gs}, and C_{gd}, but also on extrinsic parameters R_g, R_s, C_{ds}, R_p, and C_p. Figure 4-1-4 shows the cross section of a metal-semiconductor field-effect transistor and its equivalent circuit.

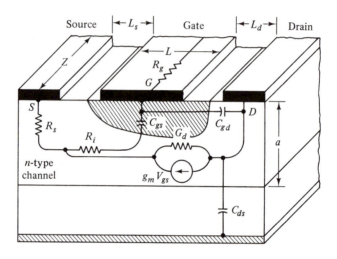

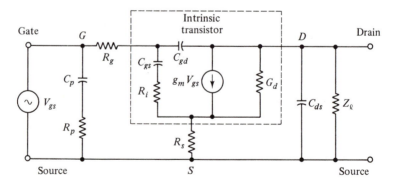

Figure 4-1-4   Cross section and equivalent circuit of a GaAs MESFET

Microwave properties of the metal-semiconductor field-effect transistor were investigated and analyzed by many scientists and engineers [6–12]. The noise behavior of metal-semiconductor field-effect transistors at microwave frequencies has been investigated and measured by van der Ziel and others [13–17].

Intrinsic Elements:

g_m is the transconductance of the FET.
G_d is the drain conductance.
R_i is the input resistance.
C_{gs} is the gate-source capacitance.
C_{gd} is the gate-drain (or feedback) capacitance.

Extrinsic Elements:

R_g is the gate metallization resistance.
R_s is the source-gate resistance.
C_{ds} is the drain-source capacitance.
R_p is the gate bonding-pad parasitic resistance.
C_p is the gate bonding-pad parasitic capacitance.
Z_l is the load impedance.

The values of these intrinsic and extrinsic elements depend on the channel type, material, structure, and dimensions of the metal semiconductor-gate FET. The large values of the extrinsic resistances will seriously decrease power gain and efficiency and increase the noise figure of the MESFET. It is advantageous to increase the channel doping N as high as possible in order to decrease the relative influence of the feedback capacitance C_{gd} and to increase the transconductance g_{m0} and the dc open-circuit voltage gain. An increase in concentration N decreases the breakdown voltage of the gate, however. A doping of 10^{18} per cubic centimeter might be the upper limit.

4-1-4 Drain Current I_{ds}

The drain current I_{ds} of a metal semiconductor-gate FET is expressed [7] as

$$I_{ds} = I_{dss} \frac{3(u^2 - \rho^2) - 2(u^3 - \rho^3)}{1 + \eta(u^2 - \rho^2)} \quad \text{amperes} \tag{4-1-6}$$

where $I_{dss} = \dfrac{qN\mu aZV_p}{3L}$ is the saturation drain current for the Shockley case at $V_{gs} = 0$

$\mu =$ low-field mobility in square meters per volt-second
$q = 1.6 \times 10^{-19}$ coul is the electron charge
$N =$ doping concentration in electrons per cubic meter
$a =$ channel height
$Z =$ channel depth or width
$L =$ gate length

$V_p =$ pinch-off voltage as defined in Eq. (4-1-5)

$u = \left(\dfrac{V_{ds} + |V_{gs}|}{|V_p|} \right)^{1/2}$ is the normalized sum of the drain and gate voltages with respect to the pinch-off voltage

$\rho = \left(\dfrac{|V_{gs}|}{|V_p|} \right)^{1/2}$ is the normalized voltage with respect to the pinch-off voltage

$\eta = \dfrac{\mu |V_p|}{v_s L} = \dfrac{v}{v_s}$ is the normalized drift velocity with respect to the saturation drift velocity

$v_s =$ saturation drift velocity

$v = \dfrac{\mu E_x}{1 + \mu E_x / v_s}$ is the drift velocity in the channel

$E_x =$ absolute value of the electric field in the channel

Figure 4-1-3 shows a plot of drain current I_{ds} versus drain voltage V_{ds} with gate voltage V_{gs} as parameter for a typical n-channel GaAs MESFET. The drain current has a maximum at $u = u_m$, given by

$$u_m^3 + 3u_m \left(\frac{1}{z} - \rho^2 \right) + 2\rho^3 - \frac{3}{z} = 0 \qquad (4\text{-}1\text{-}7)$$

where $z = \mu |V_p| / (v_x L)$ is the ratio between the small-field velocity extrapolated linearly to the field $|V_p|/L$ and the saturation velocity.

Substitution of Eq. (4-1-7) into Eq. (4-1-6) yields the saturation drain current as

$$I_{ds(\text{sat})} = \frac{3 I_{dss} (1 - u_m)}{z} \qquad (4\text{-}1\text{-}8)$$

The transconductance in the saturation region is then given by

$$g_m = \left. \frac{\partial I_{ds(\text{sat})}}{\partial V_{gs}} \right|_{V_{ds} = \text{constant}} = \frac{g_{\max} (u_m - \rho)}{1 + z(u_m^2 - \rho^2)} \qquad (4\text{-}1\text{-}9)$$

In practice, the drain current and mutual conductance of a GaAs MESFET can be expressed by

$$I_{ds} = I_{dss} \left(1 - \frac{|V_{gs}|}{|V_p|} \right)^2 \qquad (4\text{-}1\text{-}9a)$$

and

$$g_m = - \frac{2 I_{dss}}{|V_p|} \left(1 - \frac{|V_{gs}|}{|V_p|} \right) \qquad (4\text{-}1\text{-}9b)$$

The velocity-field curves of a GaAs MESFET are very complicated. Figure 4-1-5 shows these curves in the saturation region [2].

The narrowest channel cross section is located under the drain end of the gate. The peak electric field appears near the drain. The drift velocity rises to a peak at

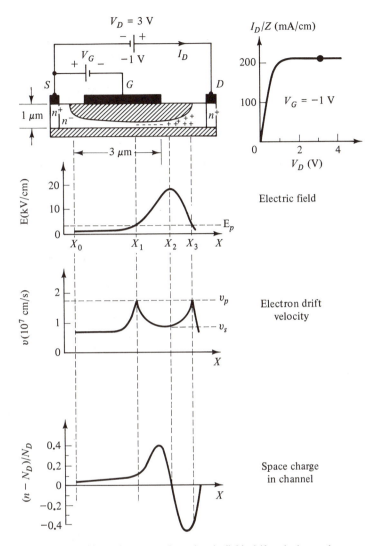

Figure 4-1-5   Channel cross section, electric field, drift velocity, and space charge distribution in the channel of a GaAs MESFET in saturation region. (From C. A. Liechti [2]; reprinted by permission of the IEEE, Inc.)

x_1, close to the center of the channel, and falls to the low saturated value under the gate edge. To preserve current continuity, heavy electron accumulation must form in this region because the channel cross section is narrowing. In addition, the electrons are moving progressively slower with increasing x. Exactly the opposite occurs between x_2 and x_3. The channel widens and the electrons move faster, causing a strong depletion layer. The charges in the accumulation and depletion layers are nearly equal, and most of the drain voltage drops in this stationary dipole layer.

The drain current for a Si MESFET is expressed as [2]

$$I_{ds} = Zqn(x)v(x)\,d(x) \tag{4-1-10}$$

where Z = channel depth or width

$n(x) = N_d$ is the density of conductance electrons

$v(x)$ = drift velocity

$d(x)$ = conductive layer thickness

x = coordinate in the direction of the electron drift

The current-voltage characteristics of a Si MESFET are shown in Fig. 4-1-6 [2]. In Fig. 4-1-6(a), there is no metal gate electrode. At the surface of the conducting layer,

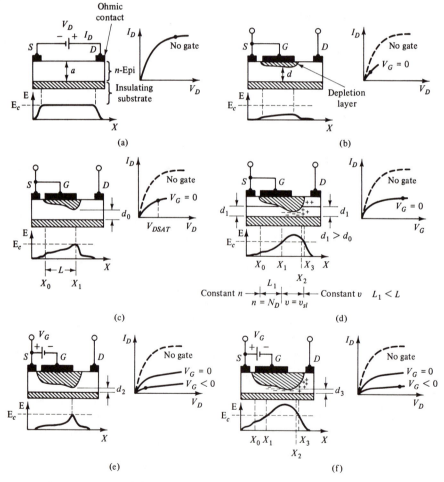

Figure 4-1-6   Electric-field diagrams and voltage-current characteristics of a Si MESFET. (After C. A. Liechti [2]; reprinted by permission of the IEEE, Inc.)

the source and drain contacts are made. When a positive voltage V_{ds} is applied to the drain, electrons flow from source to drain. In Fig. 4-1-6(b), a metal gate is added and shorted to the source. When a small drain voltage is applied, a depletion layer is created. The current I_{ds} flowing from drain to source is indicated by Eq. (4-1-10). When drain voltage V_{ds} is increased, the depletion layer becomes wider. The resulting decrease in conductive cross section d must be compensated by an increase of electron velocity v to maintain constant current through the channel. As drain voltage is increased further, the electrons reach their saturation velocity v_s, as shown in Fig. 4-1-6(c). When the drain voltage is increased beyond $V_{d\text{sat}}$, the depletion layer widens toward the drain, as shown in Fig. 4-1-6(d). When a negative voltage is applied to the gate, as shown in Fig. 4-1-6(e), the gate-to-channel junction is reverse biased and the depletion layer becomes wider. As the gate voltage V_{gs} is more negative, the channel is almost pinched off, as shown in Fig. 4-1-6(f), and the drain current I_{ds} is nearly cut off.

Example 4-1-4   *V-I* Characteristics of a GaAs MESFET

A typical *n*-channel GaAs MESFET has the following parameters:

Electron concentration	$N = 9.52 \times 10^{17} \, \text{cm}^{-3}$
Channel height	$a = 0.1 \, \mu\text{m}$
Relative dielectric constant	$\epsilon_r = 13.10$
Channel length	$L = 14 \, \mu\text{m}$
Channel width	$Z = 36 \, \mu\text{m}$
Electron mobility	$\mu = 0.4 \, \text{m}^2/\text{V-s}$
	$= 4000 \, \text{cm}^2/\text{V-s}$
Drain voltage	$V_{ds} = 5 \, \text{volts}$
Gate voltage	$V_{gs} = -2 \, \text{volts}$
Saturation drift velocity	$v_s = 2 \times 10^5 \, \text{m/s}$

(a) Calculate the pinch-off voltage.
(b) Compute the velocity ratio.
(c) Determine the saturation current at $V_g = 0$.
(d) Find the drain current I_{ds}.

Solution.   (a) From Eq. (4-1-5) the pinch-off voltage is

$$V_p = -\frac{1.6 \times 10^{-19} \times 9.52 \times 10^{23} \times 10^{-14}}{2 \times 8.854 \times 10^{-12} \times 13.10}$$

$$= -6.57 \text{ volts}$$

(b) From Eq. (4-1-6) the velocity ratio is

$$\eta = \frac{0.4 \times 6.57}{2 \times 10^5 \times 14 \times 10^{-6}}$$

$$= 0.94$$

(c) From Eq. (4-1-6) the saturation current at $V_g = 0$ is

$$I_{dss} = \frac{qN\mu aZV_p}{3L}$$

$$= \frac{1.6 \times 10^{-19} \times 9.52 \times 10^{23} \times 0.4 \times 10^{-7} \times 36 \times 10^{-6} \times 6.57}{3 \times 14 \times 10^{-6}}$$

$$= 34.31 \text{ mA}$$

(d) From Eq. (4-1-6) the u and ρ factors are

$$u = \left(\frac{5+2}{6.57}\right)^{1/2} = 1.03 \qquad u^2 = 1.07 \qquad u^3 = 1.10$$

$$\rho = \left(\frac{2}{6.57}\right)^{1/2} = 0.55 \qquad \rho^2 = 0.30 \qquad \rho^3 = 0.17$$

Then the drain current is

$$I_{ds} = 34.31 \times 10^{-3} \times \frac{3(1.07 - 0.30) - 2(1.10 - 0.17)}{1 + 0.94(1.07 - 0.30)}$$

$$= 34.31 \times 10^{-3} \times 0.26$$
$$= 8.92 \text{ mA}$$

4-1-5 Cutoff Frequency f_{co} and Maximum Oscillation Frequency f_{max}

The cutoff frequency of a metal semiconductor-gate FET in a circuit depends on the way in which the transistor is being made. In a wideband lumped circuit the cutoff frequency is expressed [9] as

$$f_{co} = \frac{g_m}{2\pi C_{gs}} = \frac{v_s}{4\pi L} \quad \text{Hz} \tag{4-1-11}$$

where g_m = transconductance

C_{gs} = gate capacitance = $\left.\dfrac{dQ}{dV_{gs}}\right|_{V_{gd}=\text{constant}}$

L = gate length

v_s = saturation drift velocity

It is interesting to note that the cutoff frequency of the lumped circuit analysis shown in Eq. (4-1-11) is different from the charge-carrier transit-time cutoff frequency shown in Eq. (3-1-45) by a factor of one-half.

The maximum frequency of oscillation depends on the device transconductance and the drain resistance in a distributed circuit. It is expressed [7] as

$$f_{max} = \frac{f_{co}}{2}(g_m R_d)^{1/2} = \frac{f_{co}}{2}\left[\frac{\mu E_p(u_m - \rho)}{v_s(1 - u_m)}\right]^{1/2} \quad \text{Hz} \tag{4-1-12}$$

where R_d = drain resistance
g_m = device transconductance
E_p = electric field at the pinch-off region in the channel
u_m = saturation normalization of u
and v_s, μ, and ρ are defined previously.

For $\rho = 0$ (i.e., $V_{gs} = 0$) and $\eta \gg 1$ so that

$$f_{co} = \frac{v_s}{4\pi L} \quad \text{and} \quad u_m \simeq \left(\frac{3}{\eta}\right)^{1/3} \ll 1$$

we have
$$f_{max} = \gamma \frac{v_s}{L}\left(\frac{3}{\eta}\right)^{1/6} \quad \text{Hz} \qquad (4\text{-}1\text{-}13)$$

where $\gamma = 0.14$ for $\mu E_p/v_s = 13$ and $\gamma = 0.18$ for $\mu E_p/v_s = 20$ in case of GaAs

It has been found experimentally [7] that the maximum frequency of oscillation for a GaAs MESFET with gate length less than 10 μm is

$$f_{max} = \frac{33 \times 10^3}{L} \quad \text{Hz} \qquad (4\text{-}1\text{-}14)$$

where L is the gate length in meters. The best value of L is 0.5 μm.

The maximum frequency of oscillation f_{max} is similar to the cutoff frequency f_{co} determined by the transit time. From Eq. (3-1-45) the charge-carrier transit-time cutoff frequency is

$$f_{co} = \frac{1}{2\pi\tau} = \frac{v_s}{2\pi L} \quad \text{Hz} \qquad (4\text{-}1\text{-}15)$$

where $\tau = \dfrac{L}{v_s}$ is the transit time in seconds

$L =$ gate length in meters
$v_s =$ saturation drift velocity in meters per second

Clearly the GaAs MESFET has a better figure of merit than the silicon MESFET for an X-band amplifier because the saturation drift velocity v_s is 2×10^7 cm/s for GaAs at an electric field of 3 kV/cm and 8×10^6 cm/s for silicon at 15 kV/cm. Comparing Eq. (4-1-15) with Eq. (4-1-11), we find that the difference is a factor of one-half.

The highest oscillation frequency for maximum power gain with the input and output networks matched is given [10] as

$$f_{max} = \frac{f_{co}}{2}\left(\frac{R_d}{R_s + R_g + R_i}\right)^{1/2} \quad \text{Hz} \qquad (4\text{-}1\text{-}16)$$

where R_d = drain resistance
R_s = source-to-gate resistance
R_g = gate metallization resistance
R_i = input resistance

Example 4-1-5   Cutoff Frequency of a GaAS MESFET

A certain GaAs MESFET has the following parameters:

$$R_g = 3 \, \Omega \qquad R_i = 2.5 \, \Omega \qquad g_m = 50 \text{ m}\mho$$
$$R_d = 450 \, \Omega \qquad R_s = 2.5 \, \Omega \qquad C_{gs} = 0.60 \text{ pF}$$

(a) Determine the cutoff frequency.
(b) Find the maximum operating frequency.

Solution.   (a) From Eq. (4-1-11) the cutoff frequency is

$$f_{co} = \frac{g_m}{2\pi C_{gs}} = \frac{0.05}{2\pi \times 0.6 \times 10^{-12}}$$

$$= 13.26 \text{ GHz}$$

(b) From Eq. (4-1-16) the maximum frequency is

$$f_{max} = \frac{f_{co}}{2}\left(\frac{R_d}{R_s + R_g + R_i}\right)^{1/2}$$

$$= \frac{13.26 \times 10^9}{2}\left(\frac{450}{2.5 + 3 + 2.5}\right)^{1/2}$$

$$= 49.73 \text{ GHz}$$

4-1-6   Power Gain

The measurable circuit parameters for a two-port metal-semiconductor FET in the microwave frequency range are the scattering parameters (S parameters), described in Chapter 3 for the design of microwave transistors. Once the values of the **S** parameters are obtained, the power gain G_p, the maximum available power gain G_{max}, and the unilateral power gain G_u are easily calculated by using the equations derived in Section 3-1-4.

Power gain G_p.   The power gain G_p of a microwave metal-semiconductor FET amplifier is defined as the ratio of the output power P_ℓ delivered to the load Z_ℓ over the input power P_{avs} available from the source to the FET as shown in Eq. (3-1-16). Then applying Mason's rules as described in Section 3-1-4 the power gain for a MESFET amplifier as shown in Eq. (3-1-22) is given by

$$G_p = \frac{|S_{21}|^2(1 - |\Gamma_S|^2)(1 - |\Gamma_\ell|^2)}{|(1 - S_{11}\Gamma_S)(1 - S_{22}\Gamma_\ell) - S_{21}S_{12}\Gamma_S\Gamma_\ell|^2} \qquad (4\text{-}1\text{-}17)$$

Maximum available power gain G_{max}.   To maximize the forward power gain G_{max}, the input and output networks for a MESFET amplifier, must be conjugately matched. The maximum available power as shown in Eq. (3-1-36) is

$$G_{max} = \frac{|S_{21}|}{|S_{12}|}\left|K \pm (K^2 - 1)^{1/2}\right| \qquad (4\text{-}1\text{-}18)$$

Here K is the MESFET's inherent stability factor and it is expressed in Eq. (3-1-23) as

$$K = \frac{1 + |\Delta|^2 - |S_{11}|^2 - |S_{22}|^2}{2|S_{12}S_{21}|} > 1 \qquad (4\text{-}1\text{-}19)$$

where

$$\Delta \equiv S_{11}S_{22} - S_{12}S_{21}$$

In order for a MESFET amplifier to be unconditionally stable, the stability factor must be greater than positive unity.

Unilateral power gain G_u.   The unilateral power gain G_u is the forward power gain in a feedback amplifier that has its reverse power gain set to zero (i.e., $|S_{12}|^2 = 0$) by adjusting a lossless reciprocal feedback network connected around the microwave MESFET amplifier as shown in Eq. (3-1-37). That is,

$$G_u = \frac{|S_{21}|^2(1 - |\Gamma_s|^2)(1 - |\Gamma_\ell|^2)}{|1 - S_{11}\Gamma_s|^2|1 - S_{22}\Gamma_\ell|^2} \qquad (4\text{-}1\text{-}20)$$

The maximum unilateral power gain is achieved with $\Gamma_S = S_{11}^*$ and $\Gamma_\ell = S_{22}^*$. Then

$$G_{u\,max} = \frac{|S_{21}|^2}{(1 - |S_{11}|^2)(1 - |S_{22}|^2)} \qquad (4\text{-}1\text{-}21)$$

Example 4-1-6   Maximum Unilateral Power Gain of a GaAs MESFET

A certain GaAs MESFET has the following S parameters measured at 10 GHz:

$$S_{11} = 0.72 \underline{/157°} \qquad S_{22} = 0.61 \underline{/-151°}$$
$$S_{21} = 1.32 \underline{/-16°} \qquad S_{12} = 0.07 \underline{/-29°}$$

Compute its maximum unilateral power gain.

Solution.   The maximum unilateral power gain is

$$G_{u\,max} = \frac{|1.32|^2}{(1 - |0.72|^2)(1 - |0.61|^2)}$$

$$= 5.80 = 7.6 \text{ dB}$$

4-1-7 Noise Figure F

The overall effect of many noise sources in an electronic circuit is frequently specified by means of the noise figure of the circuit. The noise figure F of any linear two-port network can be defined in terms of its performance with a standard noise source connected to its input terminals. That is,

$$F \equiv \frac{\text{available noise power at output}}{\text{available noise power at input}} = \frac{N_0}{GkTB} \qquad (4\text{-}1\text{-}22)$$

where $GkTB$ = the available noise power of the standard source in a bandwidth B
at temperature $T = 300°K$ and Boltzmann constant $k = 1.381 \times 10^{-23}$ J/°K

G = the available power gain of the network at the frequency of the band considered

N_0 = the available noise power at outputs

Noise N_0 at the output of the network within the same frequency band arises from amplification of the input noise and from noise N_n generated within the network. Then the output noise N_0 can be expressed as

$$N_0 = N_n + GkTB \qquad (4\text{-}1\text{-}23)$$

Thus the noise figure results in

$$F = 1 + \frac{N_n}{GkTB} \qquad (4\text{-}1\text{-}24)$$

In general, when the noise powers are expressed by their noise temperatures, Eq. (4-1-24) becomes

$$F = 1 + \frac{T_n}{T_0} \qquad (4\text{-}1\text{-}25)$$

where T_n = noise temperature of the network in degrees Kelvin
T_0 = ambient noise temperature at 300°K

There are three types of noise figures for a MESFET amplifier.

1. Intrinsic Noise Figure. The intrinsic noise figure of a GaAs MESFET is given [13, 15] as

$$F = 2 + \gamma \left(\frac{E}{E_{\text{sat}}} \right)^3 \qquad (4\text{-}1\text{-}26)$$

where E = electric field in volts per meter
E_{sat} = saturation electric field at 300 kV/m
$\gamma = 6$

At a frequency of 10 GHz, a noise figure of 6.6 dB for a GaAs MESFET has been measured [14], which is much better than the noise figure of any other FET or bipolar transistor. At 5 GHz, the noise figure for a GaAs FET is approximately 3 dB. For silicon MESFET, the noise figure could be approximately expressed [13, 15] as

$$F = 2 + \gamma \left(\frac{E}{E_{\text{sat}}} \right)^2 \qquad (4\text{-}1\text{-}27)$$

where $\gamma = 2.3$
E_{sat} = saturation electric field at 1500 kV/m

2. Intervalley Scattering Noise Figure in GaAs. According to the energy-band theory of the n-type GaAs, the low-mobility upper valley is separated by an energy gap of 0.36 eV from the high-mobility lower valley and the lower valley is separated by an energy gap of 1.43 eV from the valence band. The saturation drift velocity of electrons is at an electric field of 300 kV/m. The intervalley scattering noise figure is given [14] as

$$F = \frac{T_{nv}}{T_0}(1-p) + \frac{T_{ni}}{T_0} + 1 \qquad (4\text{-}1\text{-}28)$$

where $T_0 = 300°\text{K}$ is the ambient noise temperature
T_{nv} = noise temperature of electrons in the valence band
T_{ni} = noise temperature of intervalley scattering
$1 - p$ = ratio of the number of electrons in the valence band
p = population probability of electrons in the lower and upper valleys (satellite valleys)

3. Extrinsic Noise Figure. Several sources of extrinsic noise for a metal semiconductor field-effect transistor (MESFET) are as follows:
   (a) Gate metallization resistance R_g. If the gate is made of relatively long, narrow, thin aluminum film, the resistance between the gate bonding pad and the active transistor region is quite large.
   (b) Source resistance R_s. The region between the source and gate contributes the source resistance.
   (c) Drain resistance R_d. The region between the drain contact and the drain end of the channel causes resistance.
   (d) Bonding pad resistance R_p. The gate bonding pads lie on the n-channel epitaxial end and may be represented by an impedance of a resistance R_p in series with a capacitance C_p.

The extrinsic noise figure of a MESFET contributed by these noise-source resistances is expressed [13] as

$$F = F_0 + \frac{R_n}{G_s}\left[(G_s - G_{\text{on}})^2 + (B_s - B_{\text{on}})^2\right] \qquad (4\text{-}1\text{-}29)$$

where $F_0 =$ optimum noise figure
$R_n =$ noise resistance
$Y_s = G_s + jB_s$ is the source admittance
$Y_{\text{on}} = G_{\text{on}} + jB_{\text{on}}$ is the optimum source admittance with respect to noise

Noise parameters F_0, G_{on}, B_{on}, and R_n can be determined by calculating noise figure F for four different source admittances $Y_s = G_s + jB_s$ and applying Eq. (4-1-29). It has been shown that the optimum extrinsic noise figure does not depend on the drain and gate voltages and that it is about 4 dB at 2 GHz and 8 dB at 8 GHz.

If the values of R_g, R_p, and R_s are reduced to a minimum, the noise figure could be much improved. This situation is achieved by using heavier metallization of the gate, by removing the low-resistivity epitaxial layer below the gate bonding pad, and by reducing the distance between the source and gate. The best noise figure and power gain can be obtained for 0-volt gate voltage and 4-volt drain voltage for a GaAs MESFET.

4-1-8 Power-Frequency Limitations

In Section 3-1-5 the power-frequency limitations for microwave bipolar transistors were described. Again the same question occurs as to whether microwave unipolar metal semiconductor field-effect transistors have any limitations on their frequency and power gain. The answer is yes. The basic principle of constant gain-bandwidth product is still valid. When the frequency is increased, the power gain is decreased. Figure 4-1-7 shows the measured power gain for a silicon metal-semiconductor field-effect transistor [6].

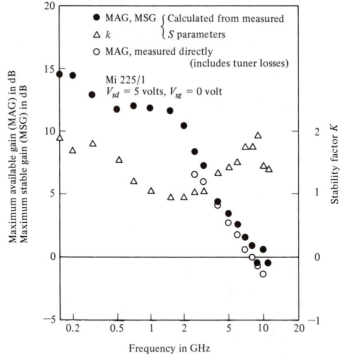

Figure 4-1-7 Measured power gain for a silicon MESFET (From P. Wolf [6]; reprinted by permission of IBM Corp.)

Here the power gain near the X band is very low. The GaAs MESFET, however, has about a 4-dB power gain higher than the counterpart of silicon

MESFET. So the best figure of a GaAs MESFET may be about 6 dB at the frequency range of the X band. The output power may reach a range of 1 watt in the X band.

Recent reports of GaAs MESFET performance at microwave frequencies such as about 5 dBW at X band and 10 dBm at 60 GHz have made the GaAs MESFET device attractive in local oscillator and phase array modules applications. Its noise figure is about 1 dB at 10 GHz and 3 dB at 30 GHz [42].

Electron carriers in GaAs will slow down at an electric field greater than 3 kV/cm. The electrons move from a high-mobility state to a low-mobility state in about 1 picosecond and thus the velocity of the carriers reaches a peak and slows down in the middle of the channel. The velocity versus electric field of GaAs and Si is shown in the appendix. The change in velocity of the carriers in GaAs causes the Gunn effect in Gunn diodes or TEOs (Transfer Electron Oscillators).

A nonequilibrium velocity-field characteristic must be considered in short-channel devices (less than 3 μm). When electrons enter the high-field domain, they are accelerated to a high velocity. This effect can cause peak velocities of about 4×10^7 cm/s, which relax to 1×10^7 cm/s after traveling about 0.5 μm. The overshoot in velocity reduces the transit time and shifts the dipole charge to the right of the channel.

Example 4-1-8   Power-Frequency Limitation of a GaAs MESFET

The output power of a GaAs MESFET is inversely proportional to its frequency squared. A certain GaAs MESFET has the following parameters:

$$f_T = 10 \text{ GHz} \qquad E_m = 2 \times 10^5 \text{ V/cm}$$
$$X_c = 1 \ \Omega \qquad v_s = 1 \times 10^6 \text{ cm/s}$$

Determine the maximum output power of the device.

Solution.   From Eq. (3-1-47) we have

$$P_m = \frac{1}{X_c f_T^2} \left(\frac{E_m v_s}{2\pi} \right)^2$$

$$= \frac{1}{1 \times (10 \times 10^9)^2} \left(\frac{2 \times 10^5 \times 10^6}{2\pi} \right)^2$$

$$= 10.13 \text{ W}$$

4-1-9 High-Gain and Wideband Balanced MESFET Amplifiers

The principle of the constant gain-bandwidth product limits the high gain and wide bandwidth of a single MESFET amplifier. However, a balanced GaAs MESFET amplifier (two MESFETs in parallel) incorporated with two 3-dB and 90-degree hybrid couplers or two Wilkinson power dividers and a pair of 90-degree phase shifters can operate with both high gain and wide bandwidth [37–39]. Figure 4-1-8 shows two circuit diagrams for a balanced GaAs MESFET amplifier.

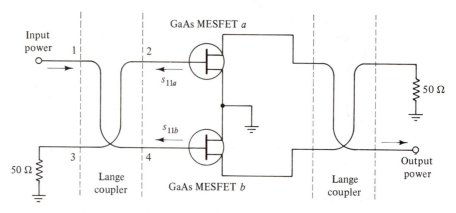

(a) Balanced amplifier with Lange couplers

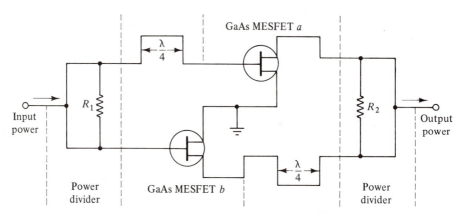

(b) Balanced amplifier with Wilkinson power dividers and phase shifters

Figure 4-1-8 Circuit diagrams for a balanced amplifier

Two GaAs MESFETs are connected at the ports of a 90-degree and 3-dB hybrid, usually a Lange coupler. The Lange coupler is an interdigitated microstrip quadrature hybrid and consists of three or more parallel striplines with alternate lines tied together. A signal wave incident in port 1 couples equal power into ports 2 and 4 but none into port 3. The input and output signals are out of phase by 90°. Thus the quadrature, or 90-degree hybrid, is also called 3-dB directional coupler. This design allows the input and output amplifier ports to be mismatched, but the combined amplifier appears to be matched if the two parallel stages are balanced. The total reflection and power transmission can be expressed as

$$\mathbf{S}_{11} = \tfrac{1}{2}(\mathbf{S}_{11a} - \mathbf{S}_{11b}) \tag{4-1-30}$$

$$\mathbf{S}_{22} = \tfrac{1}{2}(\mathbf{S}_{22a} - \mathbf{S}_{22b}) \tag{4-1-31}$$

and
$$\text{Gain} = |S_{21}|^2 = \tfrac{1}{4}|S_{21a} + S_{21b}|^2 \qquad (4\text{-}1\text{-}32)$$

where a and b indicate the two GaAs MESFETs. The 1 and 2 refer to the input and output ports, respectively. The input and output reflections are reduced to half of the corresponding different reflection of the two MESFETs, and the total power gain is equal to the individual power gain if the two MESFETs are identical ($|S_{21}|^2 = |S_{21a}|^2 = |S_{21b}|^2$).

If one MESFET fails, the loss for the balanced amplifier would be only 6 dB. The balanced amplifier configuration is by far the most common design in modern microwave integrated circuits. Its advantages are:

1. good input and output VSWRs
2. good stability
3. high reliability
4. low tuning work
5. linear output power increases by 3 dB
6. up to 20 dB power gain with 10% bandwidth at X band.

As indicated by Eqs. (4-1-30) and (4-1-31), the amplifier will be balanced if the two GaAs MESFETs are identical ($S_{11a} = S_{11b}$ and $S_{22a} = S_{22b}$). In other words, the VSWR at the output terminal will be unity. However, the characteristics of the two MESFETs are not actually measured in practice, and they are not the same. When their characteristics are different, the amplifier will not be balanced, and manual tuning work is needed to bring the amplifier on balance. For mass production it is required to characterize the GaAs MESFET chips before placing them in the circuits in order to increase the hybrid reproducibility, minimize tuning work, reduce production costs, and to improve the module performance [41].

There are three types of characterization for GaAs MESFET chips:

1. DC and Low-Frequency Tests.   The dc and low-frequency tests are individually measured for each chip on a wafer by a standard test fixture such as made by the Rucker-Kolls Company.

2. RF Packaged-chip Test.   Packaged chips are those which are ribbon-bonded to bring the source, drain, and gate contacts out to bonding pads over the semi-insulating substrate after the chips are removed from the wafer. In practice, the S parameters at microwave frequencies are measured on a sample packaged chip; it is then assumed that the hundred chips around the sample chip would have the same S parameters. This assumption is not accurate because the wafer yield is often low—about from 6 to 10%—due to lattice imperfections and fabrication defects. The test fixture for packaged chip test is available from Hughes model Z1010H-1000 and Maury Microwave MT950 series.

3. **RF Chip Test.**   There are two types of RF chip characterization—destructive and nondestructive tests. After the chip is removed from a wafer and wire-bonded a RF test is made on the chip. This test is called destructive because the tested chip cannot be used again. The test fixture for destructive testing is the same as for the packaged-chip test. The nondestructive chip test is made to each chip after the chip is removed from the wafer without wire bonding. So far the nondestructive chip test has not been successful because a test fixture for nondestructive testing is not yet available. Cascade Microwave is developing a nondestructive test fixture and reports that it will be available in the near future.

Example 4-1-9   GaAs MESFET Balanced Amplifier

A GaAs MESFET has the following parameters:

$$\text{MESFET a: Reflection coefficients} \quad S_{11a} = 0.81 \underline{/53^\circ}$$

$$S_{22a} = 0.55 \underline{/20^\circ}$$

$$\text{Forward transmission} \quad S_{21a} = 6.22 \underline{/180^\circ}$$
$$\text{coefficient}$$

$$\text{MESFET b: Reflection coefficients} \quad S_{11b} = 0.79 \underline{/55^\circ}$$

$$S_{22b} = 0.50 \underline{/22^\circ}$$

$$\text{Forward transmission} \quad S_{21b} = 6.10 \underline{/178^\circ}$$
$$\text{coefficient}$$

a. Calculate the input and output reflection coefficients of the balanced amplifier.
b. Compute the input and output VSWRs.
c. Determine the power gain in dB for the balanced amplifier.
d. Find the power loss in dB if one MESFET fails.
e. Calculate the linear output power capability in comparison with two amplifiers in series.

Solution.   a. Using Eqs. (4-1-30) and (4-1-31) the input and output reflection coefficients are

$$S_{11} = \tfrac{1}{2}\left(0.81 \underline{/53^\circ} - 0.79 \underline{/55^\circ}\right) = 0.02$$

$$S_{22} = \tfrac{1}{2}\left(0.55 \underline{/20^\circ} - 0.50 \underline{/22^\circ}\right) = 0.03$$

b. The input and output VSWRs are

$$\text{VSWR(input)} = \frac{1 + |S_{11}|}{1 - |S_{11}|} = \frac{1 + 0.02}{1 - 0.02} = 1.04$$

$$\text{VSWR(output)} = \frac{1 + |S_{22}|}{1 - |S_{22}|} = \frac{1 + 0.03}{1 - 0.03} = 1.06$$

c. Using Eq. (4-1-32) the power gain is

$$\text{Gain} = \tfrac{1}{4} \left| 6.22 \underline{/180°} + 6.10 \underline{/178°} \right|^2 = \tfrac{1}{4} \times 151.82$$
$$= 37.96 = 15.8 \text{ dB}$$

d. If one MESFET fails, the power loss is

$$\text{Loss} = 10 \log\left(\tfrac{1}{4}\right) = -6 \text{ dB}$$

e. The linear output power is increased by

$$P_{\text{out}} = 10 \log(2) = 3 \text{ dB}$$

4-1-10 Design Example

Both Si and GaAs MESFETs can be designed as an amplifier or an oscillator, depending on whether the stability factor K is greater or less than unity. Common-source configuration is preferred for amplifier design but common-gate configuration for oscillator design. The stability factor K is greater than unity for an amplifier, but the input and output reflection coefficients (Γ_{in} and Γ_{out}) are less than unity. For an oscillator, however, the stability factor K is less than unity, but the input and output reflection coefficients are equal to or greater than unity. In amplifier design two matching networks are required to match the input and output ports of the device to give a maximum transducer power gain, a minimum noise figure, and low input and output reflection coefficients. In a small-signal oscillator design the same two-port (or one that is properly modified with a feedback path) can be designed to deliver nearly the same output power to the same 50-Ω load. Figure 4-1-9 shows a diagram of two matching networks for the design of a GaAs MESFET amplifier.

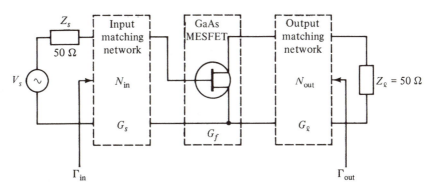

Figure 4-1-9 Matching networks to be designed for a GaAs MESFET amplifier

Design Procedures for Amplifier:

1. Calculate the stability factor K.
2. Compute the maximum unilateral transducer power gain in decibels.

3. Design an input matching network N_{in}.
4. Design an output matching network N_{out}.
5. Sketch the complete matching networks.

Example 4-1-10

A certain GaAs MESFET has the following parameters measured at 5 GHz:

$$\mathbf{S}_{11} = 0.64\underline{/-170°} \qquad \mathbf{S}_{22} = 0.56\underline{/-100°}$$

$$\mathbf{S}_{21} = 2.10\underline{/30°} \qquad \Gamma_{in} = 0.62\underline{/165°}$$

$$\mathbf{S}_{12} = 0.06\underline{/160°} \qquad \Gamma_{out} = 0.59\underline{/100°}$$

where Γ_{in} and Γ_{out} indicate the values of the optimum input and output reflection coefficients for maximum available power gain and minimum noise figure, respectively.

Solution. 1. From Eq. (4-1-19) the stability factor K is

$$K = \frac{1 + |\Delta|^2 - |\mathbf{S}_{11}|^2 - |\mathbf{S}_{22}|^2}{2|\mathbf{S}_{12}\mathbf{S}_{21}|}$$

where $|\Delta|^2 = |\mathbf{S}_{11}\mathbf{S}_{22} - \mathbf{S}_{12}\mathbf{S}_{21}|^2$

$$= \left|0.64\underline{/-170°} \times 0.56\underline{/-100°} - 0.06\underline{/160°} \times 2.10\underline{/30°}\right|^2$$

$$= 0.16$$

$$K = \frac{1 + 0.16 - |0.64|^2 - |0.56|^2}{2\left|0.06\underline{/160°} \times 2.10\underline{/30°}\right|}$$

$$= 1.70$$

2. From Eq. (4-1-21) the maximum unilateral transducer power gain is

$$G_{u\,max} = 10\log\left(\frac{1}{1 - |\mathbf{S}_{11}|^2}\right) + 10\log|\mathbf{S}_{21}|^2 + 10\log\left(\frac{1}{1 - |\mathbf{S}_{22}|^2}\right)$$

$$= 10\log\left(\frac{1}{1 - |0.64|^2}\right) + 10\log|2.10|^2 + 10\log\left(\frac{1}{1 - |0.56|^2}\right)$$

$$= 2.29 \text{ dB} + 6.44 \text{ dB} + 1.61 \text{ dB}$$

$$= 10.34 \text{ dB}$$

3. Input matching network: From the basic reflection coefficient equation of the transmission line the input impedance $\mathbf{Z}_{in}$ is

$$\mathbf{Z}_{in} = \frac{\mathbf{Z}_0(1 + \Gamma_{in})}{1 - \Gamma_{in}}$$

$$= \frac{\mathbf{Z}_0\left(1 + |\Gamma_{in}|\cos\underline{/\Gamma_{in}} + j|\Gamma_{in}|\sin\underline{/\Gamma_{in}}\right)}{1 - |\Gamma_{in}|\cos\underline{/\Gamma_{in}} - j|\Gamma_{in}|\sin\underline{/\Gamma_{in}}}$$

$$= \frac{\mathbf{Z}_0\left(1 - |\Gamma_{in}|^2\right) + j2\mathbf{Z}_0|\Gamma_{in}|\sin\left/\Gamma_{in}\right.}{1 + |\Gamma_{in}|^2 - 2|\Gamma_{in}|\cos\left/\Gamma_{in}\right.}$$

$$= \frac{50\left(1 - |0.62|^2\right) + j100|0.62|\sin(165°)}{1 + |0.62|^2 - 2|0.62|\cos(165°)}$$

$$= \frac{31 + j16.05}{2.58}$$

$$= 12.02 + j6.22 \ \Omega$$

$$\mathbf{Y}_{in} = \frac{1}{\mathbf{Z}_{in}} = \frac{1}{12.02 + j6.22} = 0.066 - j0.034 \ \text{mho}$$

Realization:

(a) Using a quarter-wave transformer of characteristic impedance $\mathbf{Z}_{0\,in}$ to match the source impedance 50 Ω to 0.066 mho or 15.15 Ω,

$$\mathbf{Z}_{0\,in} = \sqrt{\frac{50}{0.066}} = 27.52 \ \Omega \qquad \text{for } \ell = 1.5 \ \text{cm}$$

(b) Using an inductor with a reactance,

$$jX_\ell = \frac{1}{-jX_c} = \frac{1}{-j0.034} = j29.41 \ \Omega$$

$$L = \frac{29.41}{\omega} = \frac{29.41}{2\pi \times 5 \times 10^9} = 0.94 \ \text{nH}$$

(c) Alternatively, an open-circuited stub ($\mathbf{Y} = \mathbf{Y}_0\tan \beta\ell$) looks like a shunt admittance. Therefore an open-circuited stub of three-eighths wavelength long with a characteristic impedance $\mathbf{Z}_{0\,in}$ looks like a shunt inductor of an admittance $-j\mathbf{Y}_{0\,in}$.

$$\mathbf{Z}_{0\,in} = \frac{1}{0.034} = 29.41 \ \Omega \qquad \text{for } \ell = 2.25 \ \text{cm}$$

Similarly, a short-circuited stub that is one-eighth wavelength long looks like a shunt inductor of impedance $j\mathbf{Z}_{0\,in}$, where $\mathbf{Z}_{0\,in}$ is the characteristic impedance of the stub.

4. Output matching network: Similarly, the impedance of the output matching network is

$$\mathbf{Z}_{out} = \frac{50\left(1 - |0.59|^2\right) + j100|0.59|\sin(100°)}{1 + |0.59|^2 - 2|0.59|\cos(100°)}$$

$$= \frac{32.60 + j58.10}{1.55}$$

$$= 21.03 + j37.48 \ \Omega$$

$$\mathbf{Y}_{out} = \frac{1}{\mathbf{Z}_{out}} = \frac{1}{21.03 + j37.48} = 0.0114 - j0.020 \ \text{mho}$$

Realization:

(a) Using a quarter-wave transformer of characteristic impedance $\mathbf{Z}_{0\,\text{out}}$ to match the load impedance 50 Ω to 0.0114 mho or 87.72 Ω,

$$\mathbf{Z}_{0\,\text{out}} = \sqrt{\frac{50}{0.0114}} = 66.23 \ \Omega \qquad \text{for } \ell = 1.5 \text{ cm}$$

(b) Using an inductor with a reactance,

$$X_\ell = \frac{1}{X_c} = \frac{1}{0.020} = 50 \ \Omega$$

$$L = \frac{50}{2\pi \times 5 \times 10^9} = 1.60 \text{ nH}$$

(c) Alternatively, an open-circuited stub of three-eighths wavelength long with a characteristic impedance $\mathbf{Z}_{0\,\text{out}}$ will also serve the matching purpose.

$$\mathbf{Z}_{0\,\text{out}} = \frac{1}{0.020} = 50 \ \Omega \qquad \text{for } \ell = \frac{3\lambda}{8} = 2.25 \text{ cm}$$

5. The complete matching networks with a GaAs MESFET are shown in Fig. 4-1-10.

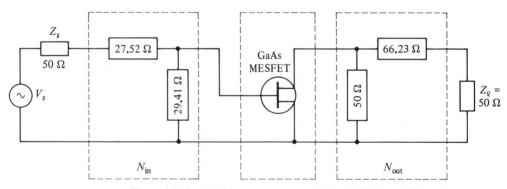

Figure 4-1-10   Matching networks for a GaAs MESFET amplifier

4-2 *METAL-OXIDE-SEMICONDUCTOR FIELD-EFFECT TRANSISTORS (MOSFETs)*

The metal-insulator-semiconductor field-effect transistor (MOSFET) may be formed by a metal, such as Al, and a semiconductor, such as Ge, Si, or GaAs, with an insulator, such as SiO_2, Si_3N_4, or Al_2O_3 sandwiched in between. If the structure is in the form Al-SiO_2-Si, it is called a MOSFET. This device is very useful in VLSI microwave circuits. In the near future a microwave device chip of 1-μm dimension containing a million or more devices will be commercially available. The basic component of a MOSFET is the MIS diode, described in Section 3-4. Here we consider MOSFETs.

4-2-1 Physical Structure

The metal-oxide-semiconductor field-effect transistor (MOSFET) is a four-terminal device and it has two types: n-channel MOSFET and p-channel MOSFET. The n-channel MOSFET consists of a slightly doped p-type semiconductor substrate into which two highly doped n^+ sections are diffused (see Fig. 4-2-1). These n^+ sections, which will act as the source and the drain, are separated by about 0.5 μm. A thin layer of insulating silicon dioxide (SiO_2) is grown over the surface of the structure. The metal contact on the insulator is called the gate.

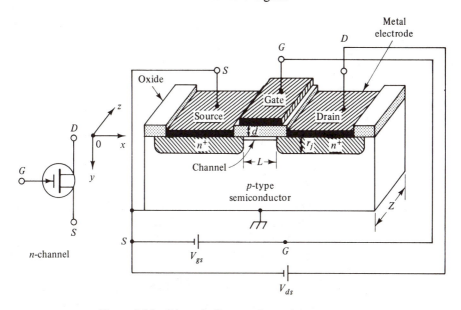

Figure 4-2-1 Schematic diagram of an n-channel MOSFET

Similarly, the p-channel MOSFET is made of a slightly doped n-type semiconductor substrate with two highly doped p^+-type regions for the source and drain. The heavily doped polysilicon or a combination of silicide and polysilicon can also be used as the gate electrode. In practice, a MOSFET is commonly surrounded by a thick oxide to isolate it from adjacent devices in a microwave integrated circuit. The basic device parameters of a MOSFET are as follows:

$L =$ the channel length which is the distance between the two n^+-p junctions just beneath the insulator (say 0.5 μm)

$Z =$ the channel depth (say 5 μm)

$d =$ the insulator thickness (say 0.1 μm)

$r_j =$ the junction thickness of the n^+ section (say 0.2 μm)

4-2-2 Electronic Mechanism

When no voltage is applied to the gate of an n-channel MOSFET, the connection between the source electrode and the drain electrode corresponds to a link of two p-n junctions connected back to back. The only current that can flow from the source to the drain is the reverse leakage current. When a positive voltage is applied to the gate relative to the source (the semiconductor substrate is grounded or connected to the source), positive charges are deposited on the gate metal. As a result, negative charges are induced in the p-type semiconductor at the insulator-semiconductor interface. A depletion layer with a thin surface region containing mobile electrons is formed. These induced electrons form the n channel of the MOSFET and allow the current to flow from the drain electrode to the source electrode. For a given value of the gate voltage V_g, the drain current I_d will be saturated for some drain voltages V_d.

A minimum gate voltage is required to induce the channel and it is called the threshold voltage V_{th}. For an n-channel MOSFET, the positive gate voltage V_g must be larger than the threshold voltage V_{th} before a conducting n channel (mobile electrons) is induced. Similarly, for a p-channel MOSFET, the gate voltage V_g must be more negative than the threshold voltage V_{th} before the p channel (mobile holes) is formed.

4-2-3 Modes of Operation

Basically there are four modes of operation for n-channel and p-channel MOSFETs.

1. *n-Channel Enhancement Mode (normally OFF).* When the gate voltage is zero, the channel conductance is very low and it is not conducting. A positive voltage must be applied to the gate to form an n channel for conduction. The drain current is enhanced by the positive voltage. This type is called the enhancement-mode (normally OFF) n-channel MOSFET.

2. *n-Channel Depletion Mode (normally ON).* If an n channel exists at equilibrium (i.e., at zero bias), a negative gate voltage must be applied to deplete the carriers in channel. In effect, the channel conductance is reduced and the device is turned OFF. This type is called the depletion-mode (normally ON) n-channel MOSFET.

3. *p-Channel Enhancement Mode (normally OFF).* A negative voltage must be applied to the gate to induce a p channel for conduction. This type is called the enhancement-mode (normally OFF) p-channel MOSFET.

4. *p-Channel Depletion Mode (normally ON).* A positive voltage must be applied to the gate to deplete the carriers in the channel for nonconducting. This type is called the depletion-mode (normally ON) p-channel MOSFET.

Figure 4-2-2 shows the four modes of MOSFETs and Fig. 4-2-3 illustrates their electric symbols, output V-I characteristics, and transfer characteristics [21].

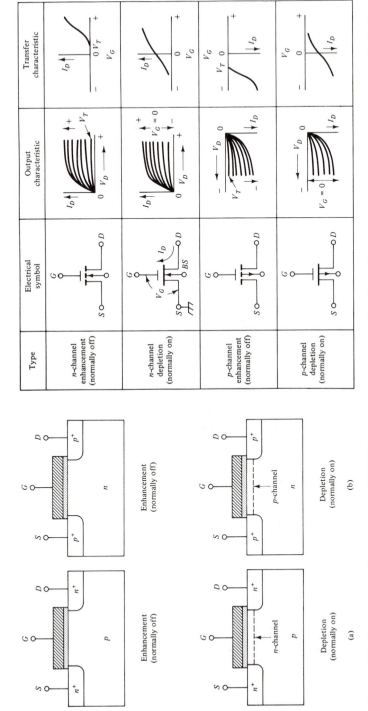

Figure 4-2-3 Electric symbols, output, and transfer characteristics of the four modes of MOSFETs. (After R. C. Gallagher [21]; reprinted by permission of Pergamon Press, Great Britain)

Figure 4-2-2 Four modes of operation for MOSFET: (a) n-channel, (b) p-channel. (After R. C. Gallagher [21]; reprinted by permission of Pergamon Press, Great Britain)

4-2-4 Drain Current and Transconductance

Drain current.   The drain current I_d of the MOSFET depends on the drain voltage V_d. It first increases linearly with the drain voltage in the linear region and then gradually levels off to a saturated value in the saturation region. Figure 4-2-4 shows the voltage-current characteristic curves of an n-channel MOSFET [22].

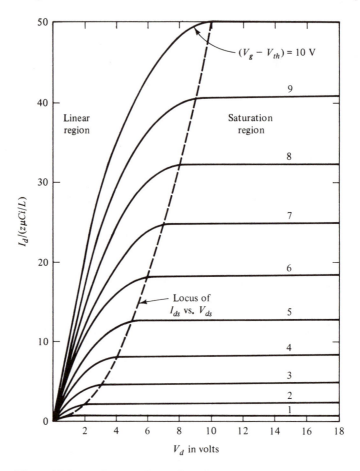

Figure 4-2-4 $V - I$ curves of an n-channel MOSFET. (From S. M. Sze [22]; reprinted by permission of John Wiley & Sons, Ltd.)

The drain current I_d is given by [22]

$$I_d = \frac{Z}{L}\mu_n C_i \left\{ \left(V_g - 2\psi_b - \frac{V_d}{2} \right) V_d - \frac{2}{3C_i}(2\epsilon_s q N_a)^{1/2} \left[(V_d + 2\psi_b)^{3/2} - (2\psi_b)^{3/2} \right] \right\}$$

$$(4\text{-}2\text{-}1)$$

where $\mu_n =$ electron carrier mobility
$C_i = \dfrac{\epsilon_i}{d}$ is the insulator capacitance per unit area
$\epsilon_i =$ insulator permittivity
$V_g =$ gate voltage
$\psi_b = \dfrac{E_i - E_F}{q}$ is the potential difference between the Fermi level E_F and the intrinsic Fermi level E_i

$V_d =$ drain voltage
$\epsilon_s =$ semiconductor permittivity
$q =$ carrier charge
$N_a =$ acceptor concentration

The drain voltage in the linear region is small and Eq. (4-2-1) becomes

$$I_d \simeq \frac{Z}{L}\mu_n C_i \left[(V_g - V_{\text{th}})V_d - \left(\frac{1}{2} + \frac{\sqrt{\epsilon_s q N_a / \psi_b}}{4C_i} \right) V_d^2 \right] \tag{4-2-2}$$

or $$I_d \simeq \frac{Z}{L}\mu_n C_i (V_g - V_{\text{th}})V_d \qquad \text{for } V_d \ll (V_g - V_{\text{th}}) \tag{4-2-3}$$

where $$V_{\text{th}} = 2\psi_b + \frac{2}{C_i}(\epsilon_s q N_a \psi_b)^{1/2}$$

The drain current in the saturation region is given by

$$I_{d(\text{sat})} \simeq \frac{mZ}{L}\mu_n C_i (V_g - V_{\text{th}})^2 \tag{4-2-4}$$

where $m = 0.5$ is the low doping factor
$V_{d(\text{sat})} - V_g - V_{\text{th}}$ may be assumed

Transconductance.   The transconductance g_m, also called the mutual conductance, in the linear region can be found from Eq. (4-2-3) as

$$g_m = \left. \frac{\partial I_d}{\partial V_g} \right|_{V_d = \text{constant}} = \frac{Z}{L}\mu_n C_i V_d \tag{4-2-5}$$

The transconductance in the saturation region becomes

$$g_{m(\text{sat})} = \frac{2mZ}{L}\mu_n C_i (V_g - V_{\text{th}}) \tag{4-2-6}$$

The channel conductance g_d is given by

$$g_d = \left. \frac{\partial I_d}{\partial V_d} \right|_{V_g = \text{constant}} = \frac{Z}{L} \mu_n C_i (V_g - V_{\text{th}}) \qquad (4\text{-}2\text{-}7)$$

Here all equations derived so far are based on the idealized n-channel MOSFET. For an idealized p-channel MOSFET, all voltages V_g, V_d, and V_{th} are negative and the drain current I_d flows from the source to the drain. For a real n-channel MOSFET (say Al-Oxide-Si structure), the saturation drain current is

$$I_{d(\text{sat})} = ZC_i (V_g - V_{\text{th}}) v_s \qquad (4\text{-}2\text{-}8)$$

and the transconductance becomes $g_m = ZC_i v_s$, where $v_s = L/\tau$ is the carrier drift velocity.

Threshold voltage V_{th} is given by

$$V_{\text{th}} = \frac{\Phi_{ms}}{q} - \frac{Q_f}{C_i} + 2\psi_b + \frac{2}{C_i} (\epsilon_s q N_a \psi_b)^{1/2} \qquad (4\text{-}2\text{-}9)$$

where $\Phi_{ms} = \Phi_m - \Phi_s$ is the work function difference (in eV) between the metal work function Φ_m and the semiconductor work function Φ_s

$Q_f =$ fixed oxide charges

Example 4-2-4 Threshold Voltage of an Ideal MOSFET

A certain p-channel MOSFET has the following parameters:

Doping concentration	$N_a = 3 \times 10^{17}\,\text{cm}^{-3}$
Relative dielectric constant	$\epsilon_r = 11.8$
Relative dielectric constant of SiO_2	$\epsilon_{ir} = 4$
Insulator depth	$d = 0.01\ \mu\text{m}$
Operating temperature	$T = 300°\text{K}$

(a) Calculate the surface potential $\psi_s(\text{inv})$ for strong inversion.
(b) Compute the insulator capacitance.
(c) Determine the threshold voltage.

Solution. (a) From Eq. (3-4-4) the surface potential for strong inversion is

$$\psi_s(\text{inv}) = 2 \times 26 \times 10^{-3} \ell\text{n} \left(\frac{3 \times 10^{17}}{1.5 \times 10^{10}} \right) = 0.874 \text{ volt}$$

(b) From Eq. (3-4-9) the insulator capacitance is

$$C_i = \frac{\epsilon_i}{d} = \frac{4 \times 8.854 \times 10^{-12}}{0.01 \times 10^{-6}} = 3.54 \text{ mF/m}^2$$

(c) From Eq. (4-2-3) the threshold voltage is

$$V_{th} = 0.874 + \frac{2}{3.54 \times 10^{-3}} \times (8.854 \times 10^{-12} \times 11.8$$

$$\times 1.6 \times 10^{-19} \times 3 \times 10^{23} \times 0.437)^{1/2}$$

$$= 0.874 + 0.56 \times 10^{3} \times 14.80 \times 10^{-4} = 1.70 \text{ volts}$$

4-2-5 Maximum Operating Frequency

The maximum operating frequency of a MOSFET is determined by its circuit parameters. A common-source equivalent circuit of the MOSFET is shown in Fig. 4-2-5 [23].

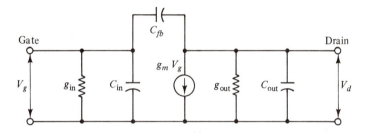

g_{in} = input conductance due to the leakage current (because the leakage current is very small, say 10^{-10} A/cm^2, g_{in} is negligible)

$C_{in} = ZLC_i$ is the input capacitance

C_{fb} = feedback capacitance

$g_{out} = g_d$ = output conductance

$C_{out} = C_iC_s/(C_i + C_s)$ is the sum of the two p-n junction capacitances in series with the semiconductor capacitance per unit area

Figure 4-2-5 Equivalent circuit of a common-source MOSFET. (After H. K. J. Ihantola and J. L. Moll [23]; reprinted with permission of Pergamon Press, Great Britain)

The maximum operating frequency of a MOSFET in the linear region can be expressed as

$$f_m = \frac{\omega_m}{2\pi} = \frac{g_m}{2\pi C_{in}} = \frac{\mu_n V_d}{2\pi L^2} \tag{4-2-10}$$

In the saturation region, $V_g \gg V_{th}$, the transconductance is reduced to

$$g_{m(sat)} \simeq \frac{Z}{L} \mu_n C_i V_g = ZC_i v_s \tag{4-2-11}$$

where $v_s = \mu_n V_g/L$ is the carrier drift velocity.

The maximum operating frequency in the saturation region is

$$f_{m(\text{sat})} = \frac{v_s}{2\pi L} = \frac{1}{2\pi\tau} \qquad (4\text{-}2\text{-}12)$$

where $\tau = L/v_s$ is the carrier transit time.

It is interesting to note that Eq. (4-2-12) for a MOSFET is identical to Eq. (4-1-53) for a MESFET.

Example 4-2-5   Transit Time and Operating Frequency of a Silicon MOSFET

A Silicon MOSFET has the following parameters:

Channel length $L = 5\ \mu\text{m}$
Drift velocity $v_s = 10^7\ \text{cm/s}$

(a) Calculate the carrier transit time.
(b) Compute the maximum operating frequency.

Solution.   (a) The transit time is

$$\tau = \frac{5.0 \times 10^{-6}}{10^5} = 50\ \text{ps}$$

(b) From Eq. (4-2-13) the maximum operating frequency is

$$f = \frac{1}{2 \times 3.1416 \times 50 \times 10^{-12}} = 3.183\ \text{GHz}$$

4-2-6 Microwave Applications

The MOSFETs are often used as power amplifiers because they offer two advantages over MESFETs and JFETs.

1. In the active region of an enhancement-mode MOSFET the input capacitance and the transconductance are almost independent of gate voltage and the output capacitance is independent of the drain voltage. This leads to very linear (class A) power amplification.
2. The active gate-voltage range can be larger because n-channel depletion-type MOSFETs can be operated from the depletion-mode region $(-V_g)$ to the enhancement-mode region $(+V_g)$.

Example 4-2-6   Characteristics of a MOSFET

A certain n-channel MOSFET has the following parameters:

Channel length $L = 4\ \mu\text{m}$
Channel depth $Z = 12\ \mu\text{m}$
Insulator thickness $d = 0.05\ \mu\text{m}$
Gate voltage $V_g = 5\ \text{V}$
Doping factor $m = 1$
Threshold voltage $V_{\text{th}} = 0.10\ \text{V}$
Electron mobility $\mu_n = 1350 \times 10^{-4}\ \text{m}^2/\text{V-s}$

Electron velocity $\quad\quad\quad v_s = 1.70 \times 10^7\ cm/s$

Relative dielectric
constant of SiO$_2$ $\quad\quad\quad \epsilon_{ir} = 3.9$

(a) Compute the insulator capacitance in farads per square meter.
(b) Calculate the saturation drain current in milliamperes.
(c) Determine the transconductance in the saturation region in millimhos.
(d) Estimate the maximum operating frequency in the saturation region in gigahertz.

Solution.   (a) The capacitance of the insulator SiO$_2$ is

$$C_i = \frac{\epsilon_i}{d} = \frac{3.9 \times 8.854 \times 10^{-12}}{0.05 \times 10^{-6}} = 6.91 \times 10^{-4}\ F/m^2$$

(b) From Eq. (4-2-8) the saturation drain current is

$$I_{d(sat)} = ZC_i\left(V_g - V_{th}\right)v_s$$
$$= 12 \times 10^{-6} \times 6.91 \times 10^{-4} \times (5 - 0.1) \times 1.7 \times 10^5$$
$$= 6.91\ mA$$

(c) From Eq. (4-2-9) the transconductance in the saturation region is

$$g_{m(sat)} = ZC_iv_s$$
$$= 12 \times 10^{-6} \times 6.91 \times 10^{-4} \times 1.7 \times 10^5$$
$$= 1.41\ mmhos$$

(d) From Eq. (4-2-13) the maximum operating frequency is

$$f_m = \frac{v_s}{2\pi L} = \frac{1.7 \times 10^5}{2\pi \times 4 \times 10^{-6}} = 6.76\ GHz$$

4-3 SHORT-CHANNEL MOS FIELD-EFFECT TRANSISTORS (VMOSFETs, UMOSFETs, DMOSFETs, AND VDMOSFETs)

The MOSFET was described earlier. It is a typical small-signal and low-power device, however, for it has a small channel area, low breakdown voltage, and a long channel length. Conversely, the V-channel MOS field-effect transistor (VMOSFET), the U-channel MOSFET (UMOSFET), the double-diffused MOSFET (DMOSFET), and the vertical double-diffused MOSFET (VDMOSFET) are the modified MOSFETs with short channels. Because their channel length L is much shorter than the MOSFET's, these devices have high transconductance and high amplification, and can handle high current and high power at the microwave frequency range.

4-3-1 Physical Structures

Figures 4-3-1 and 4-3-2, respectively, show the schematic diagrams for one VMOS-FET (V-channel, or Vertical channel, or V-shaped grooved MOSFET) structure, one

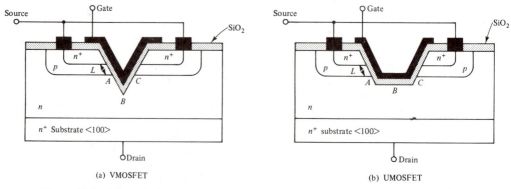

(a) VMOSFET   (b) UMOSFET

Figure 4-3-1 Schematic diagrams of VMOSFET and UMOSFET (After F. E. Holmes and C. A. T. Salama [24]; reprinted with permission of Pergamon Press, Great Britain)

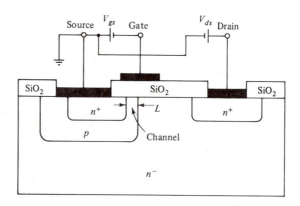

(a) DMOSFET

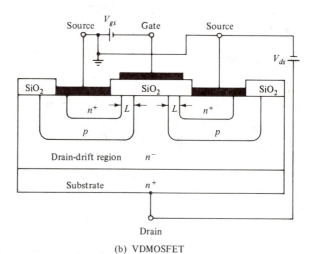

(b) VDMOSFET

Figure 4-3-2 Schematics diagrams of DMOSFET and VDMOSFET (orig Oxner)

UMOSFET structure, one DMOSFET structure, and one VDMOSFET structure [24, 25, 40].

Each type of device has two channels in parallel, one on each side of the etched groove, so the device can handle high current and high power. The two channels are in common at the drain, and the channel length is L. The source is grounded with respect to the gate and the drain.

4-3-2 Drain Current and Transconductance

Assuming negligible penetration of the source and drain space regions into the channel, the threshold voltage is given by [25]

$$V_{th} = \frac{\Phi_{ms}}{q} + 2\psi_b - \frac{Q_i}{C_i} - \frac{Q_b}{C_b} \tag{4-3-1}$$

where $\Phi_{ms} = \Phi_m - \Phi_s$ is the work function difference in electronvolts between the metal and semiconductor

$\psi_b = \dfrac{E_i - E_F}{q}$ is the surface potential between the Fermi level E_F and the intrinsic Fermi level E_i

$Q_i =$ insulator charge per unit area
$Q_b =$ bulk charge per unit area
$C_i = \epsilon_i/d$ is the insulator capacitance per unit area
$C_b = \epsilon_b/L$ is the bulk-semiconductor capacitance per unit area

Assuming that the gradual channel approximation is valid at low drain voltages (linear region) for a short channel device, the drain current can be expressed as

$$I_d = Z\mu_n C_i \left(V_g - V_{th} - V_d \right) \frac{dV}{dy} \tag{4-3-2}$$

where $\mu_n =$ electron carrier mobility
$Z =$ channel depth
$\dfrac{dV}{dy} = E_t$ is the transverse electric field along the channel
$V_{eg} = V_g - V_{th}$ is the effective gate voltage

Integration of Eq. (4-3-2) from 0 to L yields

$$I_d = \frac{Z}{L}\mu_n C_i \left[\left(V_g - V_{th} \right) V_d - \frac{V_d^2}{2} \right] \tag{4-3-3}$$

The drain conductance in the linear region is

$$g_d = \frac{Z}{L}\mu_n C_i \left(V_g - V_{th} \right) \tag{4-3-4}$$

In the saturation region ($dV/dy = E_t =$ constant) the drain current is expressed as

$$I_d = \frac{Z}{L}\mu_n C_i \left[\left(V_g - V_{th} \right) V_d - \frac{V_d^2}{2} \right] \tag{4-3-5}$$

where $V_d = (V_g - V_{th}) + V_{cr} - \sqrt{(V_g - V_{th})^2 + V_{cr}^2}$ is the drain voltage at the saturation region and $V_{cr} = E_t L$ is the critical voltage.

The transconductance in the saturation region is

$$g_m = \frac{dI_d}{dV_g}\bigg|_{V_d = \text{constant}} = \frac{Z}{L} \mu_n C_i (V_g - V_{th}) V_{cr} \left[(V_g - V_{th})^2 + V_{cr}^2 \right]^{-1/2} \quad (4\text{-}3\text{-}6)$$

Then for $(V_g - V_{th}) \gg V_{cr}$, Eq. (4-3-6) becomes

$$g_{m(\text{sat})} \simeq \frac{Z}{L} \mu_n C_i V_{cr} = ZC_i v_s \quad (4\text{-}3\text{-}7)$$

where $v_s = \mu_n V_{cr}/L$ is the carrier saturation drift velocity.

4-3-3 Microwave Applications

The maximum operating frequency in the saturation region is

$$f_m = \frac{v_s}{2\pi L} = \frac{1}{2\pi\tau} = \frac{\mu_n V_{cr}}{2\pi L^2} \quad (4\text{-}3\text{-}8)$$

Equation (4-3-8) shows that the VMOSFET and UMOSFET can operate at much higher frequencies than the standard MOSFET because its channel lengths L are much shorter. Also, the VMOSFET and UMOSFET have large transconductance and are capable of carrying high current. Therefore VMOSFET and UMOSFET are widely used as high-power amplifiers at the microwave frequency range in digital integrated circuits.

Example 4-3-3 Characteristics of a VMOSFET

A certain VMOSFET has a short channel length L of 0.4 μm and the transverse electric field along the channel E_t is 2×10^6 V/m. The other parameters are the same as in Example 4-2-6.
(a) Compute the critical voltage in volts.
(b) Calculate the drain voltage at the saturation region in volts.
(c) Determine the saturation drain current in milliamperes.
(d) Compute the carrier saturation drift velocity in meters per second.
(e) Find the transconductance in millimhos.
(f) Estimate the maximum operating frequency in gigahertz.

Solution. (a) From Eq. (4-3-5) the critical voltage is

$$V_{cr} = E_t L = 2 \times 10^6 \times 0.4 \times 10^{-6} = 0.80 \text{ volt}$$

(b) From Eq. (4-3-5) the drain voltage is

$$V_d = \left(V_g - V_{th}\right) + V_{cr} - \sqrt{\left(V_g - V_{th}\right)^2 + V_{cr}^2}$$

$$= (5 - 0.1) + 0.8 - \sqrt{(5 - 0.1)^2 + 0.8^2}$$

$$= 0.74 \text{ volt}$$

(c) From Eq. (4-3-5) the saturation drain current is

$$I_{d(sat)} = \frac{Z}{L}\mu_n C_i \left[\left(V_g - V_{th}\right)V_d - \frac{V_d^2}{2}\right]$$

$$= \frac{12 \times 10^{-6}}{0.4 \times 10^{-6}} \times 0.135 \times 6.91 \times 10^{-4}$$

$$\times \left[(5 - 0.1) \times 0.74 - \frac{0.74^2}{2}\right]$$

$$= 9.40 \text{ mA}$$

(d) From Eq. (4-3-7) the carrier drift velocity is

$$v_s = \frac{\mu_n V_{cr}}{L} = \frac{0.135 \times 0.8}{0.4 \times 10^{-6}} = 2.7 \times 10^5 \text{ m/s}$$

(e) From Eq. (4-3-7) the transconductance is

$$g_m = ZC_i v_s = 12 \times 10^{-6} \times 6.91 \times 10^{-4} \times 2.7 \times 10^5$$

$$= 2.24 \text{ m}\mho$$

(f) From Eq. (4-3-8) the maximum operating frequency is

$$f_m = \frac{v_s}{2\pi L} = \frac{2.7 \times 10^5}{2\pi \times 0.4 \times 10^{-6}}$$

$$= 107.4 \text{ GHz}$$

From the results of Examples 4-2-6 and 4-3-3 it is clear that the VMOSFET with a shorter channel length than that of the standard MOSFET can have higher power amplification and higher operating frequency.

4-4 CHARGE-COUPLED DEVICES (CCDs)

The charge-coupled device (CCD) is a metal-oxide-semiconductor (MOS) diode structure that was proposed in 1969 by Boyle and Smith [26, 27]. The CCD can move the charges in the MOS diode along a predetermined path under the control of clock pulses and so it is also called the charge-transfer device (CTD). CCDs have many microwave applications, such as in infrared detection and imaging and digital signal processing. There are three basic types of CCDs: surface-channel CCD (SCCD), buried-channel CCD (BCCD), and junction CCD (JCCD). In the SCCD or BCCD the charge is stored and transferred at the semiconductor surface or in the semiconductor interior, respectively; whereas in the JCCD the store and transfer of the charge packet occur at the *p-n* junction.

The motion of the charge packets in a charge-coupled device is transversely controlled by the applied gate voltages. This phenomenon is similar to the carrier motion in a microwave field-effect transistor like MESFET or MOSFET. In effect, the CCD can be referred to as the field-effect CCD.

4-4-1 *Operational Mechanism*

A charge-coupled device (CCD) is an array of many MOS or MIS diodes. In operation, the information (or signal) is stored in the form of electrical charge packets in the potential wells created in a MOS diode. Under the control of externally applied voltage (i.e., gate voltage), the potential wells and the charge packets can be shifted from one well to an adjacent one rapidly through the entire CCD structure.

Three separate mechanisms allow the charge packets to move from one well to another: self-induced drift, thermal diffusion, and fringing field drift. Thermal diffusion results in an exponential decay of the remaining charge under the transferring electrode. The fringing field is the electric field in the direction of charge flow and it can help speed the charge-transfer process considerably. Self-induced drift (or a charge-repulsion effect) is only important at relatively large signal-charge densities. It is the dominant mechanism in the transfer of the first 99% of the charge signal.

Energy band of MIS diode.   A single MIS structure on an *n*-type semiconductor (or *p*-type semiconductor) is the basic element of the CCD. Figure 4-4-1 shows the energy band diagrams for a MIS structure [26].

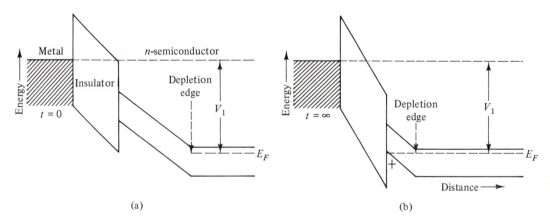

Figure 4-4-1 Energy-band diagrams of MIS structure (From W. S. Boyle and G. E. Smith [26]; reprinted with permission from *The Bell System*, AT&T)

The voltage applied to the metal electrode is negative with respect to the semiconductor and large enough to cause depletion. When the voltage is first applied at $t = 0$, there are no holes at the insulator-semiconductor interface [see Fig. 4-4-1(a)]. As holes are introduced into the depletion region, they will accumulate at the interface and cause the surface potential to be more positive [see Fig. 4-4-1(b)].

Three-phase structure.   The CCD can be constructed in the form of a typical three-phase structure as shown in Fig. 4-4-2 [27].

$V_1 = -5$ V $V_2 = -10$ V $V_3 = -5$ V

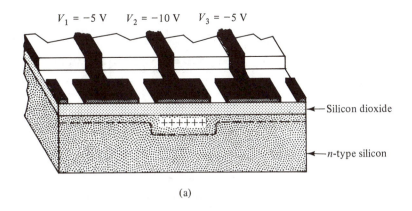

(a)

$V_1 = -5$ V $V_2 = -10$ V $V_3 = -15$ V

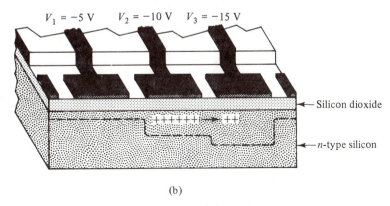

(b)

Figure 4-4-2 Cutaway of CCD (From W. S. Boyle and G. E. Smith [27]; reprinted with permission of the IEEE, Inc.)

The CCD consists of a closely spaced array of MIS diodes on an n-type semiconductor substrate with a large negative gate voltage applied. Its basic function is to store and transfer the charge packets from one potential well to an adjacent one. As shown in Fig. 4-4-2(a), $V_1 = V_3$ and V_2 is more negative. In effect, a potential well with stored holes is created at gate electrode 2. The stored charge is temporary because a thermal effect will diffuse the holes out of the wells. Therefore the switching time of the voltage clock must be fast enough to move all charges out of the occupied well to the next empty one. When the voltage V_3 is pulsed to be more negative than the other two voltages V_1 and V_2, the charge begins to transfer to the potential well at gate electrode 3 as shown in Fig. 4-4-2(b).

Store and transfer of charge packets.   A linear array of MIS diodes on an n-type semiconductor is shown in Fig. 4-4-3 [26].

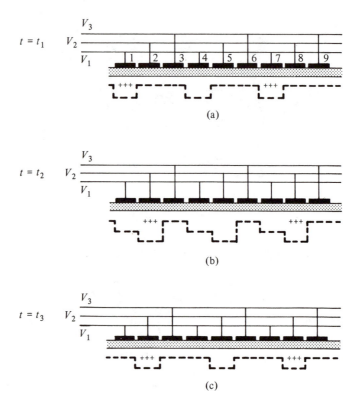

(a)

(b)

(c)

Figure 4-4-3   Store and transfer of charge packets for a three-phase CCD (From W. S. Boyle and G. E. Smith [26]; reprinted with permission from *The Bell System*, AT&T)

For a three-phase CCD, every third gate electrode is connected to a common line (see Fig. 4-4-3). At $t = t_1$, a more negative voltage V_1 is applied to gate electrodes 1, 4, 7, and so on and less negative voltages V_2 and V_3 ($V_2 = V_3$) are applied to the other gate electrodes. It is assumed that the semiconductor substrate is grounded and that the magnitude of V_1 is larger than the threshold voltage V_{th} for the production of inversion under steady-state conditions. As a result, positive charges are stored in the potential wells under electrodes 1, 4, 7,... as shown in Fig. 4-4-3(a).

At $t = t_2$, when voltage V_2 at the gate electrode 2, 5, 8,... is pulsed to be more negative than V_1 and V_3, the charge packets will be transferred from the potential wells at gate electrodes 1, 4, 7,... to the potential minimum under gate electrodes 2, 5, 8,... as shown in Fig. 4-4-3(b).

At $t = t_3$, when voltages $V_1 = V_3$ and voltage V_2 remains more negative, the charge packets have been transferred one spatial position and the sequence is ready to continue as shown in Fig. 4-4-3(c).

4-4-2 *Surface-Channel Charge-Coupled Devices (SCCDs)*

A surface-channel charge-coupled device (SCCD) is basically an MOS-diode structure as shown in Fig. 4-4-4(a) [28]. Its energy band diagrams are illustrated in parts (b) and (c).

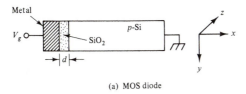

(a) MOS diode

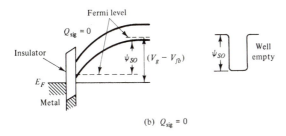

(b) $Q_{\text{sig}} = 0$

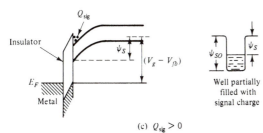

(c) $Q_{\text{sig}} > 0$

Figure 4-4-4 Physical structure and energy-band diagrams of a BCCD (From D. F. Barle [28]; reprinted with permission of the IEEE, Inc.)

There are two cases for the creation of a charge packet under deep depletion.

1. Zero Signal-Charge ($Q_{sig} = 0$). When the signal charge is zero, an empty well is formed by the potential minimum at the semiconductor surface as shown in Fig. 4-4-4(b). Gate voltage V_g and surface potential ψ_s are related by

$$V_g - V_{fb} = V_i + \psi_s = \frac{qN_aW}{C_i} + \psi_s \qquad (4\text{-}4\text{-}1)$$

where V_{fb} = flatband voltage
        V_i = voltage across the insulator

Substitution of Eq. (3-4-5) into Eq. (4-4-1) yields

$$V_g - V_{fb} = \frac{1}{C_i}(2\epsilon_s qN_a\psi_s)^{1/2} + \psi_s \qquad (4\text{-}4\text{-}2)$$

2. Stored Signal-Charge ($Q_{sig} > 0$). When a signal-charge packet is stored at the semiconductor surface, the surface potential decreases and the potential well is partially filled as shown in Fig. 4-4-4(c). The surface potential equation of Eq. (4-4-2) becomes

$$V_g - V_{fb} = \frac{Q_{sig}}{C_i} + \frac{1}{C_i}(2\epsilon_s qN_a\psi_s)^{1/2} + \psi_s \qquad (4\text{-}4\text{-}3)$$

The maximum charge density (electron or hole) that can be stored on an MOS capacitor is approximately equal to

$$N_{max} \simeq \frac{C_iV_g}{q} \quad \text{for } V_g \gg 1 \qquad (4\text{-}4\text{-}4)$$

Example 4-4-2   Maximum Charge Density of a SCCD

A certain surface-channel charge-coupled device has the following parameters:

$$V_g = 10 \text{ volts}$$
$$d = 0.1 \ \mu\text{m}$$
$$\epsilon_{ir} = 3.9$$

Determine the maximum charge density.

Solution.

$$C_i = \frac{\epsilon_i}{d} = \frac{8.854 \times 10^{-12} \times 3.9}{0.1 \times 10^{-6}}$$

$$= 34.53 \times 10^{-5} \text{ F/m}^2$$

$$= 34.53 \times 10^{-9} \text{ F/cm}^2$$

$$N_{max} = \frac{C_iV_g}{q} = \frac{34.53 \times 10^{-9} \times 10}{1.6 \times 10^{-19}}$$

$$= 2.16 \times 10^{12} \text{ charges/cm}^2$$

4-4-3 *Buried-Channel Charge-Coupled Devices (BCCDs)*

The potential wells of an SCCD are formed at the insulator-semiconductor interface and the minority-carrier charge packets are moved along the surface. The effects of interface traps limit the performance of the SCCD. Conversely, the buried-channel CCD (BCCD) forms a buried channel with potential wells below the semiconductor surface to avoid charge trapping by interface states (see Fig. 4-4-5) [29, 30]. The device consists of an *n*-type semiconductor layer on a *p*-type substrate with n^+ contacts at either end of the channel.

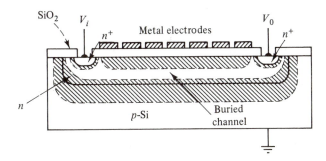

Figure 4-4-5 Cross-sectional view of a BCCD (From R. H. Walden et al. [30]; reprinted with permission from *The Bell System*, AT&T)

Figure 4-4-6 shows the diagrams of a BCCD structure and its energy bands [31].

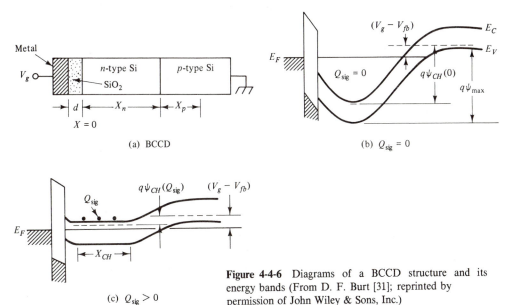

(a) BCCD

(b) $Q_{sig} = 0$

(c) $Q_{sig} > 0$

Figure 4-4-6 Diagrams of a BCCD structure and its energy bands (From D. F. Burt [31]; reprinted by permission of John Wiley & Sons, Inc.)

The silicon n-type semiconductor has an additional thin layer whose conductivity type is opposite to that of the semiconductor as shown in Fig. 4-4-6(a). During the operation of a BCCD and with no signal charge ($Q_{sig} = 0$), the top layer is fully depleted of the mobile charge as indicated in part (b). So the potential minimum forms below the silicon surface. When signal charges are introduced to the buried channel, they are stored in the channel as shown in part (c).

The potential distributions for part (b) of Fig. 4-4-6 can be obtained by solving the following Poisson equations [32]:

$$\frac{d^2\psi}{dx^2} = 0 \qquad\qquad -d < x < 0 \qquad\qquad (4\text{-}4\text{-}5)$$

$$\frac{d^2\psi}{dx^2} = -\frac{qN_d}{\epsilon_s} \qquad\qquad 0 < x < x_n \qquad\qquad (4\text{-}4\text{-}6)$$

$$\frac{d^2\psi}{dx^2} = \frac{qN_a}{\epsilon_s} \qquad\qquad x_n < x < (x_n + x_p) \qquad\qquad (4\text{-}4\text{-}7)$$

with the boundary conditions as

1. $\psi = V_g - V_{fb}$ at $x = -d$
2. $\psi = 0$ at $x = x_n + x_p$
3. Continuity of potential and the electric displacement at $x = 0$ and $x = x_n$

Then the maximum potential depth ψ_{max} is given by

$$\psi_{max} = \frac{qN_a}{2\epsilon_s}\left(1 + \frac{N_a}{N_d}\right)x^2 \qquad\qquad (4\text{-}4\text{-}8)$$

and the applied voltage (gate voltage) becomes

$$V_g - V_{fb} = \psi_j + \sqrt{V_{0x}\psi_j} - V_1 \qquad\qquad (4\text{-}4\text{-}9)$$

where

$$\psi_j = \frac{qN_a}{2\epsilon_s}x_p^2$$

$$V_{0x} = \frac{2qN_a}{\epsilon_s}\left(1 + \frac{\epsilon_s d}{\epsilon_i x_n}\right)^2 x_p^2$$

$$V_1 = \frac{qN_d}{2\epsilon_s}\left(1 + \frac{2\epsilon_s d}{\epsilon_i x_n}\right)x_n^2$$

When charge packets are introduced to the buried channel, potential distribution can be determined from Eqs. (4-4-4) through (4-4-7) with $[N_d - n(x)]$ replacing N_d in Eq. (4-4-6), and $[-p(x) + n(x) + N_a]$ replacing N_a in Eq. (4-4-7), where $n(x)$ and $p(x)$ are the free electron and free-hole density, respectively.

Example 4-4-3   Maximum Potential Depth of a BCCD

A certain Si BCCD has the following parameters:

Electron density	$N_d = 2 \times 10^{15} \text{ cm}^{-3}$
Hole density	$N_a = 5 \times 10^{14} \text{ cm}^{-3}$
Thickness of n-type Si	$x_n = 3 \ \mu\text{m}$
Thickness of p-type Si	$x_p = 2 \ \mu\text{m}$
Relative dielectric constant	$\epsilon_r = 11.8$

Compute the maximum potential-well depth.

Solution. From Eq. (4-4-8), the maximum potential-well depth is

$$\psi_{max} = \frac{1.6 \times 10^{-19} \times 5 \times 10^{20}}{2 \times 8.854 \times 10^{-12} \times 11.8} \left(1 + \frac{5 \times 10^{14}}{2 \times 10^{15}} \right) (5 \times 10^{-6})^2$$

$$= 11.96 \text{ V}$$

4-4-4 Junction Charge-Coupled Devices (JCCDs)

A junction charge-coupled device (JCCD) can be constructed by replacing the upper MOS capacitor in a buried-channel CCD by reverse-biased p-n junction diodes [33, 34]. Through the p-n diodes, charges can be injected into the CCD transport channel and can be extracted from the CCD. Substitution of a junction diode for the MOS capacitor does not alter the principle of CCD operation. The JCCDs do have some advantages in image-sensing applications over the SCCDs and BCCDs because the latter two have no light-absorbing conducting layers on top of the oxide layer. JCCDs, however, may have disadvantages, such as low transfer efficiency (about 0.999) and low charge-handling capability (about 2×10^{11} electrons/cm^2).

When a positive voltage, large enough to deplete the n-type layer completely, is applied to the drain the fixed positive space charge between the gates causes potential wells or barriers for electrons in the underlying parts of the n-type layer. These potential wells will impair the charge-transfer efficiency. This problem can be solved by increasing the space-charge density of the n-type layer under the gates with respect to the space-charge density between the gates. Figure 4-4-7 shows the cross section and top view of a JCCD [35].

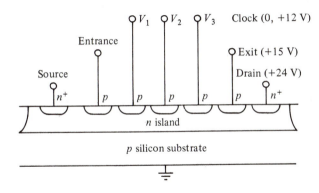

Figure 4-4-7(a) Cross section of a JCCD (From C. D. Hartgring and M. Kleefstra [35]; reprinted with permission of the IEEE, Inc.)

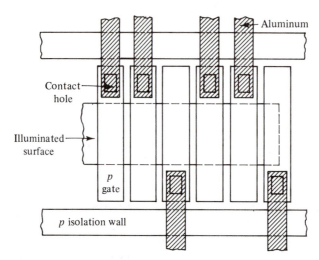

Figure 4-4-7(b) Top view of a JCCD (From C. D. Hartgring and M. Kleefstra [35]; reprinted with permission of the IEEE, Inc.)

It should be noted that only one cell of a three-phase JCCD is shown in part (a). When voltages on the source and drain are sufficiently positive with respect to voltages on the substrate and the *p* gates, the whole *n* island can be depleted of electrons. The potential in the *n* region between the *p* gates and the substrate can then be modulated by the gate voltage. The presence of electrons under one gate lowers the potential there compared to the potential when electrons are absent. By making the source less positive for a short time, electrons can be injected and transported as charge packets by applying a suitable sequence of clock pulses to the *p*-type gates. Electron-hole pairs can be generated in the semiconductor by light absorption. The holes will be collected in the *p* region and the electrons can be collected in the most positive parts of the *n* layer.

As shown in Fig. 4-4-7(b), only that part of the JCCD is illuminated where no contact holes to the *p* gates and no metallization layers are present. For such a *p-n-p* structure, the fraction of electrons collected under the positive gates can be close to unity if the surface recombination rate is lower than $10^3/\text{cm}^3$ per second and the diffusion length of the minority carriers in the *p* gates is higher than two times the thickness of the neutral *p* gate.

The channel-potential diagram of a three-phase JCCD is shown in Fig. 4-4-8 with different voltages applied to the transfer gates [34].

It is assumed that no charges are present in the channel when no voltages are applied. When the potential applied to gate 1 is 0 V, then gate 2 is switched from 10 V to 2 V and gate 3 is kept at 10 V. The potential well under gate 3 is 90% filled ($\psi_{ch} = 11$ V).

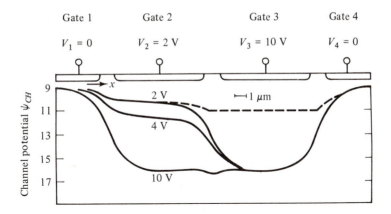

Figure 4-4-8   Channel potential profile of a JCCD (From E. A. Wolsheimer [34]; reprinted with permission of the IEEE Inc.)

The signal carrier transit time t_r is given by

$$t_r = \frac{1}{\mu_n} \int_0^L \frac{dx}{E_x(x)} = \frac{L}{\mu_n E_x} \tag{4-4-10}$$

Example 4-4-4   Signal Carrier Transit Time in a JCCD

A three-phase JCCD has a length L of 10 μm, an electric field E_x of 500 V/cm, and an electron mobility μ_n of 10^3 cm^2/V-s. Determine the signal carrier transit time.

Solution.   From Eq. (4-4-10), the signal transit time is

$$t_r = \frac{10 \times 10^{-4}}{10^3 \times 500} = 2 \text{ ns}$$

Although not all charges are transferred within a period of 2 ns, clock frequencies up to 100 MHz hardly impair the transfer efficiency. The maximum charge-handling capability is measured at about 10^{12} electrons/cm^2.

4-4-5 Dynamic Characteristics

The dynamic characteristics can be described in terms of charge-transfer efficiency η, frequency response, and power dissipation.

Charge-transfer efficiency η.   The charge-transfer efficiency is defined as the fraction of charge transferred from one well to the next in a CCD. The fraction left behind is the transfer loss and it is denoted by ϵ. Therefore the charge-transfer efficiency is given by

$$\eta = 1 - \epsilon \tag{4-4-11}$$

If a single charge pulse with an initial amplitude P_0 transfers down a CCD register, after n transfers, the amplitude P_n becomes

$$P_n = P_0\eta^n = P_0(1 - n\epsilon) \quad \text{for } \epsilon \ll 1 \tag{4-4-12}$$

where n equals the number of transfers or phases.

If many transfers are required, the transfer loss ϵ must be very small. For example, if a transfer efficiency of 99.99% is required for a three-phase, 330-stage shift register, the transfer loss must be less than 0.01%. The maximum achievable transfer efficiency depends on two factors: how fast the free charge can be transferred between adjacent gates and how much of the charge gets trapped at each gate location by stationary states.

Frequency response.   There are, in fact, upper and lower frequency limitations for CCDs. The potential well will not remain indefinitely and thermally generated electrons (or holes) eventually fill the well completely. Also, the time stored by the charge must be much shorter than the thermal relaxation time of the CCD's capacitor. So the maximum frequency is limited by the channel length L.

Power dissipation.   The power dissipation per bit is given by

$$P = nfVQ_{\max} \tag{4-4-13}$$

Example 4-4-5   Power Dissipation of a Three-Phase CCD

A three-phase CCD is operating under the following conditions:

Applied voltage	$V = 10$ volts
Number of phases	$n = 3$
Maximum stored charges	$Q_{\max} = 0.04$ pC
Clock frequency	$f = 10$ MHz

Determine the power dissipation per bit.

Solution.   From Eq. (4-4-13) the power dissipation per bit is

$$P = nfVQ_{\max} = 3 \times 10^7 \times 10 \times 0.04 \times 10^{-12}$$
$$= 12 \ \mu W$$

4-4-6 Microwave Applications

Charge-coupled devices (CCDs) have many microwave applications in electronic components and systems, such as in infrared systems and signal processing.

Infrared detection and imaging.   Because varying amounts of charge corresponding to information can be introduced into the potential wells at one end of the CCD structure to emerge after some delay at the other end, the CCD is capable of detecting and imaging the infrared light from a target [36]. Ten years ago, when the CCD was first developed, only an array of 12×12 photodetectors could be used to detect the infrared images of a target. Today, due to the availability of greatly improved CCDs, an array of photodetectors, such as Indium Antimonide (InSb), 128×128, are used to form charge packets that are proportional to the light intensity of a target; and these packets are shifted to a detector point for detection, readout, multiplexing, and time delay and integration (TDI). In scanned IR systems TDI is one of the most important functions performed by CCDs. In such a system the scene is mechanically scanned across an array of detector elements. By using CCD columns to shift the detector output signals (in the form of charge packets) along the focal plane with the same speed as the mechanical scan moves the scene across the array, the signal-to-noise ratio can be improved by the square root of the number of detector elements in the TDI column.

Signal processing.   The CCD can perform several analog and digital signal processing functions, such as delay, multiplexing, demultiplexing, transversal filtering, recursive filtering, integration, analog memory, digital memory, and digital logic. Thus CCDs are being used widely in special applications for the very large scale integration (VLSI) circuits.

4-4-7 Design Example

The project here is to design an n-type three-phase surface-channel CCD with the following specifications:

Electron density	$N_{\max} = 2 \times 10^{12}$ cm^{-2}
Insulator relative dielectric constant	$\epsilon_{ir} = 3.9$
Insulator thickness	$d = 0.15 \ \mu$m
Insulator cross section	$A = 0.5 \times 10^{-4}$ cm^2
Power dissipation allowable per bit	$P = 0.67$ mW

Procedures:
(a) Compute the insulator capacitance in farads per square centimeter.
(b) Determine the maximum stored charges per well in coulombs.
(c) Find the required applied gate voltage in volts.
(d) Choose the clock frequency in megahertz.

Solution.   (a) The insulator capacitance is

$$C_i = \frac{\epsilon_i}{d} = \frac{3.9 \times 8.854 \times 10^{-12}}{0.15 \times 10^{-6}} = 23 \text{ nF/cm}^2$$

(b) The maximum stored charges per well is

$$Q_{max} = N_{max}qA = 2 \times 10^{12} \times 1.6 \times 10^{-19} \times 0.5 \times 10^{-4}$$

$$= 16 \text{ pC}$$

(c) From Eq. (4-4-4) the required applied gate voltage is

$$V_g = \frac{N_{max}q}{C_i} = \frac{2 \times 10^{12} \times 1.6 \times 10^{-19}}{23 \times 10^{-9}} = 14 \text{ V}$$

(d) From Eq. (4-4-13) the clock frequency is

$$f = \frac{P}{nVQ_{max}} = \frac{0.67 \times 10^{-3}}{3 \times 14 \times 16 \times 10^{-12}} = 1 \text{ MHz}$$

REFERENCES

[1] Shockley, W. A unipolar field-effect transistor. *Proc. IRE*, **40**, no. 11 (November 1952), 1365–1376.

[2] Liechti, C. A. Microwave field-effect transistors—1976. *IEEE Trans. on Microwave Theory and Techniques*, **MTT-24**, no. 6 (June 1976), 279–300.

[3] Schottky, W. *Natarwiss.*, **26**, (1938), 843.

[4] Mead, C. A. Schottky Barrier Gate Field-Effect Transistor. *Proc. IEEE*, **54**, no. 2 (February 1966), 307–308.

[5] Hooper, W. W., et al. An Epitaxial GaAs Field-Effect Transistor. *Proc. IEEE*, **55**, no. 7 (July 1967), 1237–1238.

[6] Wolf, P. Microwave Properties of Schottky-Barrier Field-Effect Transistors. *IBM J. Res. Develop.* **14** (March 1970), 125–141.

[7] Zuleeg, R. and K. Lehovec. High Frequency and Temperature Characteristics of GaAs Junction Field-Effect Transistors in the Hot Electron Range. *Proc. Symp. GaAs*, Institute of Physics Conf. Series no. 9, pp. 240–245, 1970.

[8] Hower, P. L., and N. G. Bechtel. Current Saturation and Small-Signal Characteristics of GaAs Field-Effect Transistors. *IEEE Trans. on Electron Devices*, **ED-20**, no. 3 (March 1973), 213–220.

[9] Lehovec, K., and R. Zuleeg. Voltage-Current Characteristics of GaAs J-FET's in the Hot Electron Range. *Solid-State Electronics*, vol. 13, pp. 1415–1426. Great Britian: Pergamon Press, 1970.

[10] Drangeid, K. E., and R. Sommerhalder. Dynamic Performance of Schottky-Barrier Field-Effect Transistors. *IBM J. Res. Develop.* **14** (March 1970), 82–94.

[11] Hower, P. L., et al. The Schottky Barrier Gallium Ansenide Field-Effect Transistor. *Proc. Symp. GaAs*, Institute of Physics and Physical Society Conf. Series no. 7, pp. 187–194, 1968.

[12] Das, M. B., and P. Schmidt. High-Frequency Limitations of Abrupt-Junction FET's. *IEEE Trans. on Electron Devices*, **ED-20**, no. 9 (September 1973), 779–792.

[13] Baechtold, W. Noise Behavior of Schottky Barrier Gate Field-Effect Transistors at Microwave Frequencies. *IEEE Trans. on Electron Devices*, **ED-18**, no. 2 (February 1971), 97–104.

[14] Baechtold, W. Noise Behavior of GaAs Field-Effect Transistors with Short Gate Lengths. *IEEE Trans. on Electron Devices*, **ED-19**, no. 5 (May 1972), 674–680.

[15] van der Ziel, A. Gate Noise in Field-Effect Transistors at Moderately High Frequencies. *Proc. IEEE*, **51**, no. 3 (March 1963), 461–467.

[16] van der Ziel, A. and J. W. Ero. Small-Signal, High-Frequency Theory of Field-Effect Transistors. *IEEE Trans. on Electron Devices*, **ED-11**, no. 2 (February 1964), 128–135.

[17] van der Ziel, A. Thermal Noise in Field-Effect Transistors. *Proc. IRE*, **50** (January 1962), 108–112.

[21] Gallagher, R. C., and W. S. Corak. A metal-oxide-semiconductor (MOS) Hall element. *Solid-state Electronics*, **9** (1966), 571.

[22] Sze, S. M. *Physics of Semiconductor Devices*, 2nd ed, p. 445, 455, 441. New York: John Wiley & Sons, 1981.

[23] Ihantola, H. K. J., and J. L. Moll. Design theory of a surface field-effect transistor. *Solid-state Electronics*, **7** (1964), 423.

[24] Holmes, F. E., and C. A. T. Salama. VMOS—a new MOS integrated circuit technology. *Solid-state Electronics*, **17** (1974), 791.

[25] Salama, C. A. T. A new short channel MOSFET structure (UMOST). *Solid-state Electronics*, **20** (1977), 1003.

[26] Boyle, W. S., and G. E. Smith. Charge couple semiconductor devices. *Bell Syst. Tech. J.*, **49** (April 1970), 587–593.

[27] Boyle, W. S., and G. E. Smith. Charge-coupled devices—a new approach to MIS device structures. *IEEE spectrum*, **8**, no. 7 (July 1971), 18–27.

[28] Barle, D. F. Imaging devices using the charge-coupled concept. *Proc. IEEE*, **63**, no. 1 (January 1975), 38–67.

[29] Boyle, W. S., and G. E. Smith. U.S. patent 3,792,322 (1974).

[30] Walden, R. H., et al. The buried-channel charge-coupled devices. *Bell Syst. Tech. J.*, **51** (1972), 1635.

[31] Burt, D. J. Basic operation of the charge-coupled device. *Int. Conf. Technol. Appl. CCD*, University of Edinburgh, p. 1, 1974.

[32] Kim, C. K. The physics of charge-coupled devices. In M. J. Howes and D. V. Morgan, eds., *Charge-coupled Devices and Systems*. New York: John Wiley & Sons, 1979.

[33] Wolsheimer, E. A., and M. Kleefstra. Experimental results on junction charge-coupled devices. *IEEE Trans. on Electron Devices*, **ED-29**, no. 12 (December 1982), 1930–1936.

[34] Wolsheimer, E. A. Optimization of potential profiles in junction charge-coupled devices. *IEEE Trans. on Electron Devices*, **ED-28**, no. 7 (July 1981), 811–817.

[35] Hartgring, C. D., and M. Kleefstra. Quantum efficiency and blooming suppression in junction charge-coupled devices. *IEEE J. Solid-State Circuits*, **SC-13**, no. 5 (October 1978), 728–730.

[36] Steckl, A. J., et al. Application of charge-coupled devices to infrared detection and imaging. *Proc. IEEE*, **63**, no. 1 (January 1975), 67–74.

[37] Kurokawa, K. Design theory of balanced transistor amplifier. *Bell Syst. Tech. J.*, **44** Oct. 1965, 1675–1698.

[38] Lange, Julius. Interdigitated stripline quadrature hybrid. *IEEE Trans. on MTT*, vol. MTT-17, December 1969, 1150–1151.

[39] Engelbrecht, R. S. et al., A wide-band low noise L-band balanced transistor amplifier. *Proc. IEEE*, **53**, March 1965, 237–247.

[40] Oxner, Edwin S. *Power FETs and Their Applications*, Englewood Cliffs, NJ, Prentice-Hall, Inc., 1982, 41, 47.

[41] Liao, Samuel Y., *RF Characterization of GaAs MESFETs*. Report to Hughes Aircraft Company, El Segundo, CA. August 1983.

[42] Hieslmair, Hans, et al., State of the art of solid-state and tube transistors. *Microwave Journal*, Vol. 26, no. 10. pp. 46–48. October 1983.

SUGGESTED READINGS

1. Barbe, D. F., ed. Charge-coupled devices. In *Topics in Applied Physics*, vol. 38. Berlin: Springer-Verlag, 1980.

2. Dilorenzo, James V., ed. *GaAs FET Principles and Technology*. Dedham, Mass.: Artech House, 1982.

3. Hobson, G. S. *Charge-Transfer Devices*. New York: John Wiley & Sons, 1978.

4. *IEEE Transactions on Electron Devices*. Special issues on microwave solid-state devices.
   vol. ED-27, no. 2, February 1980.
   vol. ED-27, no. 6, June 1980.
   vol. ED-28, no. 2, February 1981.
   vol. ED-28, no. 8, August 1981.

5. *IEEE Transactions on Microwave Theory and Techniques*. Special issues on microwave solid-state devices.
   vol. MTT-21, no. 11, November 1973.
   vol. MTT-24, no. 11, November 1976.
   vol. MTT-27, no. 5, May 1979.
   vol. MTT-28, no. 12, December 1980.
   vol. MTT-30, no. 4, April 1982.
   vol. MTT-30, no. 10, October 1982.

6. Liao, Samuel Y. *Microwave Devices and Circuits*, Chapter 6. Englewood Cliffs, N.J.: Prentice-Hall, Inc., 1980.

7. Melen, Roger, and Dennis Buss, eds., *Charge-Coupled Devices*. New York: IEEE Press, 1977.

8. Milnes, A. G. *Semiconductor Devices and Integrated Electronics*, Chapters 6, 7, and 10. New York: Van Nostrand Reinhold Company, 1980.

9. Seymour, J. *Electronic Devices and Components*, Chapter 6. New York: John Wiley & Sons, 1981.

10. Sze, S. M. *Physics of Semiconductor Devices*, 2nd ed., Chapters 6, 7, and 8. New York: John Wiley & Sons, 1981.

11. Vendelin, George D., *Design of Amplifiers and Oscillators by the S-Parameter Method.* New York: John Wiley & Sons, 1982.

PROBLEMS

4-1 MESFETs

4-1-1. A certain microwave GaAs MESFET has a gate length 6 μm. Calculate the maximum frequency of oscillation for the device.

4-1-2. The S parameters of a certain GaAs MESFET measured at 5 GHz with a 50-Ω resistance matching the input and output are as follows:

$$\mathbf{S}_{11} = 0.55 \underline{/25°} \qquad\qquad \mathbf{S}_{21} = 5.0 \underline{/180°}$$

$$\mathbf{S}_{22} = 0.35 \underline{/-20°} \qquad\qquad \mathbf{S}_{12} = 0.6 \underline{/-180°}$$

   (a) Calculate the maximum available power gain $G_{\max}$.
   (b) Compute the unilateral power gain G_u for $\Gamma_s = 0.1$ and $\Gamma_\ell = 0.1$.
   (c) Determine the maximum unilateral power gain $G_{u\max}$.
   (d) Find the intrinsic noise figure in decibels for $E = 2$ kV/m.

4-1-3. The S parameters of a GaAs MESFET are as follows:

$$\mathbf{S}_{11} = 0.6 \underline{/-150°} \qquad \mathbf{S}_{21} = 5.5 \underline{/180°} \qquad\qquad f = 10 \text{ GHz}$$

$$\mathbf{S}_{22} = 0.4 \underline{/-15°} \qquad \mathbf{S}_{12} = 0.5 \underline{/-180°} \qquad R_0 = 50 \ \Omega$$

   (a) Calculate and plot the input and output constant power gain circles on a Smith chart.
   (b) Overlap two Smith charts to facilitate the design procedures. Use the original chart to read the impedance and the overlaid chart to read the admittance.
   (c) Determine the capacitance and inductance of the output matching network.
   (d) Find the capacitance and inductance of the input matching network.

4-1-4. A GaAs MESFET has a thickness of 0.30 μm and a doping concentration N of 10^{17} cm^{-3}. The relative dielectric constant ϵ_r of GaAs is 13.10. Calculate the pinch-off voltage.

4-1-5. The S parameters of a certain GaAs MESFET are measured with a 50-Ω line at 5 GHz as

$$\mathbf{S}_{11} = 0.45 \underline{/-150°} \quad \mathbf{S}_{22} = 0.35 \underline{/-25°} \quad \mathbf{S}_{21} = 7 \underline{/180°}$$

The MESFET is to be used as an amplifier in an amplifier circuit. The problem is to design two networks to match the input and output impedances of the MESFET. Load resistance and output impedance of the signal source are considered to be 50 Ω.

(a) Determine the capacitance in picofarads and the inductance in millihenries of the output matching network.

(b) Find the capacitance in picofarads and the inductance in millihenries of the input matching network.

(c) Draw the complete circuit diagram for the designed amplifier.

4-1-6. A certain n-channel GaAs MESFET has the following parameters:

Electron concentration	$N = 2.38 \times 10^{23} \text{ m}^{-3}$
Channel height	$a = 0.2 \ \mu\text{m}$
Relative dielectric constant	$\epsilon_r = 13.10$
Channel length	$L = 10 \ \mu\text{m}$
Channel width	$Z = 60 \ \mu\text{m}$
Electron mobility	$\mu = 0.3 \text{ m}^2/\text{V-s}$
Drain voltage	$V_{ds} = 6 \text{ volts}$
Gate voltage	$V_{gs} = -3 \text{ volts}$
Saturation drift velocity	$v_s = 10^5 \text{ m/s}$

(a) Calculate the pinch-off voltage.

(b) Compute the velocity ratio.

(c) Determine the saturation drain current at $V_g = 0$.

(d) Find the drain current I_d.

4-1-7. A certain GaAs MESFET has the following parameters:

$$S_{11} = 0.66 \underline{/-160°} \qquad S_{22} = 0.55 \underline{/-110°} \qquad f = 6 \text{ GHz}$$

$$S_{21} = 2.20 \underline{/25°} \qquad \Gamma_{\text{in}} = 0.65 \underline{/160°} \qquad R_0 = 50 \ \Omega$$

$$S_{12} = 0.05 \underline{/150°} \qquad \Gamma_{\text{out}} = 0.60 \underline{/110°}$$

(a) Calculate the stability factor K.

(b) Compute the maximum unilateral transducer power gain in decibels.

(c) Design an input matching network N_{in}.

(d) Design an output matching network N_{out}.

(e) Sketch the complete matching networks.

4-1-8. Write a FORTRAN program to compute the drain current of an n-type GaAs MESFET. The device parameters are as follows:

Electron concentration	$N = 2 \times 10^{23} \text{ m}^{-3}$
Channel height	$a = 0.2 \times 10^{-6} \text{ m}$
Relative dielectric constant	$\epsilon_r = 13.10$
Channel length	$L = 10 \times 10^{-6} \text{ m}$
Channel width	$Z = 60 \times 10^{-6} \text{ m}$
Drain voltage	$V_{ds} = 0 \text{ to } 6 \text{ volts}$
Gate voltage	$V_{gs} = 0 \text{ to } -3 \text{ volts}$
Saturation drift velocity	$v_s = 10^5 \text{ m/s}$

Here are the program specifications.
(a) The drain voltage V_{ds} varies from 0 to 6 volts with an increment of 1 volt per step.
(b) The gate voltage V_{gs} varies from 0 to -3 volts with a decrease of -0.5 volt per step.
(c) The electron mobility μ varies from 0.9 to 0.3 m²/V-s with a decrease of 0.1 per step.
(d) Use F10.5 format for numerical outputs and Hollerith format for character outputs.
(e) Print the outputs in three columns, such as drain voltage V_{ds} (volts), gate voltage V_{gs} (volts), and drain current I_{ds} (mA).

4-1-9. A GaAs MESFET balanced amplifier has the following parameters:

MESFET a: Reflection coefficients $S_{11a} = 0.57 \underline{/178°}$

$S_{22a} = 0.62 \underline{/180°}$

Forward transmission $S_{21a} = 5.6 \underline{/50°}$
coefficient

MESFET b: Reflection coefficients $S_{11b} = 0.50 \underline{/160°}$

$S_{22b} = 0.64 \underline{/170°}$

Forward transmission $S_{21b} = 6.0 \underline{/60°}$
coefficient

(a) Calculate the input and output reflection coefficients of the balanced amplifier.
(b) Compute the input and output VSWRs.
(c) Determine the power gain in dB for the balanced amplifier.
(d) Find the power loss in dB if one MESFET fails.
(e) Calculate the linear output power capability as compared with two MESFETs in series.

4-2 MOSFETs

4-2-1. A basic MOSFET is formed of Al metal, a SiO_2 insulator, and a Si semiconductor. The insulator capacitance is 4 pF, channel length L is 12 μm, and the electron mobility of Si is 1350 cm²/V-s.
(a) Determine the drain current $I_{ds(sat)}$ in the saturation region with $V_{th} = 2$ volts and $V_{gs} = 4$ volts for the enhancement mode.
(b) Compute the transconductance g_m for the same mode.
(c) Calculate the drain current $I_{ds(sat)}$ in the saturation region with $V_{th} = -1.5$ volts and $V_{gs} = 0$ volts for the depletion mode.
(d) Find the transconductance g_m for the depletion mode.

4-2-2. An Al-Oxide-Si MOSFET has an insulator capacitance of 3 pF and a channel length L of 10 μm. The gate voltage V_{gs} is 10 volts and the threshold voltage is 1.5 volts.

(a) Determine the carrier drift velocity in the real case.

(b) Calculate the drain current.

(c) Compute the carrier transit time.

(d) Find the maximum operating frequency in gigahertz.

4-2-3. The insulator SiO_2 in a Si MOSFET has a relative dielectric constant ϵ_{ir} of 4.0 and a depth d of 0.08 μm. The channel length L is 15 μm and the channel depth Z is 150 μm. Calculate the insulator capacitance C_{in}.

4-2-4. Compare the advantages and disadvantages of a GaAs MOSFET with those of a Si MOSFET.

4-2-5. A certain Si p-channel MOSFET has the following parameters:

Doping concentration	$N_a = 2 \times 10^{17}$ cm^{-3}
Relative dielectric constant	$\epsilon_r = 11.8$
Insulator relative dielectric constant	$\epsilon_{ir} = 4$
Insulator thickness	$d = 0.01$ μm
Operating temperature	$T = 320°$F

(a) Calculate the surface potential ψ_s(inv) for strong inversion.

(b) Compute the insulator capacitance.

(c) Determine the threshold voltage.

4-2-6. A certain Si n-channel MOSFET has the following parameters:

Channel length	$L = 5$ μm
Channel depth	$Z = 10$ μm
Insulator thickness	$d = 0.02$ μm
Gate voltage	$V_{gs} = 8$ volts
Threshold voltage	$V_{th} = 1.5$ volts
Electron velocity	$v_s = 2 \times 10^7$ cm/s
Insulator relative dielectric constant	$\epsilon_{ir} = 4$

(a) Compute the insulator capacitance in millifarads per square meter.

(b) Calculate the saturation drain current in milliamperes.

(c) Determine the saturation transconductance in millimhos.

(d) Estimate the maximum saturation operating frequency in gigahertz.

4-3 VMOSFETs and UMOSFETs

4-3-1. Compare a VMOSFET with a MOSFET and describe their advantages and disadvantages.

4-3-2. A Si VMOSFET has a channel length L of 1.8 μm and a critical electric field E_t of 1.28×10^4 V/cm.

(a) Calculate the critical voltage.

(b) Compute the insulator capacitance C_i per unit area for a depth d of 0.1 μm.

4-3-3. A Si VMOSFET has a critical voltage of 2.5 volts, a channel length L of 2 μm, and a channel depth Z of 150 μm.

(a) Compute the carrier saturation drift velocity.

(b) Calculate the transit time of the carrier.

(c) Determine the maximum operating frequency in gigahertz.

4-3-4. A Si UMOSFET has a critical voltage of 3 volts. Its insulator capacitance is 1 pF, its channel length L is 2 μm, and its channel depth Z is 150 μm.

(a) Determine the drain current for $V_{gs} = 10$ volts and $V_{th} = 2$ volts.

(b) Calculate the transconductance g_m.

4-3-5. A certain VMOSFET has the following parameters:

Short channel length	$L = 0.2\,\mu$m
Transverse electric field	$E_t = 5 \times 10^6$ V/m
Channel depth	$Z = 10\,\mu$m
Insulator thickness	$d = 0.01\,\mu$m
Gate voltage	$V_{gs} = 6$ volts
Threshold voltage	$V_{th} = 0.5$ volts
Electron mobility	$\mu_n = 0.14$ m^2/V-s
Electron velocity	$v_s = 1.5 \times 10^7$ cm/s
Insulator relative dielectric constant	$\epsilon_{ir} = 3.9$

(a) Compute the critical voltage.

(b) Calculate the insulator capacitance.

(c) Find the saturation drain voltage.

(d) Determine the saturation drain current in milliamperes.

(e) Compute the saturation carrier drift velocity.

(f) Find the transconductance.

(g) Estimate the maximum operating frequency in gigahertz.

4-4   CCDs

4-4-1. A charge-coupled device has 484 (or 22×22 array) elements each with a transfer inefficiency of 10^{-4} and is clocked at a frequency of 100 kHz.

(a) Determine the delay time between input and output.

(b) Find the percentage of input charge appearing at the output terminal.

4-4-2. A surface-channel CCD is operated by a gate voltage V_g of 10 volts. The insulator has a relative dielectric constant ϵ_{ir} of 6 and a depth d of 0.1 μm. Determine the stored charge density on this MOS capacitor.

4-4-3. A p-type surface-channel CCD (Al-Oxide-Si) has the following parameters:

$N_a = 10^{14}$ cm^{-3}	$Q_{in} = 10^{-8}$ C/cm^2
$\Phi_{ms} = -0.85$ eV	$W = 1\,\mu$m
$\epsilon_{ir} = 4$	$d = 0.1\,\mu$m

(a) Determine the flatband potential

(b) Calculate the voltage across the insulator.

4-4-4. A three-phase CCD has $Q_{max} = 0.06$ pC operating at a clock frequency of 20 MHz with 10 volts applied. Determine the power dissipation per bit.

4-4-5. A three-phase JCCD has a length L of 8 μm, an electric field E_x of 480 V/cm, and an electron mobility μ_n of 1000 cm^2/V-s. Calculate the signal carrier transit time.

4-4-6. An n-type three-phase SCCD has the following parameters:

Insulator relative dielectric constant	$\epsilon_{ir} = 4$
Insulator thickness	$d = 0.03$ μm
Insulator cross section	$A = 0.6 \times 10^{-4}$ cm^2
Electron density	$N_{max} = 4 \times 10^{12}$ cm^{-2}
Power dissipation allowable per bit	$P = 0.8$ mW

(a) Compute the insulator capacitance in farads per square meter.
(b) Calculate the maximum stored charges per well in coulombs.
(c) Find the required applied gate voltage in volts.
(d) Choose the clock frequency in megahertz.

Chapter 5

Microwave Solid-State
Gunn-Effect Devices

5-0 INTRODUCTION

The application of two-terminal semiconductor devices at microwave frequencies has increased during the past decade. The continuous wave (CW), average, and peak power outputs of these devices at higher microwave frequencies are much larger than those obtained with the best power transistor.

Usually microwave solid-state bulk-effect devices are called *Gunn-effect diodes*. Gunn-effect diodes are used widely as radar local oscillators and as transmitters for low-power radars. Their operating frequency range is from 5 to 25 GHz and their continuous power is up to a few hundred milliwatts. If the domain formation is suppressed, these devices can be used as small-signal, broadband, reflection-type microwave amplifiers. Gunn-effect diodes produce less noise (typically 10 to 20 dB less) than IMPATT diodes; for this reason, they are more suitable as local oscillators in radar systems. LSA diodes generally produce pulsed power up to several hundred watts peak at a frequency range of 3 to 20 GHz. Pulsed LSA oscillators are commonly used as beacon transponders and moderate-power pulsed transmitters in radars.

The common characteristic of all active two-terminal solid-state devices is their negative resistance. The real part of their impedance is negative over a range of frequencies. In a positive resistance the current through the resistance and the voltage across it are in phase. The voltage drop across a positive resistance will be positive and a power of (I^2R) will be dissipated in the resistance. In a negative resistance, however, current and voltage are out of phase by $180°$. The voltage drop across a negative resistance will be negative and a power of $(-I^2R)$ will be generated by the power supply associated with the negative resistance. In other words, positive resistances absorb power (passive devices), whereas negative resistances generate power (active devices). Gunn-effect diodes are negative-resistance devices and their carriers are transferred from one end to the other in a bulk material by an electric field. Thus they are usually called *transferred electron devices* (TEDs). In this chapter transferred electron devices are analyzed.

There are fundamental differences between microwave transistors and transferred electron devices. Transistors, for instance, operate with either junctions or gates. TEDs are bulk devices that have no junctions or gates. Most transistors are fabricated from elemental semiconductors, such as silicon or germanium, whereas TEDs are fabricated from compound semiconductors, such as gallium arsenide (GaAs), indium phosphide (InP), or cadmium telluride (CdTe). Transistors operate with "warm" electrons whose energy is not much larger than the thermal energy (0.026 eV at room temperature) of electrons in the semiconductor. TEDs operate with "hot" electrons whose energy is very much greater than the thermal energy. Because of their basic differences, transistor theory and technology cannot be applied to TEDs.

5-1 GUNN-EFFECT DIODES: GaAs DIODE

Gunn-effect diodes were named after J. B. Gunn, who in 1963 discovered a periodic fluctuation of current passing through an *n*-type gallium arsenide (GaAs) specimen when the applied voltage exceeded a certain critical value. Several years later the limited space-charge-accumulation diode (LSA diode) and the indium phosphide diode (InP diode) were successfully developed. They are bulk devices in the sense that microwave amplification and oscillation are derived from the bulk negative-resistance property of uniform semiconductors rather than from the junction negative-resistance property between two different types of semiconductors, such as the tunnel diode. Gunn-effect oscillators operate at the frequency range of 4 to 80 GHz. Their power outputs vary from 10 to 700 mW. They have the advantages of broadband, low-noise, low-voltage, and simple power supply and are linear as an amplifier. Because these crystals oscillate by themselves in a microwave cavity, they are often used as local oscillators.

5-1-1 Background

In 1954 Shockley suggested that two-terminal negative-resistance devices using semiconductors may have advantages over transistors at high frequencies [1]. Then in 1961 Ridley and Watkins described a new method for obtaining negative differential mobility in semiconductors [2]. The principle was to heat carriers in a light-mass, high-mobility subband with an electric field so that the carriers can transfer to a heavy-mass, low-mobility higher-energy subband when they have a high enough temperature. Ridley and Watkins also mentioned that GeSi alloys and some

III-V compounds may have suitable subband structures in the conduction bands. Their theory for achieving negative differential mobility in bulk semiconductors by transferring electrons from high-mobility energy bands to low-mobility energy bands was taken a step further by Hilsum in 1962 [3]. Hilsum carefully calculated the transferred electron effect in several III-V compounds and was the first to use the terms transferred electron amplifiers (TEAs) and oscillators (TEOs). He predicted accurately that a TEA-bar of semi-insulating GaAs would operate at 373°K at a field of 3200 V/cm. Unfortunately, Hilsum's attempts to verify his theory experimentally failed because the GaAs diode available to him at that time was not of sufficiently high quality.

It was not until 1963 that J. B. Gunn of IBM discovered the so-called Gunn effect from thin disks of *n*-type GaAs and *n*-type InP specimens while studying the noise properties of semiconductors [4]. (The Gunn effect is discussed in detail in the next section.) Gunn did not connect his discoveries with the theories of Ridley, Watkins, and Hilsum and, in fact, immediately rejected them. In 1963 Ridley predicted [5] that the field domain is continually moving down through the crystal, disappearing at the anode and then reappearing at a favorable nucleating center, and starting the whole cycle once more. Finally, Kroemer stated [6] that the origin of the negative differential mobility was Ridley–Watkins–Hilsum's mechanism of electron transfer into the satellite valleys that occur in the conduction bands of both the *n*-type GaAs and the *n*-type InP and that the properties of the Gunn effect are the current oscillations caused by the periodic nucleation and disappearance of traveling space-charge instability domains. Thus the connection between theoretical predictions and experimental discoveries completed the theory of transferred electron devices.

5-1-2 Gunn Effect

A schematic diagram of a uniform *n*-type GaAs diode with ohmic contacts at the end surfaces is shown in Fig. 5-1-1.

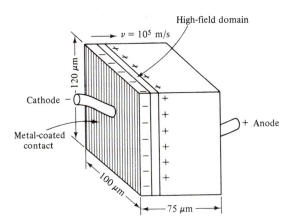

Figure 5-1-1 *n*-type GaAs diode

Gunn observed the Gunn effect in the *n*-type GaAs bulk diode in 1963 and it is perhaps best explained by Gunn himself, who published several papers about his observations [7–10]. He stated in his first paper [7] that:

> Above some critical voltage, corresponding to an electric field of 2000–4000 volts/cm, the current in every specimen became a fluctuating function of time. In GaAs specimens, this fluctuation took the form of a periodic oscillation superimposed upon the pulse current.... The frequency of oscillation was determined mainly by the specimen, and not by the external circuit.... The period of oscillation was usually inversely proportional to the specimen length and closely equal to the transit time of electrons between the electrodes, calculated from their estimated velocity of slightly over 10^7 cm/sec.... The peak pulse microwave power delivered by the GaAs specimens to a matched load was measured. Values as high as 0.5 W at 1 Gc/sec, and 0.15 W at 3 Gc/sec, were found, corresponding to 1–2% of the pulse input power. (After J. B. Gunn [7]. Copyright 1964 by International Business Machines Corporation; reprinted with permission.)

According to Gunn's observations, the carrier drift velocity increases linearly from zero to a maximum when the electric field varies from zero to a threshold value. When the electric field is beyond the threshold value of 3000 V/cm for the *n*-type GaAs, the drift velocity decreases and the diode exhibits a phenomenon of negative resistance. This situation is shown in Fig. 5-1-2. The current fluctuations are shown in Fig. 5-1-3.

The current waveform was produced by applying a voltage pulse of 16-V amplitude and 10-nsec duration to a specimen of *n*-type GaAs 2.5×10^{-3} cm in

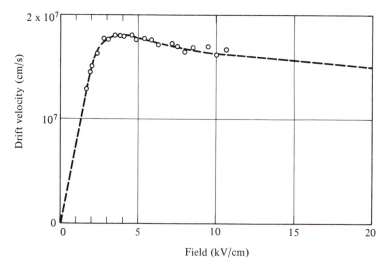

Figure 5-1-2   Drift velocity of electrons in *n*-type GaAs versus electric field

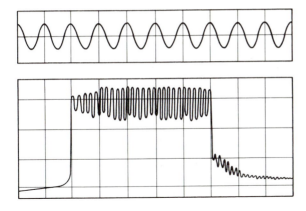

Figure 5-1-3 Current waveform of *n*-type GaAs reported by Gunn [9]; reprinted with permission of IBM Corp.

length. The oscillation frequency was 4.5 GHz. The lower trace had 2 nsec/cm in the horizontal axis and 0.23 A/cm in the vertical axis. The upper trace was the expanded view of the lower trace. Gunn found that the period of these oscillations was equal to the transit time of the electrons through the specimen calculated from the threshold current.

Gunn also discovered that the threshold electric field E_{th} varied with the length and type of material. He developed an elaborate capacitive probe for plotting the electric field distribution within a specimen of *n*-type GaAs of length $L = 210$ μm and cross-sectional area 3.5×10^{-3} cm^2 with a low-field resistance of 16 Ω. Current instabilities occurred at specimen voltages above 59 V, which means that the threshold field is

$$E_{th} = \frac{V}{L} = \frac{59}{210 \times 10^{-6} \times 10^2} = 2810 \text{ volts/cm} \tag{5-1-1}$$

5-2 RIDLEY–WATKINS–HILSUM (RWH) THEORY

Many explanations have been offered for the Gunn effect. In 1964 Kroemer [6] suggested that Gunn's observations were in complete agreement with the Ridley–Watkins–Hilsum (RWH) theory.

5-2-1 Differential Negative Resistance

The fundamental concept of the Ridley–Watkins–Hilsum (RWH) theory is the differential negative resistance developed in a bulk solid-state III-V compound when either a voltage (or electric field) or a current is applied to the terminals of the sample. There are two modes of negative-resistance devices: voltage-controlled or current-controlled modes as shown in Fig. 5-2-1(a) and (b), respectively [5].

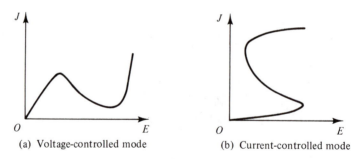

(a) Voltage-controlled mode   (b) Current-controlled mode

Figure 5-2-1 Diagram of negative resistance (From B. K. Ridley [5]; reprinted by permission of the Institute of Physics)

In the voltage-controlled mode the current density can be multivalued whereas in the current-controlled mode the voltage can be multivalued. The major effect of the appearance of a differential negative-resistance region in the current-density-field curve is to render the sample electrically unstable. As a result, the initially homogeneous sample becomes electrically heterogeneous in an attempt to reach stability. In the voltage-controlled negative-resistance mode high-field domains are formed, separating two low-field regions. The interfaces separating low- and high-field domains lie along equipotentials; thus they are in planes perpendicular to the current direction as shown in Fig. 5-2-2(a). In the current-controlled negative-resistance mode splitting the sample results in high-current filaments running along the field direction as shown in Fig. 5-2-2(b).

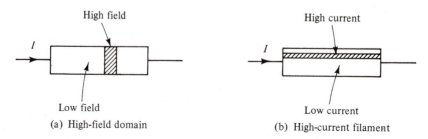

(a) High-field domain   (b) High-current filament

Figure 5-2-2 Diagrams of high field domain and high current filament (From B. K. Ridley [5]; reprinted by permission of the Institute of Physics)

Expressed mathematically, the negative resistance of the sample at a particular region is

$$\frac{dI}{dV} = \frac{dJ}{dE} = \text{negative resistance} \tag{5-2-1}$$

If an electric field E_0 (or voltage V_0) is applied to the sample, for example, the current density J_0 is generated. As the applied field (or voltage) is increased to E_2 (or V_2), the current density is decreased to J_2. When the field (or voltage) is decreased to E_1 (or V_1), the current density is increased to J_1. These phenomena of the voltage-controlled negative resistance are shown in Fig. 5-2-3(a). Similarly, for the current-controlled mode, the negative-resistance profile is as shown in Fig. 5-2-3(b).

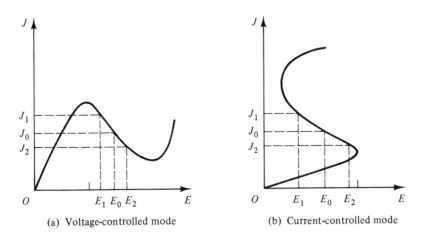

(a) Voltage-controlled mode   (b) Current-controlled mode

Figure 5-2-3 Multiple values of current density for negative resistance. (From B. K. Ridley [5]; reprinted by permission of the Institute of Physics)

5-2-2 Two-Valley Model Theory

A few years before the Gunn effect was discovered, Kroemer proposed a negative-mass microwave amplifier in 1958 [11] and 1959 [12]. According to the energy band theory of the n-type GaAs, a high-mobility lower valley is separated by an energy of 0.36 eV from a low-mobility upper valley as shown in Fig. 5-2-4. Table 5-2-1 lists

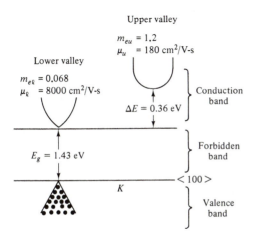

Figure 5-2-4 Two-valley model of electron energy versus wave number for n-type GaAs

TABLE 5-2-1   DATA FOR TWO VALLEYS IN GaAs

Valley	Effective Mass M_e	Mobility μ	Separation ΔE
Lower	$M_{e\ell} = 0.068$	$\mu_\ell = 8000 \text{ cm}^2/\text{V-sec}$	$\Delta E = 0.36 \text{ eV}$
Upper	$M_{eu} = 1.2$	$\mu_u = 180 \text{ cm}^2/\text{V-sec}$	$\Delta E = 0.36 \text{ eV}$

the data for the two valleys in the *n*-type GaAs and Table 5-2-2 shows the data for two-valley semiconductors.

TABLE 5-2-2   DATA FOR TWO-VALLEY SEMICONDUCTORS

Semiconductor	Gap energy (at 300 °K) E_g (eV)	Separation energy between two valleys ΔE (eV)	Threshold field E_{th} (kV/cm)	Peak velocity v_p (10^7 cm/s)
Ge	0.80	0.18	2.3	1.4
GaAs	1.43	0.36	3.2	2.2
InP	1.33	0.60	10.5	2.5
		0.80		
CdTe	1.44	0.51	13.0	1.5
InAs	0.33	1.28	1.60	3.6
InSb	0.16	0.41	0.6	5.0

Note: InP is a three-valley semiconductor: 0.60 eV is the separation energy between the middle and lower valleys, 0.8 eV that between the upper and lower valleys.

Electron densities in the lower and upper valleys remain the same under an equilibrium condition. When the applied electric field is lower than the electric field of the lower valley ($E < E_\ell$), no electrons will transfer to the upper valley as shown in Fig. 5-2-5(a). When the applied electric field is higher than that of the lower

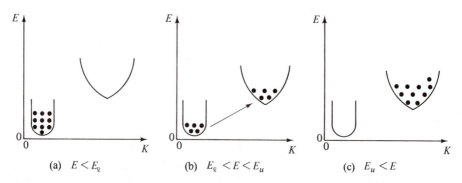

(a) $E < E_\varrho$ (b) $E_\varrho < E < E_u$ (c) $E_u < E$

Figure 5-2-5   Transfer of electron densities

valley and lower than that of the upper valley ($E_\ell < E < E_u$), electrons will begin to transfer to the upper valley as shown in Fig. 5-2-5(b). And when the applied electric field is higher than that of the upper valley ($E_u < E$), all electrons will transfer to the upper valley as shown in Fig. 5-2-5(c).

If electron densities in the lower and upper valleys are n_ℓ and n_u, the conductivity of the n-type GaAs is

$$\sigma = e(\mu_\ell n_\ell + \mu_u n_u) \tag{5-2-2}$$

where e = the electron charge

μ = the electron mobility, and

$n = n_\ell + n_u$ is the electron density

Example 5-2-2   Conductivity of an n-type GaAs Gunn Diode

Electron density	$n = 10^{18}\,\text{cm}^{-3}$
Electron density at lower valley	$n_\ell = 10^{10}\,\text{cm}^{-3}$
Electron density at upper valley	$n_u = 10^8\,\text{cm}^{-3}$
Temperature	$T = 300\,°\text{K}$

Determine the conductivity of the diode.

Solution.   From Eq. (5-2-2) the conductivity is

$$\sigma = e(\mu_\ell n_\ell + \mu_u n_u)$$
$$= 1.6 \times 10^{-19}(8000 \times 10^{-4} \times 10^{16} + 180 \times 10^{-4} \times 10^{14})$$
$$\simeq 1.6 \times 10^{-19} \times 8000 \times 10^{-4} \times 10^{16} \qquad \text{for } n_\ell \gg n_u$$
$$= 1.28 \text{ mmhos}$$

When a sufficiently high field E is applied to the specimen, electrons are accelerated and their effective temperature rises above the lattice temperature. Furthermore, the lattice temperature also increases. Thus electron density n and mobility μ are both functions of electric field E. Differentiation of Eq. (5-2-2) with respect to E yields

$$\frac{d\sigma}{dE} = e\left(\mu_\ell \frac{dn_\ell}{dE} + \mu_u \frac{dn_u}{dE}\right) + e\left(n_\ell \frac{d\mu_\ell}{dE} + n_u \frac{d\mu_u}{dE}\right) \tag{5-2-3}$$

If the total electron density is given by $n = n_\ell + n_u$ and it is assumed that μ_ℓ and μ_u are proportional to E^p, where p is a constant, then

$$\frac{d}{dE}(n_\ell + n_u) = \frac{dn}{dE} = 0 \tag{5-2-4}$$

$$\frac{dn_\ell}{dE} = -\frac{dn_u}{dE} \tag{5-2-5}$$

and

$$\frac{d\mu}{dE} \propto \frac{dE^p}{dE} = pE^{p-1} = p\frac{E^p}{E} \propto p\frac{\mu}{E} = \mu\frac{p}{E} \tag{5-2-6}$$

Substitution of Eqs. (5-2-4) to (5-2-6) into Eq. (5-2-3) results in

$$\frac{d\sigma}{dE} = e(\mu_\ell - \mu_u)\frac{dn_\ell}{dE} + e(n_\ell\mu_\ell + n_u\mu_u)\frac{p}{E} \qquad (5\text{-}2\text{-}7)$$

Then differentiation of Ohm's law $J = \sigma E$ with respect to E yields

$$\frac{dJ}{dE} = \sigma + \frac{d\sigma}{dE}E \qquad (5\text{-}2\text{-}8)$$

Equation (5-2-8) can be rewritten

$$\frac{1}{\sigma}\frac{dJ}{dE} = 1 + \frac{d\sigma/dE}{\sigma/E} \qquad (5\text{-}2\text{-}9)$$

Clearly, for negative resistance, the current density J must decrease with increasing field E or the ratio of dJ/dE must be negative. Such would be the case only if the right-hand term of Eq. (5-2-9) is less than zero. In other words, the condition for negative resistance is

$$-\frac{d\sigma/dE}{\sigma/E} > 1 \qquad (5\text{-}2\text{-}10)$$

Substitution of Eqs. (5-2-2) and (5-2-7) with $f = n_u/n_\ell$ results in [41]

$$\left[\left(\frac{\mu_\ell - \mu_u}{\mu_\ell + \mu_u f}\right)\left(-\frac{E}{n_\ell}\frac{dn_\ell}{dE}\right) - p\right] > 1 \qquad (5\text{-}2\text{-}11)$$

Note that the field exponent p is a function of the scattering mechanism and should be negative and large. This factor makes impurity scattering quite undesirable because when it is dominant, the mobility rises with an increasing field and thus p is positive. When lattice scattering is dominant, however, p is negative and must depend on the lattice and carrier temperature. The first bracket in Eq. (5-2-11) must be positive in order to satisfy unequality. This means that $\mu_\ell > \mu_u$. Electrons must begin in a low-mass valley and transfer to a high-mass valley when they are heated by the electric field. The maximum value of this term is unity—that is, $\mu_\ell \gg \mu_u$. The factor dn_ℓ/dE in the second bracket must be negative. This quantity represents the rate with field at which electrons transfer to the upper valley; this rate depends upon differences between electron densities, electron temperature, and gap energies in the two valleys.

On the basis of the Ridley–Watkins–Hilsum theory as described earlier, the band structure of a semiconductor must satisfy three criteria in order to exhibit negative resistance [3].

1. The separation energy between the bottom of the lower valley and the bottom of the upper valley must be several times larger than the thermal energy (about 0.026 eV) at room temperature. This means that $\Delta E > kT$ or $\Delta E > 0.026$ eV.
2. The separation energy between the valleys must be smaller than the gap energy between the conduction and valence bands. This means that $\Delta E < E_g$. Otherwise the semiconductor will break down and become highly conductive before

the electrons begin to transfer to the upper valleys because hole-electron pair formation is created.

3. Electrons in the lower valley must have high mobility, small effective mass, and a low density of state whereas those in the upper valley must have low mobility, large effective mass, and a high density of state. In other words, electron velocities (dE/dk) must be much larger in the lower valleys than in the upper valleys.

The two most useful semiconductors—silicon and germanium—do not meet all these criteria. Some compound semiconductors, such as gallium arsenide (GaAs), indium phosphide (InP), and cadmium telluride (CdTe) do satisfy these criteria. Others such as indium arsenide (InAs), gallium phosphide (GaP), and indium antimonide (InSb) do not. Figure 5-2-6 shows a possible current versus field characteristic of a two-valley semiconductor.

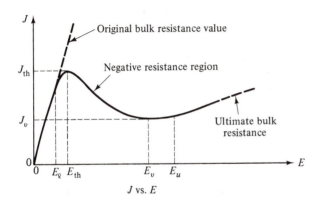

Figure 5-2-6 Current versus field characteristic of a two-valley semiconductor

A mathematical analysis of differential negative resistance requires a detailed analysis of high-field carrier transports [14–15]. From electric field theory the magnitude of the current density in a semiconductor is given by

$$J = qnv \qquad (5\text{-}2\text{-}12)$$

where q = electric charge
$\quad\quad\ n$ = electron density, and
$\quad\quad\ v$ = average electron velocity.

Differentiation of Eq. (5-2-12) with respect to electric field E yields

$$\frac{dJ}{dE} = qn\frac{dv}{dE} \qquad (5\text{-}2\text{-}13)$$

The condition for negative differential conductance may then be written

$$\frac{dv_d}{dE} = \mu_n < 0 \qquad (5\text{-}2\text{-}14)$$

where μ_n denotes the negative mobility, which is shown in Fig. 5-2-7.

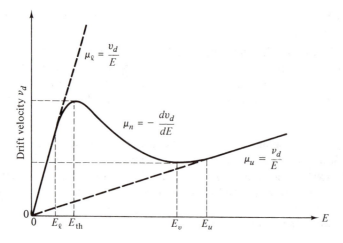

Figure 5-2-7 Electron drift velocity versus field

The direct measurement of the dependence of drift velocity on the electric field and direct evidence for the existence of the negative differential mobility were made by Ruch and Kino [16]. Experimental results, along with the theoretical results of analysis by Butcher and Fawcett [14], are shown in Fig. 5-2-8.

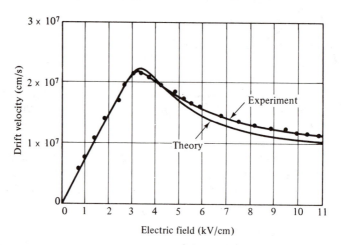

Figure 5-2-8 Theoretical and experimental velocity-field characteristics of a GaAs diode

Example 5-2-2A   Characteristics of a GaAs Gunn Diode

A typical n-type GaAs Gunn diode has the following parameters:

Threshold field	$E_{th} = 2800 \text{ V/cm}$
Applied field	$E = 3200 \text{ V/cm}$
Device length	$L = 10 \ \mu\text{m}$
Doping concentration	$n_0 = 2 \times 10^{14} \text{ cm}^{-3}$
Operating frequency	$f = 10 \text{ GHz}$

(a) Compute the electron drift velocity.
(b) Calculate the current density.
(c) Estimate the negative electron mobility.

Solution.   (a) The electron drift velocity is

$$v_d = 10 \times 10^9 \times 10 \times 10^{-6} = 10^5 \text{ m/sec} = 10^7 \text{ cm/sec}$$

(b) From Eq. (5-2-12) the current density is

$$J = qnv = 1.6 \times 10^{-19} \times 2 \times 10^{20} \times 10 \times 10^9 \times 10^{-5}$$

$$= 3.2 \times 10^6 \text{ A/m}^2$$

$$= 320 \text{ A/cm}^2$$

(c) The negative electron mobility is

$$\mu_n = -\frac{v_d}{E} = -\frac{10^7}{3200} = -3100 \text{ cm}^2/\text{V-sec}$$

5-2-3 High-Field Domain

In the last section we described how differential resistance can occur when an electric field of a certain range is applied to a multivalley semiconductor compound, such as the *n*-type GaAs. Here we show how a decrease in drift velocity with increasing electric field can lead to the formation of a high-field domain for microwave generation and amplification.

In the *n*-type GaAs diode, the majority carriers are electrons. When a small voltage is applied to the diode, the electric field and conduction current density are uniform throughout the diode. At low voltage the GaAs is ohmic because the drift velocity of the electrons is proportional to the electric field. This situation was shown in Fig. 5-1-2. The conduction current density in the diode is given by

$$\mathbf{J} = \sigma \mathbf{E}_x = \frac{\sigma V}{L}\mathbf{U}_x = \rho v_x \mathbf{U}_x \qquad (5\text{-}2\text{-}15)$$

where $\mathbf{J}$ = conduction current density

σ = conductivity

$\mathbf{E}_x$ = electric field in the *x* direction

L = length of the diode

V = applied voltage

ρ = charge density

v = drift velocity, and

$\mathbf{U}$ = unit vector.

The current is carried by free electrons that are drifting through a background of fixed positive charge. The positive charge, which is due to impurity atoms that have donated an electron (donors), is sometimes reduced by impurity atoms that have accepted an electron (acceptors). As long as the fixed charge is positive, the semiconductor is *n* type because the principal carriers are the negative charges. The density of donors less the density of acceptors is a condition called *doping*. When the space charge is zero, the carrier density is equal to the doping.

When the applied voltage is above the threshold value, which was measured at about 3000 V/cm times the thickness of the GaAs diode, a high-field domain is formed near the cathode that reduces the electric field in the rest of the material and causes the current to drop to about two-thirds of its maximum value. This situation occurs because the applied voltage given by

$$V = -\int_0^L E\,dx \qquad (5\text{-}2\text{-}16)$$

which is constant, and an increase in electric field within the specimen must be accompanied by a decrease in the electric field in the rest of the diode. The high-field domain then drifts with the carrier stream across the electrodes and disappears at the anode contact. When the electric field increases, the electron drift velocity decreases and the GaAs exhibits negative resistance.

More specifically, it is assumed that at point A on the J-E plot [see Fig. 5-2-9(a)] there exists an excess (or accumulation) of negative charge, which could be

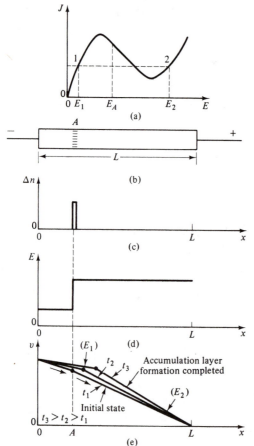

Figure 5-2-9 Formation of an electron accumulation layer in GaAs diodes (From Herbert Kroemer [17]; reprinted with permission of the IEEE, Inc.)

due to a random noise fluctuation or possibly a permanent nonuniformity in doping in the n-type GaAs diode. An electric field is then created by the accumulated charges [see Fig. 5-2-9(d)]. The field to the left of point A is lower than that to the right of A. If the diode is biased at point E_A on the J-E curve, this situation would imply that the carriers (or current) flowing into point A are greater than those flowing out of point A, thereby increasing the excess negative space charge at A. Furthermore, when the electric field to the left of point A is lower than it was before, the field to the right is then greater than the original one, resulting in an even greater space-charge accumulation. This process continues until both the low and the high fields reach values outside the differential negative-resistance region and settle at points 1 and 2 in Fig. 5-2-9(a), where the currents in the two field regions are equal. As a result of this process, a traveling space-charge accumulation is formed. This process, of course, depends on the condition that the number of electrons inside the crystal is large enough to allow the necessary amount of space charge to be built up during the transit time of the space-charge layer.

The pure accumulation layer just described is the simplest form of space-charge instability. When positive and negative charges are separated by a small distance, then a dipole domain is formed as shown in Fig. 5-2-10. The electric field inside the dipole domain would be greater than the fields on either side of the dipole in Fig. 5-2-10(c). Because of the negative differential resistance, the current in the low-field side would be greater than that in the high-field side. The two field values will tend toward an equilibrium condition outside the differential negative-resistance region where the low and high currents are the same, as described earlier. Then the dipole field reaches a stable condition and moves through the specimen toward the anode.

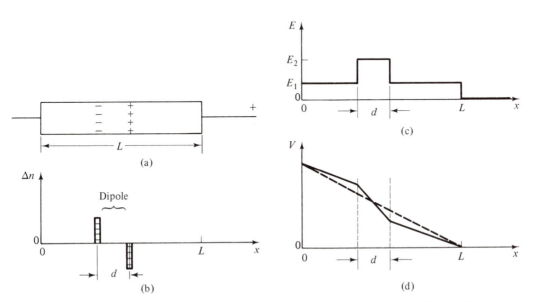

Figure 5-2-10   Formation of an electron dipole layer in GaAs diodes (From Herbert Kroemer [17]; reprinted with permission of the IEEE, Inc.)

When the high-field domain disappears at the anode, a new dipole field starts forming at the cathode and the process is repeated.

In general, the high-field domain has the following properties [3]:

1. A domain starts to form when the electric field in a region of the sample increases above the threshold electric field and drifts with the carrier stream through the device. When the electric field increases, the electron drift velocity decreases and the GaAs diode exhibits negative resistance.

2. When additional voltage is applied to a device containing a domain, the domain increases in size and absorbs more voltage than was added and the current decreases.

3. A domain does not disappear before reaching the anode unless the voltage drops appreciably below threshold (for a diode with uniform doping and area).

4. Formation of a new domain can be prevented by decreasing the voltage slightly below threshold (in a nonresonant circuit).

5. As the domain passes through regions of different doping and cross-sectional area, it may modulate the current through a device, or it may disappear. The effective doping may vary in regions along the drift path by additional contacts.

6. Generally the domain's length is inversely proportional to the doping; so devices with the same product of doping multiplied by length will behave similarly in terms of frequency multiplied by length, voltage–length, and efficiency.

7. As a domain passes a point in the device, it can be detected by a capacitive contact, for the voltage changes suddenly as the domain passes. The presence of a domain anywhere in a device can be detected by the decreased current or by the change in differential impedance.

Note that properties (3) and (6) are valid only when the length of the domain is much longer than the thermal diffusion length for carriers, which for GaAs is about 1μm for a doping of 10^{16} per cubic centimeter and about 10 μm for a doping of 10^{14} per cubic centimeter.

5-3 MODES OF OPERATION

Since J. B. Gunn first announced his observation of microwave oscillation in the n-type GaAs and n-type InP diodes in 1963, various modes of operation have been developed, depending on the material parameters and operating conditions. As noted, the formation of a strong space-charge instability depends on the conditions that enough charge is available in the crystal and that the specimen is long enough for the necessary amount of space charge to be built up within the transit time of the electrons. This requirement sets up a criterion for the various modes of operation of bulk negative-differential-resistance devices. Copeland suggested that there are four basic modes of operation of uniformly doped bulk diodes with low-resistance contacts [18] as shown in Fig. 5-3-1.

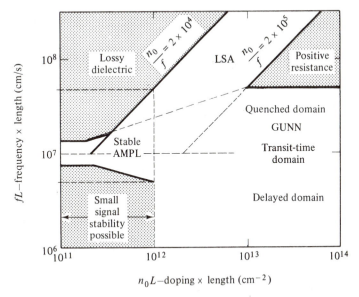

Figure 5-3-1   Modes of operation for Gunn diodes (From John A. Copeland [18]; reprinted with permission of the IEEE, Inc.)

First type: Gunn oscillation mode. This mode is defined in the region where the product of frequency multiplied by length is about 10^7 cm/sec and the product of doping multiplied by length is greater than 10^{12} per cm^2. In this region the device is unstable because of the cyclic formation of either the accumulation layer or the high-field domain. In a circuit with relatively low impedance the device operates in the high-field domain mode and the frequency of oscillation is near the intrinsic frequency. When the device is operated in a relatively high-Q cavity and coupled properly to the load, the domain is quenched and/or delayed before nucleating. In this case, the oscillation frequency is almost entirely determined by the resonant frequency of the cavity and has a value several times greater than the intrinsic frequency.

Second type: Stable amplification mode. This mode is defined in the region where the product of frequency times length is about 10^7 cm/sec and the product of doping times length is between 10^{11} and 10^{12} per cm^2.

Third type: LSA oscillation mode. This mode is defined in the region where the product of frequency times length is above 10^7 cm/sec and the quotient of doping divided by frequency is between 2×10^4 and 2×10^5.

Fourth type: Bias-circuit oscillation mode. This mode occurs only when either Gunn or LSA oscillation exists and is usually at the region where the product of frequency times length is too small to appear in the figure. When a bulk diode is biased to threshold, the average current suddenly drops as Gunn oscillation begins. The drop in current at the threshold can lead to oscillations in the bias circuit that are typically 1 kHz to 100 MHz [19].

The first three modes are discussed in detail in this section. First, however, let us consider the criterion for classifying the modes of operation.

5-3-1 Criterion of Mode Classification

Gunn-effect diodes are basically made from an n-type GaAs, with the concentrations of free electrons ranging from 10^{14} to 10^{17} per cm^3 at room temperature. Typical dimensions are 150×150 μm in cross section and 30 μm in length. During the early stages of space-charge accumulation the time rate of growth of the space-charge layers is given by

$$Q(X, t) = Q(X - vt, 0)\exp\left(\frac{t}{\tau_d}\right) \qquad (5\text{-}3\text{-}1)$$

where $\tau_d = \dfrac{\epsilon}{\sigma} = \dfrac{\epsilon}{en_0|\mu_n|}$ is the magnitude of the negative dielectric relaxation time

$\epsilon =$ the permittivity

$n_0 =$ the doping concentration

$\mu_n =$ the negative mobility

$e =$ the electron charge, and

$\sigma =$ the conductivity.

Figure 5-3-2 clarifies Eq. (5-3-1).

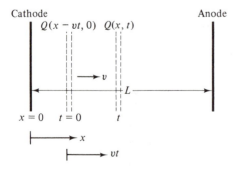

Figure 5-3-2 clarifies Eq. (5-3-1).

Figure 5-3-2 Space-charge accumulation with a velocity of v

If Eq. (5-3-1) remains valid throughout the entire transit time of the space-charge layer, the factor of maximum growth is given by

$$\text{Growth factor} = \frac{Q[L,(L/v)]}{Q(0,0)} = \exp\left(\frac{L}{v\tau_d}\right) = \exp\left(\frac{Ln_0e|\mu_n|}{\epsilon v}\right) \quad (5\text{-}3\text{-}2)$$

Here the layer is assumed to start at the cathode at $t = 0$, $X = 0$ and arrive at the anode at $t = L/v$ and $X = L$. For a large space-charge growth, this factor must be larger than unity. This means that

$$n_0L > \frac{\epsilon v}{e|\mu_n|} \quad (5\text{-}3\text{-}3)$$

This is the criterion for classifying the modes of operation for Gunn-effect diodes.

Example 5-3-1 Criterion of Mode Operation for Gunn-effect Diodes

The n-type GaAs Gunn-effect diode has the following parameters:

Electron drift velocity	$v_d = 1.7 \times 10^7$ cm/sec
Negative electron mobility	$|\mu_n| = 100$ cm^2/V-sec
Relative dielectric constant	$\epsilon_r = 13.10$

Determine the criterion for classifying the modes of operation.

Solution.   From Eq. (5-3-3) the criterion is

$$\frac{\epsilon v_d}{e|\mu_n|} = \frac{8.854 \times 10^{-14} \times 13.10 \times 1.7 \times 10^7}{1.6 \times 10^{-19} \times 100}$$

$$= 1.23 \times 10^{12} \text{ cm}^{-2}$$

Then the product of the doping concentration and the device length must be

$$n_0L > 1.23 \times 10^{12} \text{ cm}^{-2}$$

5-3-2 *Gunn Oscillation Modes* (10^{12}/cm$^2 \lesssim (n_0L) < 10^{14}$/cm^2)

Most Gunn-effect diodes have the product n_0L of doping and length greater than 10^{12}/cm^2. Yet the mode that Gunn himself observed had the product n_0L, which is much less than 10^{12}/cm^2. When the product of n_0L is greater than 10^{12}/cm^2 in GaAs, the space-charge perturbations in the specimen increase exponentially in space and time according to Eq. (5-3-1). Thus a high-field domain is formed and moves from the cathode to the anode as described in the previous section. The frequency of oscillation is given by the relation [20]

$$f = \frac{v_{\text{dom}}}{L_{\text{eff}}} \quad (5\text{-}3\text{-}4)$$

where v_{dom} is the domain velocity and L_{eff} is the effective length that the domain travels from the time it is formed until the time a new domain begins to form.

Gunn has described the behavior of Gunn oscillators under several circuit configurations [21]. When the circuit is mainly resistive or the voltage across the diode is constant, the period of oscillation is the time required for the domain to drift from the cathode to the anode. This mode is not really typical of microwave applications. Negative conductivity devices are usually operated in resonant circuits, such as high-Q resonant microwave cavities. When the diode is in a resonant circuit, the frequency can be tuned to a range of about an octave without loss of efficiency [22].

As noted, the normal Gunn domain mode (or Gunn oscillation mode) is operated with the electric field greater than the threshold field ($E > E_{th}$). The high-field domain drifts along the specimen until it reaches the anode or until the low-field value drops below the sustaining field E_s required to maintain v_s (see Fig. 5-3-3). The sustaining drift velocity for GaAs is $v_s = 10^7$ cm/sec. Because the electron drift velocity v varies with the electric field, several domain modes are possible for the Gunn domain mode.

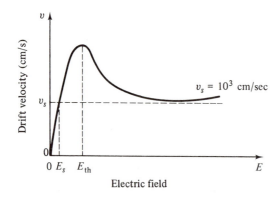

Figure 5-3-3   Electron drift velocity versus electric field

Transit-time domain mode ($fL \simeq 10^7$ cm/sec).   When the electron drift velocity v_d is equal to the sustaining velocity v_s, the high-field domain is stable. Consequently, the electron drift velocity is given by

$$v_d = v_s = fL \simeq 10^7 \text{cm/sec} \qquad (5\text{-}3\text{-}5)$$

Then the oscillation period is equal to the transit time, that is, $\tau_0 = \tau_t$. This situation is shown in Fig. 5-3-4(a). The efficiency is below 10% because the current is collected only when the domain arrives at the anode.

Delayed domain mode (10^6 cm/sec $< fL < 10^7$ cm/sec).   When the transit time is chosen so that the domain is collected while $E < E_{th}$ [see Fig. 5-3-4(b)], a new domain cannot form until the field rises above threshold again. In this case, the oscillation period is greater than the transit time—that is, $\tau_0 > \tau_t$. This delayed mode is also called *inhibited mode*. The efficiency of this mode is about 20%.

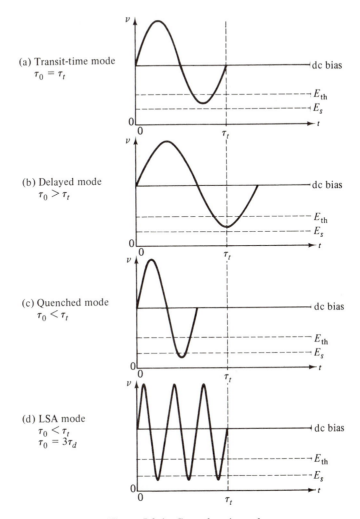

Figure 5-3-4   Gunn domain modes

Quenched domain mode ($fL > 2 \times 10^7$ cm/sec).   If the bias field drops below the sustaining field E_s during the negative half cycle [see Fig. 5-3-4(c)], the domain collapses before it reaches the anode. When the bias field swings back above threshold, a new domain is nucleated and the process repeats. Therefore the oscillations occur at the frequency of the resonant circuit rather than at the transit-time frequency. It has been found that the resonant frequency of the circuit is several times the transit-time frequency, since one dipole does not have enough time to readjust and absorb the voltage of the other dipoles [23, 24]. Theoretically the efficiency of quenched domain oscillators can reach 13% [23].

5-3-3 Limited Space-Charge Accumulation (LSA) Mode ($fL > 2 \times 10^7$ cm/sec)

When the frequency is very high, the domains do not have sufficient time to form while the field is above threshold. As a result, most of the domain is maintained in the negative conductance state during a large fraction of the voltage cycle. Any accumulation of electrons near the cathode has time to collapse while the signal is below threshold. Thus the LSA mode is the simplest mode of operation and it consists of a uniformly doped semiconductor without any internal space charges. In this instance, the internal electric field would be uniform and proportional to the applied voltage. The current in the device is then proportional to the drift velocity at this field level. The efficiency of the LSA mode can reach 20%.

The oscillation period τ_0 should be no more than several times larger than the magnitude of the dielectric relaxation time in the negative conductance region τ_d. The oscillation indicated in Fig. 5-3-4(d) is $\tau_0 = 3\tau_d$. It is appropriate here to define LSA boundaries. As described in the preceding section, the sustaining drift velocity is 10^7 cm/sec [see Eq. (5-3-5) and Fig. 5-3-3]. For the n-type GaAs, the product of doping and length $(n_0 L)$ is about $10^{12}/\text{cm}^2$. The drift velocity at the low-frequency limit is taken to be

$$v_\ell = fL = 5 \times 10^6 \text{ cm/sec} \qquad (5\text{-}3\text{-}6)$$

The ratio of $n_0 L$ to fL yields

$$\frac{n_0}{f} = 2 \times 10^5 \qquad (5\text{-}3\text{-}7)$$

It is assumed that the drift velocity at the upper-frequency limit is

$$v_u = fL = 5 \times 10^7 \text{ cm/sec} \qquad (5\text{-}3\text{-}8)$$

The ratio of $n_0 L$ to fL is

$$\frac{n_0}{f} = 2 \times 10^4 \qquad (5\text{-}3\text{-}9)$$

Both the upper and lower boundaries of the LSA mode are indicated in Fig. 5-3-1. The LSA mode is described further in Section 5-4.

5-3-4 Stable Amplification Mode ($n_0 L < 10^{12}/\text{cm}^2$)

When the $n_0 L$ product of the device is less than about $10^{12}/\text{cm}^2$, the device will exhibit amplification rather than spontaneous oscillation at the transit-time frequency. This situation occurs because the negative conductance is used without domain formation. There are too few carriers for domain formation within the transit time. Therefore the amplification of signals near the transit-time frequency can be accomplished. This type of mode was first observed by Thim and Barber [24]. Furthermore, Uenohara showed [25] that various types of amplification are possible, depending on the fL product of the device (see Fig. 5-3-5). No explanations were given, however.

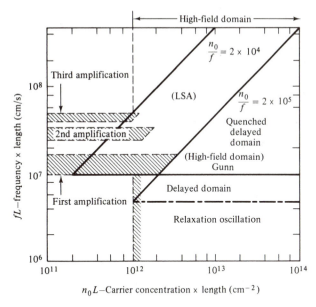

Figure 5-3-5   Mode chart (From M. Uenohara [25]; reproduced with permission of McGraw-Hill Book Company)

The various modes of operation of Gunn diodes can be classified on the basis of the times in which various processes occur. These times are defined as follows:

τ_t = domain transit time

τ_d = the dielectric relaxation time at low field

τ_g = domain growth time

τ_0 = the natural period of oscillation of a high-Q external electric circuit

The modes described previously are summarized in Table 5-3-1.

TABLE 5-3-1   MODES OF OPERATION OF GUNN OSCILLATORS

Modes	Time Relationships	Doping Level	Nature of Circuit
Stable amplifier	$\tau_0 \geq \tau_t$	$n_0 L < 10^{12}$	Nonresonant.
Gunn domain	$\tau_g \leq \tau_t$ $\tau_0 = \tau_t$	$n_0 L > 10^{12}$	Nonresonant; constant voltage
Quenched domain	$\tau_g \leq \tau_t$ $\tau_0 < \tau_t$	$n_0 L > 10^{12}$	Resonant; finite impedance
Delayed domain	$\tau_g \leq \tau_t$ $\tau_0 > \tau_t$	$n_0 L > 10^{12}$	Resonant; finite impedance
LSA	$\tau_0 < \tau_g$ $\tau_0 > \tau_d$	$2 \times 10^4 < (n_0/f) < 2 \times 10^5$	Multiply resonant; high impedance; high dc bias

5-4 LSA DIODES

The abbreviation LSA refers to the Limited Space-Charge Accumulation mode of the Gunn diode. These devices may eventually have a frequency range of 90 GHz. At present, LSA devices appear best for pulsed-power sources that have the capability of much higher peak power than IMPATT or any other Gunn-effect diodes.

As noted, if the product $n_0 L$ is larger than 10^{12} per square centimeter and if the ratio of doping n_0 to frequency f is within 2×10^5 to 2×10^4 sec/cm^3, the high-field domains and the space-charge layers do not have sufficient time to build up. The magnitude of the RF voltage must be large enough to drive the diode below threshold during each cycle to dissipate space charge. Also, the portion of each cycle during which the RF voltage is above threshold must be short enough to prevent domain formation and space-charge accumulation. Only the primary accumulation layer forms near the cathode; the rest of the sample remains fairly homogeneous. Therefore, with limited space-charge formation, the remainder of the sample appears as a series negative resistance that increases the frequency of the oscillations in the resonant circuit. Copeland discovered the LSA mode of the Gunn diode in 1966 [26]. In the LSA mode the diode is placed in a resonator tuned to an oscillation frequency of

$$f_0 = \frac{1}{\tau_0} \tag{5-4-1}$$

The device is biased to several times the threshold voltage (see Fig. 5-4-1). As the RF voltage swings beyond the threshold, the space charge starts building up at the cathode. Because the oscillation period τ_0 of the RF signal is less than the domain-

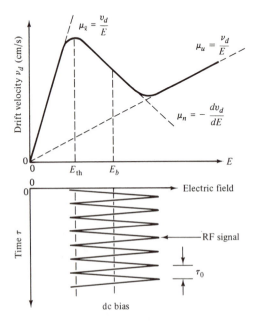

Figure 5-4-1 LSA mode operation

growth time constant τ_g, the total voltage swings below the threshold before the domain can form. Furthermore, because τ_0 is much greater than the dielectric relaxation time τ_d, the accumulated space charge is drained in a very small fraction of the RF cycle. Therefore the device spends most of the RF cycle in the negative-resistance region and the space charge is not allowed to build up. The frequency of oscillation in the LSA mode is independent of the transit time of the carriers and is determined solely by the circuit external to the device. Also, the power-impedance product does not fall off as $1/f_0^2$; thus the output power in the LSA mode can be greater than in other modes.

The LSA mode does have limitations. This mode is very sensitive to load conditions, temperatures, and doping fluctuations [27]. Also, the RF circuit must allow the field to build up very quickly to prevent domain formation. The power output of an LSA oscillator can be simply written as

$$P = \eta V_0 I_0 = \eta(ME_{th}L)(n_0 e v_0 A) \tag{5-4-2}$$

where η = the dc-to-RF conversion efficiency (primarily a function of material and circuit considerations)

V_0 = the operating voltage

I_0 = the operating current

M = the multiple of the operating voltage above negative-resistance threshold voltage

E_{th} = the threshold field (about 3400 V/cm)

L = the device length (about 10–200 μm)

n_0 = the donor concentration (about 10^{15} electrons/cm^3)

e = the electric charge (1.6×10^{-19} coulomb)

v_0 = the average carrier drift velocity (about 10^7 cm/sec)

A = the device area (about $4 \times 10^{-4} - 20 \times 10^{-4}$ cm^2)

For an LSA oscillator, n_0 is primarily determined by the desired operating frequency f_0 so that the output peak power for a properly designed circuit is directly proportional to the volume (LA) of the device length L multiplied by the area A of the active layer. Active volume cannot be increased indefinitely. Theoretically it cannot because of electrical wavelength and skin depth factors. In the practical limit, however, available bias, thermal dissipation capability, or technological problems associated with material uniformity limit device length.

Example 5-4-1   Output Power of an LSA Oscillator

An LSA oscillator has the following parameters:

Conversion efficiency	$\eta = 0.05$
Multiplication factor	$M = 3$
Threshold field	$E_{th} = 340$ kV/m
Length	$L = 10 \ \mu$m
Donor concentration	$n_0 = 10^{21}$ m^{-3}
Average carrier velocity	$v_0 = 1.66 \times 10^5$ m/sec
Area	$A = 4 \times 10^{-8}$ m^2

Determine the output power.

Solution. Using Eq. (5-4-2), we obtain the output power as

$$P = 0.05 \times (3 \times 340 \times 10^3 \times 10 \times 10^{-6})$$
$$\times (10^{21} \times 1.6 \times 10^{-19} \times 1.66 \times 10^5 \times 4 \times 10^{-8})$$
$$= 0.05 \times 10.2 \times 1.06$$
$$= 541 \text{ mW}$$

5-5 InP DIODES

When Gunn first announced his Gunn effect in 1963, the diodes that he investigated were the types of gallium arsenide (GaAs) and indium phosphide (InP). The GaAs diode was described earlier. Here the n-type InP diode is discussed. Both the GaAs diode and the InP diode operate basically the same way in a circuit with dc voltage applied at the electrodes. The two-valley model theory is the basis for explaining the electrical behavior of the Gunn effect in the n-type GaAs. Hilsum proposed, however, that indium phosphide and some alloys of indium gallium antimonide could work as three-level devices [28]. Figure 5-5-1 shows the three-valley model for indium phosphide.

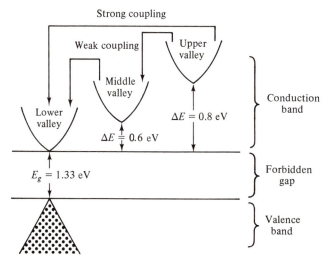

Figure 5-5-1 Three-valley model energy level for InP diode

It can be seen that besides having an upper-valley energy level and a lower-valley energy level similar to the model shown in Fig. 5-2-1 for n-type GaAs, InP also has a third middle-valley energy level. The energy difference between the upper and low valleys is 0.8 eV whereas the energy difference between the middle and lower valleys is 0.6 eV. In GaAs, the electron transfer process from the lower valley to the upper valley is comparatively slow. At a particular voltage above threshold, current flow consists of a larger contribution of electrons from the lower valley rather than from the upper valley. Because of this larger contribution from the lower energy level, a relatively low peak-to-valley current ratio results, which is shown in Fig. 5-5-2(a).

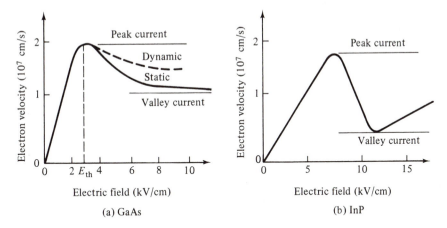

Figure 5-5-2   Peak-to-valley current ratio

The InP diode has a larger peak-to-valley current ratio [see Fig. 5-5-2(b)] because an electron transfer proceeds rapidly as the field increases. This transfer occurs because the coupling between the lower valley and upper valley in InP is weaker than in GaAs. The middle-valley energy level provides the additional energy loss mechanism required to avoid breakdown due to the high energies acquired by the lower-valley electrons from the weak coupling. It can be seen from Fig. 5-5-1 that the lower valley is weakly coupled to the middle valley but strongly coupled to the upper valley to prevent breakdown. This situation ensures that under normal operating conditions electrons concentrate in the middle valley. Because InP has a greater energy separation between the lower valley and the nearest energy levels, the thermal excitation of electrons has less effect and the degradation of its peak-to-valley current ratio is about four times less than in GaAs [29].

The mode of operation of InP is unlike the domain oscillating mode in which a high-field domain is formed that propagates with a velocity of about 10^7 cm/sec. The result is an output current waveform that is transit-time dependent. This mode reduces the peak-to-valley current ratio and so the efficiency is reduced. For this reason, an operating mode in which charge domains are not formed is usually sought. The three-valley model of InP inhibits the formation of domains because the electron diffusion coefficient is increased by the stronger coupling [29]. From experiments performed by Taylor and Colliver [30] it was determined that epitaxial InP oscillators operate via a transit-time phenomenon and do not oscillate in a bulk mode of the LSA type. From their discussion it was determined that it is not appropriate to attempt to describe the space-charge oscillations in InP in terms of modes known to exist in GaAs devices. Taylor and Colliver also determined that the frequencies obtained from a device depend on the active layer thickness. The InP oscillator could be tuned over a large frequency range, bounded only by the thickness, by adjusting the cavity size. Figure 5-5-3 shows the frequency ranges for the different active-layer thicknesses and lines of constant electron velocity [29, 30]. It can be seen that only a few InP devices operate in the domain formation area. In each case on the graph, the maximum efficiency occurs at about midband [30].

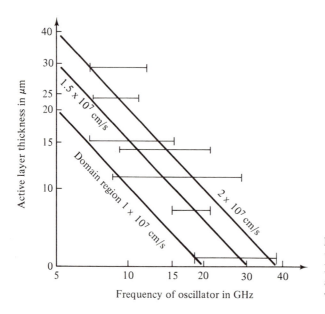

Figure 5-5-3 Active layer thickness versus frequency for InP diode (From Brian C. Taylor and D. J. Colliver [30]; reprinted with permission of the IEEE, Inc.)

5-6 CdTe DIODES

The Gunn effect, first observed by Gunn as a time variation in the current through samples of n-type GaAs when the voltage across the sample exceeded a critical value, has since been observed in n-type InP, n-type CdTe, alloys of n-GaAs and n-GaP, and in InAs. In n-type cadmium telluride (CdTe) the Gunn effect was first seen by Foyt and McWhorter [31], who observed a time variation of the current through samples 250 to 300 microns long with a carrier concentration of $5 \times 10^{14}/cm^3$ and room temperature mobility 1000 cm^2/V-s. Ludwig, Halsted, and Aven [32] confirmed the existence of current oscillations in n-CdTe and Ludwig has further reported studies of the Gunn effect in CdTe over a wider range of sample doping levels and lengths [33]. It has been confirmed that the same mechanism—the field induced transfer of electrons to a higher conduction band minimum (Gunn effect)—applies in CdTe just as it does in GaAs. From the two-valley model theory in CdTe, as in GaAs, the $\langle 000 \rangle$ minimum is the lowest in energy. The effective mass $m_{eff} = 0.11m$ and the intrinsic mobility $\mu \simeq 1100$ cm^2/V-s at room temperature. Hilsum has estimated that $\langle 111 \rangle$ minima are the next lowest in energy, being 0.51 eV higher than $\langle 000 \rangle$ minimum [34]. In comparing the Gunn effect in CdTe to that in GaAs, a major difference is the substantially higher threshold field, about 13 kV/cm for CdTe compared to 3 kV/cm for GaAs [35]. Qualitatively the higher threshold may be associated with the relatively strong coupling of the electrons to longitudinal optical phonons, a factor that limits the mobility and hence the rate of energy acquisition from the applied field and also provides an efficient mechanism for transferring energy to the lattice, thus minimizing the kinetic energy in the electron distribution.

The ratio of peak-to-valley current is another parameter of interest. In CdTe, as in GaAs, the spike amplitude can be as large as 50% of the maximum total current. A similar maximum efficiency for CdTe and GaAs can be expected. Because the domain velocities in CdTe and GaAs are approximately equal, samples of the same length operate at about the same frequency in the transit-time mode. The high threshold field of CdTe combined with its poor thermal conductivity causes a heating problem. If sufficiently short pulses are used so that the heat can be dissipated, then the high operating field of the sample can be an advantage, however.

5-7 *MICROWAVE GENERATION AND AMPLIFICATION*

The transferred electron device is one of the major low microwave power sources by either generation or amplification.

5-7-1 *Microwave Generation*

As described in high-field domain theory in Section 5-2-3, if the applied field is less than threshold, the specimen is stable. If, however, the field is greater than threshold, the sample is unstable and divides up into two domains of different conductivity and different electric field but the same drift velocity. Figure 5-7-1 shows the stable and unstable regions.

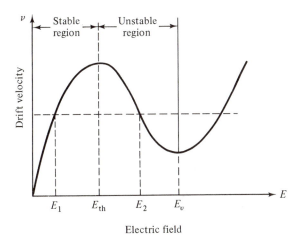

Figure 5-7-1 Electric field versus drift velocity

At the initial formation of the accumulation layer the field behind the layer decreases and the field in the front of it increases. This process continues as the layer travels from the cathode toward the anode. As the layer approaches the anode, the field behind it begins to increase again; and after the layer is collected by the anode, the field in the whole sample is higher than threshold. When the high-field domain disappears at the anode, a new dipole field starts forming again at the cathode and the process repeats itself once more. Because current density is proportional to the drift velocity of the electrons, a pulsed current output is obtained. The oscillation frequency of the pulsed current is given by

$$f = \frac{v_d}{L_{\text{eff}}} \tag{5-7-1}$$

where v_d is the velocity of the domain or approximately the drift velocity of the electrons and L_{eff} is the effective length that the domain travels.

The highest reported output power of a Gunn-effect diode has reached a 10-W pulse at a frequency of 15 GHz. Experiments have shown that n-type GaAs diodes yielded 200-W pulses at 3.05 GHz and 780-mW CW power at 8.7 GHz. Efficiencies of 29% have been obtained in pulsed operation at 3.05 GHz and 5.2% in CW operation at 24.8 GHz. Predictions have been made that 250-kW pulses from a single block of n-type GaAs are theoretically possible up to 100 GHz.

The source generation of solid-state microwave devices has many advantages over the vacuum tube devices that they are beginning to replace. It might be noted, however, that at present they also have serious drawbacks which could prevent more widespread use. The biggest disadvantages are

1. Low efficiency at frequencies above 10 GHz
2. Small tuning range
3. Large dependence of frequency on temperature
4. High noise.

These problems are common to both avalanche diodes and transferred electron devices [36].

Gunn diode oscillators have better noise performance than IMPATTs. They are used as local oscillators for receivers and as primary sources where CW powers of up to 100 mW are required. InP Gunn diodes have higher power and efficiency than GaAs Gunn diodes [43].

5-7-2 Microwave Amplification

When an RF signal is applied to a Gunn oscillator, amplification of the signal occurs, provided that the signal frequency is low enough to allow the space charge in the domain to readjust itself. There is a critical value of fL above which the device will not amplify. Below this frequency limit the sample presents an impedance with a negative real part that can be used for amplification. If n_0L becomes less than $10^{12}/cm^2$, domain formation is inhibited and the device exhibits a nonuniform field distribution that is stable with respect to time and space. Such a diode can amplify signals in the vicinity of the transit-time frequency and its harmonics without oscillating. If used in a circuit with enough positive feedback, this device will oscillate. Hakki has shown that the oscillation diode can amplify at nearby frequencies or can be used simultaneously as an amplifier and local oscillator [37]. The output power of a stable amplifier is quite low, however, because of the limitation imposed by the value of n_0L.

In contrast to the stable amplifier, the Gunn-effect diode must oscillate at the transit-time frequency while it is amplifying at some other frequency. The value of n_0L must be larger than $10^{12}/cm^2$ in order to establish traveling domain oscillations; thus substantially larger output power can be obtained. Because of the presence of high-field domains, this amplifier is called a *traveling domain amplifier* (TDA).

Although a large number of possible amplifier circuits exist, the essential feature of each is to provide both a broadband circuit at the signal frequency and a short circuit at the Gunn oscillation frequency.

In order to maintain stability with respect to the signal frequency, the Gunn diode must see a source admittance whose real part is larger than the negative conductance of the diode. The simplest circuit that satisfies this condition is shown in Fig. 5-7-2 [38]. An average gain of 3 dB was exhibited between 5.5. and 6.5 GHz.

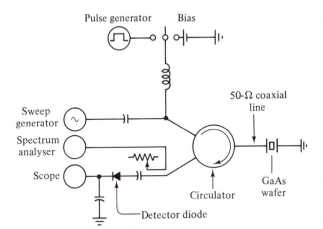

Figure 5-7-2 Gunn diode amplifier circuit (From H. W. Thim [38]; reprinted with permission of the IEEE, Inc.)

Gunn diodes have been used in conjunction with circulator-coupled networks in the design of high-level, wideband transferred electron amplifiers that have a voltage gain bandwidth product in excess of 10 dB for frequencies from 4 to 16 GHz. Linear gains of 6 to 12 dB per stage and saturated output power levels in excess of 0.5 W have been realized [39].

Example 5-7-2   Amplification of a Gunn-diode Amplifier

The negative resistance of a Gunn diode is a function of the bias electric field and is assumed to be 10 Ω. The three-port circulator and the transmission line are matched by their characteristic impedances of 50 Ω, respectively, as shown in Fig. 5-7-2. Determine the power gain in decibels if the input signal power is 1 mW.

Solution.   From the basic power gain equation we see that the output power is

$$P_{out} = P_{in} \left(\frac{R_n - R_0}{R_n + R_0} \right)^2 = 1 \times \left(\frac{-10 - 50}{-10 + 50} \right)^2 = 2.25 \text{ mW}$$

$$= 3.52 \text{ dBm}$$

The power gain is 3.52 dB.

5-8   DESIGN EXAMPLES

In many microwave solid-state source designs it is often desirable to obtain high output power and a low noise figure from the source device. Although providing greater power, IMPATT devices are usually not suitable because of higher noise. The GaAs Gunn diodes are the most common devices for high-power and low-noise oscillators at frequencies up to the W band. Two design examples for Gunn-diode oscillators are described next.

1.   Dual-Diode 73-GHz Gunn Oscillator

The design of a low-noise parametric amplifier in the X band must include a pumping source in the W band (56 to 100 GHz). A dual-diode Gunn oscillator was designed to operate at 73 GHz [40] and its configuration diagrams are shown in Fig. 5-8-1.

A dual-diode Gunn oscillator has three different structures: two diodes in parallel form, two diodes in upside-down parallel form, and two diodes in upside-down series form (see Fig. 5-8-1). The circuit used for this type of oscillator is the waveguide WR-15 with a width "a" of 0.3759 cm and a height "b" of 0.1879 cm for W-band operation. One diode is mounted on each wide wall of the waveguide. The spacing between the two diodes is $\lambda_g/2$, where λ_g is the wavelength in the waveguide. A disk is placed on top of the diode and a bias is applied through a post mounted on the disk. A movable short is inserted on one end for tuning purposes. With a bias voltage of 6.3 V, the individual output power was reported to be about 40 mW for each diode at a frequency of 73 GHz. The combined output power of the dual-diode oscillator was 80 mW at the same frequency.

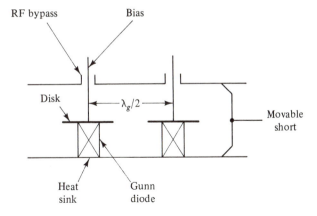

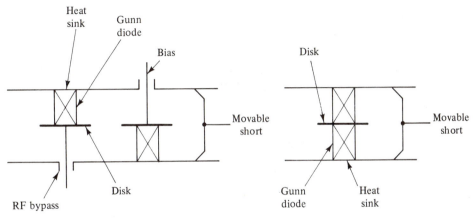

(a) Two diodes in parallel form

(b) Two diodes in upside-down parallel form   (c) Two diodes in upside-down series form

Figure 5-8-1 Schematic diagram of a dual-diode Gunn oscillator (From A. K. Talwar [40]; reprinted with permission of the IEEE, Inc.)

2. *Tunable V-Band Gunn Oscillator*

In designing a Gunn-diode oscillator, the tuning mechanism must be considered an essential part of the entire process. The tuning method may be mechanical, electronic, or both. The V band covers frequencies from 46 to 56 GHz. The circuit used in the two-diode Gunn oscillator is the reduced-height waveguide WR-15 [41] as shown in Fig. 5-8-2.

The reduced-height waveguide is the one whose height is reduced with a taper transition to a lower height at the end for tuning purposes. The waveguide used in the V-band Gunn-diode oscillator is the WR-15 type and its dimensions are 0.3759

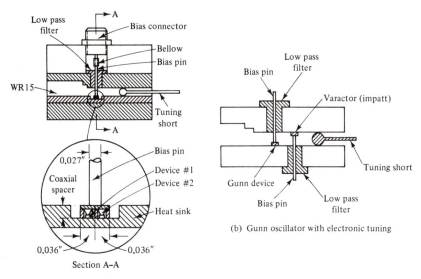

(a) Gunn oscillator with mechanical tuning

(b) Gunn oscillator with electronic tuning

Figure 5-8-2   Schematic diagram of tunable V-band Gunn oscillator (From C. Sun et al. [41]; reprinted with permission of the IEEE, Inc.)

cm by 0.1879 cm. The impedance matching to the diode is achieved by using two stages of a quarter-wave transformer at the output circuit, a tuning short, and a coaxial spacer. The diameter of the device package is 0.79 mm and the diameter for the bias pin is 0.69 mm. The equivalent circuit of the Gunn oscillator is shown in Fig. 5-8-3 [42].

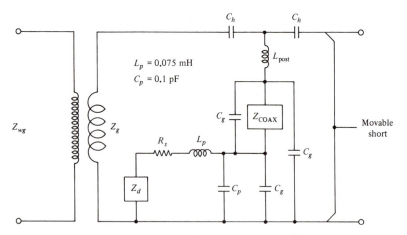

Figure 5-8-3   Equivalent circuit of Fig. 5-8-2 (From T. T. Fong et al. [42]; reprinted with permission of the IEEE, Inc.)

The subscript h in the figure refers to the lumped reactance of the post, g refers to the post gap, and p indicates the device package. Two Gunn diodes are mounted as shown. The output power at 54 GHz is 251 mW (24 dBm) for the mechanical tuning structure and 125 mW (21 dBm) for the electronic tuning configuration.

REFERENCES

[1] Shockley, W. Negative resistance arising from transit time in semiconductor diodes. *Bell Syst. Tech. J.*, **33** (July 1954), 799–826.

[2] Ridley, B. K., and T. B. Watkins. The possibility of negative resistance effects in semiconductors. *Proc. Phys. Soc.*, **78** (August 1961), 293–304.

[3] Hilsum, C. Transferred electron amplifiers and oscillators. *Proc. IEEE*, **50** (February 1962), 185–189.

[4] Gunn, J. B. Microwave oscillations of current in III-V semiconductors. *Solid-State Communications*, **1** (September 1963), 89–91.

[5] Ridley, B. K. Specific negative resistance in solids. *Proc. Phys. Soc. (London)*, **82** (December 1963), 954–966.

[6] Kroemer, Herbert. Theory of the Gunn effect. *Proc. IEEE*, **52** (1964), 1736.

[7] Gunn, J. B. Microwave oscillations of current in III-V semiconductors. *Solid-State Communications*, **1** (1963), 88–91.

[8] Gunn, J. B., and B. J. Elliott. *Phys. Lett.*, **22** (1966).

[9] Gunn, J. B. Instabilities of current in III-V semiconductors. *IBM J. Res. & Dev.*, **8** (April 1964), 141–159.

[10] Gunn, J. B. Instabilities of current and of potential distribution in GaAs and InP. *7th Int. Conf. on Physics of Semiconductors*, "Plasma Effects in Solids," pp. 199–207, Tokyo, 1964.

[11] Kroemer, Herbert. Proposed negative-mass microwave amplifier. *Phys. Rev.*, **109**, no. 5 (March 1, 1958), 1856.

[12] Kroemer, Herbert. The physical principles of a negative-mass amplifier. *Proc. IRE*, **47** (March 1959), 397–406.

[13] Copeland, John A. Bulk negative-resistance semiconductor devices. *IEEE Spectrum*, May 1967.

[14] Butcher, P. N., and W. Fawcett. Calculation of the velocity-field characteristics of gallium arsenide. *Phys. Lett.*, **21** (1966), 498.

[15] Conwell, E. M., and M. O. Vassell. High-field distribution function in GaAs. *IEEE Trans. on Electron Devices*, **ED-13** (1966), 22.

[16] Ruch, J. G., and G. S. Kino. Measurement of the velocity-field characteristics of gallium arsenide. *Appl. Phys. Lett.*, **10** (1967), 50.

[17] Kroemer, Herbert. Negative conductance in semiconductors. *IEEE Spectrum*, **5**, no. 1 (January 1968), 47.

[18] Copeland, John A. Characterization of bulk negative-resistance diode behavior. *IEEE Trans. on Electron Devices*, **ED-14**, no. 9 (September 1967).

[19] Elliott, B. J., J. G. Gunn, and J. C. McGroddy. Bulk negative differential conductivity and traveling domains in *n*-type germanium. *Appl. Phys. Lett.*, **11** (1967), 253.

[20] Copeland, J. A. Stable space-charge layers in two-valley semiconductors. *J. Appl. Phys.*, **37**, no. 9 (August 1966), 3602.

[21] Gunn, J. B. Effect of domain and circuit properties on oscillations in GaAs. *IBM J. Res. & Dev.* (July 1966), pp. 310–320.

[22] Hobson, G. S. Some properties of Gunn-effect oscillations in a biconical cavity. *IEEE Trans. on Electron Devices*, **ED-14**, no. 9 (September 1967), 526–531.

[23] Thim, H. W. Computer study of bulk GaAs devices with random one-dimensional doping fluctuations. *J. Appl. Phys.*, **39** (1968), 3897.

[24] Thim, H. W., and M. R. Barber. Observation of multiple high-field domains in *n*-GaAs. *Proc. IEEE*, **56** (1968), 110.

[25] Uenohara, M. Bulk gallium arsenide devices, Chapter 16. In H. A. Watson, ed., *Microwave Semiconductor Devices and Their Circuit Amplification*. New York: McGraw-Hill Book Company, 1969.

[26] Copeland, John A. CW operation of LSA oscillator diodes—44 to 88 GHz. *Bell Syst. Tech. J.*, **46** (January 1967), 284–287.

[27] Wilson, W. E. Pulsed LSA and TRAPATT sources for microwave systems. *Microwave J.*, **14**, no. 8 (August 1971).

[28] G. B. L. Three-level oscillator in indium phosphide. *Physics Today*, **23** (December 1970), 19–20.

[29] Colliver, D., and Brian Prew. Indium phosphide: Is it practical for solid state microwave sources? *Electronics* (April 10, 1972), pp. 110–113.

[30] Taylor, Brian C., and David J. Colliver. Indium phosphide microwave oscillators. *IEEE Trans. on Electron Devices*, **ED-18**, no. 10 (October 1971), 835–840.

[31] Foyt, A. G., and A. L. McWhorter. The Gunn effect in polar semiconductors. *IEEE Trans. on Electron Devices*, **ED-13** (January 1966), 79–87.

[32] Ludwig, G. W., R. E. Halsted, and M. Aven. Current saturation and instability in CdTe and ZnSe. *IEEE Trans. on Electron Devices*, **ED-13** (August–September 1966), 671.

[33] Ludwig, G. W. Gunn effect in CdTe. *IEEE Trans. on Electron Devices*, **ED-14**, no. 9 (September 1967), 547–551.

[34] Butcher, P. N., and W. Fawcett. *Proc. Phys. Soc. (London)*, **86** (1965), 1205.

[35] Oliver, M. R., and A. G. Foyt. The Gunn effect in *n*-CdTe. *IEEE Trans. on Electron Devices*, **ED-14**, no. 9 (September 1967), 617–618.

[36] Hilsum, C. New developments in transferred electron effects. *Proc. 3rd Conference on High Frequency Generation and Amplification*. Devices and Applications. August 17–19, 1971, Cornell University, Ithaca, N.Y.

[37] Hakki, B. W. GaAs post-threshold microwave amplifier, mixer, and oscillator. *Proc. IEEE (Letters)*, **54** (February 1966), 299–300.

[38] Thim, H. W. Linear microwave amplification with Gunn oscillators. *IEEE Trans. on Electron Devices*, **ED-14**, no. 9 (September 1967).

[39] Perlman, B. S., et al. Microwave properties and applications of negative conductance transferred electron devices. *Proc. IEEE*, **59**, no. 8 (August 1971).

[40] Talwar, A. K. A dual-diode 73-GHz Gunn oscillator. *IEEE Trans. on Microwave Theory and Techniques*, **MTT-27**, no. 5, (May 1979), 510–512.

[41] Sun, C., et al. A tunable high-power V-band Gunn oscillator. *IEEE Trans. on Microwave Theory and Techniques*, **MTT-27**, no. 5 (May 1979), 512–514.

[42] Fong, T. T., et al. Circuit characterization of V-band IMPATT oscillators and amplifiers. *IEEE Trans. on Microwave Theory and Techniques*, **MTT-24**, no. 11 (November 1976), 752–758.

[43] Hieslmair, Hans, et al. State of the art of solid-state and tube transmitters. *Microwave J.*, **26**, no. 10 (October 1983).

SUGGESTED READINGS

1. Bulman, P. J., et al. *Transferred Electron Devices*. New York: Academic Press, 1972.

2. Eastman, L. F., ed. *Gallium Arsenide Microwave Bulk and Transit-Time Devices*. Dedham, Mass.: Artech House, 1972.

3. Howes, M. J., and D. V. Morgan. *Microwave Devices*. Chapter 2. New York: John Wiley & Sons, 1976.

4. *IEEE Trans. on Electron Devices*. Special issues on microwave solid-state devices.
   Vol. ED-27, No. 2, February 1980.
   Vol. ED-27, No. 6, June 1980.
   Vol. ED-28, No. 2, February 1981.
   Vol. ED-28, No. 8, August 1981.

5. *IEEE Trans. on Microwave Theory and Techniques*. Special issues on microwave solid-state devices.
   Vol. MTT-21, No. 11, November 1973.
   Vol. MTT-24, No. 11, November 1976.
   Vol. MTT-27, No. 5, May 1979.
   Vol. MTT-28, No. 12, December 1980.
   Vol. MTT-30, No. 4, April 1982.
   Vol. MTT-30, No. 10, October 1982.

6. Liao, Samuel Y. *Microwave Devices and Circuits*. Chapter 6. Englewood Cliffs, N.J.: Prentice-Hall, Inc., 1980.

7. Milnes, A. G. *Semiconductor Devices and Integrated Electronics*. Chapter 11. New York: Van Nostrand Reinhold Company, 1980.

8. Sze, S. M. *Physics of Semiconductor Devices*, Chapter 11. 2nd ed. New York: John Wiley & Sons, 1981.

PROBLEMS

5-1 Gunn Diodes

5-1-1. Explain the operational principle of Gunn-effect diodes.

5-1-2. A Gunn diode has a length of 100 μm and a cross section of 2×10^{-3} cm^2 with a doping n_d of 10^{14} cm^{-3}. Estimate the fluctuation current.

5-1-3. An n-type GaAs Gunn diode has the following parameters:

Threshold field	$E_{th} = 2800$ V/cm
Applied field	$E = 4000$ V/cm
Device length	$L = 28$ μm
Doping concentration	$n_0 = 4 \times 10^{17}$ cm^{-3}
Operating frequency	$f = 12$ GHz

(a) Compute the electron drift velocity.
(b) Calculate the current density.
(c) Estimate the negative electron mobility.

5-1-4. A Gunn diode has a negative resistance of 30 Ω and is used with a perfect circulator as an amplifier. The three-port circulator and the transmission line are matched by their characteristic impedances of 50 Ω, respectively. Determine the power gain in decibels if the input signal power is 5 mW.

5-2 RWH Theory

5-2-1. The theory of Gunn-effect diodes was proposed independently by B. K. Ridley, T. B. Watkins, and C. Hilsum before J. B. Gunn discovered the so-called Gunn-effect diodes. Describe the RWH theory.

5-2-2. When the applied voltage across a Gunn diode is above a certain threshold level, a negative resistance will appear. Derive Eq. (5-2-11).

5-3 Modes of Operation

5-3-1. The LSA oscillation mode is defined between 2×10^4 and 2×10^5 ratios of doping over frequency as shown in Fig. 5-3-1. Derive Eqs. (5-3-7) and (5-3-9).

5-3-2. The domain velocity for a transit-time domain mode is equal to the carrier drift velocity and it is about 10^7 cm/sec. Determine the drift length of the diode at a frequency of 10 GHz.

5-4 LSA Diodes

5-4-1. The LSA diode is one type of Gunn-effect diodes. Describe the operational principle of LSA diodes.

5-4-2. A LSA diode is operating at 10 GHz with a peak output power of 300 W. Its duty cycle is 0.001 and the efficiency is 10%.
(a) Calculate the average output power in watts.
(b) Compute the input dc power in watts.

5-4-3. An LSA oscillator has the following parameters:

Conversion efficiency	$\eta = 0.05$
Multiplication factor	$M = 4$
Threshold field	$E_{th} = 320$ kV/m
Device length	$L = 12$ μm
Donor concentration	$n_0 = 1.5 \times 10^{21}$ m^{-3}
Average carrier velocity	$v_0 = 1.2 \times 10^5$ m/sec
Area	$A = 2 \times 10^{-8}$ m^2

Compute the output power in milliwatts.

5-5 InP Diodes

5-5-1. The indium phosphide (InP) diode's operating principle is the three-valley model theory. Describe this theory.

5-5-2. Compare the characteristics of an InP diode with those of a GaAs Gunn diode.

5-6 CdTe Diodes

5-6-1. Describe the characteristics of CdTe diodes.

Chapter 6

Microwave Solid-State
Avalanche-Effect Devices

6-0 INTRODUCTION

Avalanche transit-time diode oscillators rely on the effect of voltage breakdown across a reverse-biased *p-n* junction to produce a supply of holes and electrons. Ever since the development of modern semiconductor device theory, scientists have speculated as to whether it was possible to make a two-terminal negative-resistance device. The tunnel diode was the first such device to be realized in practice. Its operation depends on the properties of a forward-biased *p-n* junction in which both the *p* and *n* regions are heavily doped. Two other devices are the transferred electron devices and the avalanche transit-time devices. The latter type is investigated in this chapter.

The transferred electron device or the Gunn-effect diode operates simply by the application of a dc voltage to a bulk semiconductor. There are no *p-n* junctions in this device. Its frequency is a function of the load and of the natural frequency of the circuit. (Gunn diodes have been described in Chapter 5.) The avalanche diode oscillator uses carrier impact ionization and drift in the high-field region of a semiconductor junction to produce a negative resistance at microwave frequencies. The device was originally proposed in a theoretical paper by Read [1] in which he analyzed the negative-resistance properties of an idealized n^+-*p*-*i*-p^+ diode. Two distinct avalanche oscillator modes have been observed. One is the IMPATT mode, which stands for IMPact ionization Avalanche Transit Time operation. In this mode the typical dc to RF conversion efficiency is 5 to 10% and frequencies range as high as 100 GHz with Si diodes. The other mode is the TRAPATT mode, which represents TRApped Plasma Avalanche Triggered Transit operation. Its typical conversion efficiency is from 20 to 60%.

Another type of active microwave device is the BARITT diode (BARrier Injected Transit-Time diode) [2, 3]. It has long drift regions similar to those of IMPATT diodes; however, the carriers traversing the drift regions of BARITT diodes are generated by minority-carrier injection from forward-biased junctions rather than being extracted from the plasma of an avalanche region. Several different types of device structures have been operated as BARITT diodes, such as *p-n-p*, *p-n-v-p*, *p-n*-metal, and metal-*n*-metal. BARITT diodes have a low noise figure of 15 dB, but their bandwidth is relatively narrow with low output power.

6-1 MODES OF OPERATION

Several distinctive modes of operation occur within the broad category of avalanche transit-time devices. DeLoach suggested that the operation of avalanche diodes could be classified as static, dynamic, and parametric modes within the broad domain of space-charge mode [4]. Table 6-1-1 lists the modes of operation for avalanche transit-time devices.

TABLE 6-1-1   AVALANCHE TRANSIT-TIME DEVICES

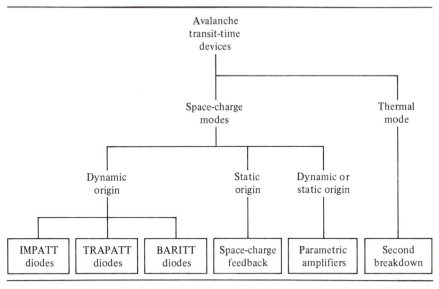

Avalanche transit-time devices have two basically different RF power-generation mechanisms. The first is the operation of the space-charge mode in which the space charge of mobile carriers changes the electric field within the device. Consequently, the produced voltage is 180° out of phase with the device current and the result leads to the generation of power. The other mechanism is the operation of thermal modes in which heat is generated by carriers from the lattice. If enough heat is generated, carriers can be thermally further generated and increase the current. As a result, the diode can now supply this current at a reduced voltage because impact avalanche fields need no longer be maintained. This type of operation is called a "second type of breakdown."

Static modes are defined as the mode in which time variations of electric field and charge density essentially follow dc solutions during oscillations. These modes are often called *quasi-static* by some authors. *Dynamic modes* are those in which the time variations of electric field and charge density are shorter than the transit time of carriers through the device. Variable reactance diodes have long been used as parametric devices. *Parametric modes* require the presence of a dynamic or static mode to exist. In a typical operation a sinusoidal pump is coupled to the diode and the output power is obtained at signal and idler frequency. Parametric operation bears a superficial resemblance to TRAPATT operation in that each presently requires an IMPATT pump.

6-2 READ DIODE

The basic principle of operation of IMPATT diodes can be most easily understood by referring to the first proposed avalanche diode, the Read diode [1]. The theory of this device was presented by Read in 1958, but the first real experimental Read diode was reported by Lee et al. in 1965 [5]. A mode of the original Read diode with a doping profile and a dc electric field distribution that exists when a large reverse bias is applied across the silicon diode is shown in Fig. 6-2-1.

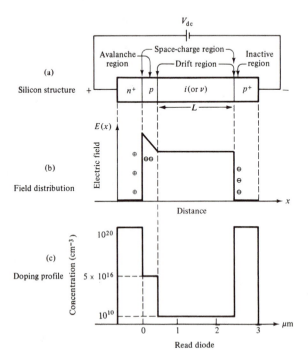

Figure 6-2-1 Read diode

6-2-1 Physical Structure

The *Read diode* is an n^+-p-i-p^+ structure in which the superscript plus sign denotes very high doping and the i or ν refers to intrinsic material. The device consists essentially of two regions. One is the narrow p region at which avalanche multiplication occurs. This region is also called the high-field region or the avalanche region. The other is the i or ν region through which the generated holes must drift in moving to the p^+ contact. This region is also called the intrinsic region or the drift region. The p region is very thin. The space between the n^+-p junction and the i-p^+ junction is called the *space-charge region*. Similar devices can be built in the p^+-n-i-n^+ structure, in which electrons generated from avalanche multiplication drift through the i region.

   The Read diode oscillator consisted of an n^+-p-i-p^+ diode biased in reverse and mounted in a microwave cavity. The impedance of the cavity is mainly inductive and is matched to the mainly capacitive impedance of the diode so as to form a resonant circuit. The device can produce a negative ac resistance that, in turn, delivers power from the dc bias to the oscillation.

6-2-2 Avalanche Multiplication

When the reverse-biased voltage is well above the punchthrough or breakdown voltage, the space-charge region always extends from the n^+-p junction through the p and i regions to the i-p^+ junction. The fixed charges in the various regions are shown in Fig. 6-2-1(b). A positive charge gives a rising field in moving from left to right. The maximum field, which occurs at the n^+-p junction, is about several hundred kilovolts per centimeter. Carriers (holes) moving in the high field near the n^+-p junction acquire energy to knock valence electrons into the conduction band, thus producing hole-electron pairs. The rate of pair production, or avalanche multiplication, is a sensitive nonlinear function of the field. By proper doping, the field can be given a relatively sharp peak so that avalanche multiplication is confined to a very narrow region at the n^+-p junction. The electrons move into the n^+ region and the holes drift through the space-charge region to the p^+ region with a constant velocity v_d of 10^7 cm/sec for silicon. The field throughout the space-charge region is above 5 kV/cm. The transit time of a hole across the drift i-region L is given by

$$\tau = \frac{L}{v_d} \qquad (6\text{-}2\text{-}1)$$

6-2-3 Carrier Current $I_0(t)$ and External Current $I_e(t)$

As noted, the Read diode is mounted in a microwave resonant circuit. An ac voltage can be maintained at a given frequency in the circuit and the total field across the diode is the sum of the dc and ac fields. This total field causes breakdown at the n^+-p junction during the positive half cycle of the ac voltage if the field is above the breakdown voltage, and the carrier current (or the hole current in this case) $I_0(t)$ generated at the n^+-p junction by the avalanche multiplication grows exponentially with time while the field is above the critical value. During the negative half cycle, when the field is below the breakdown voltage, the carrier current $I_0(t)$ decays exponentially to a small steady-state value. The carrier current $I_0(t)$ is the current at the junction only and is in the form of a pulse of very short duration as shown in Fig. 6-2-2(d). Therefore the carrier current $I_0(t)$ reaches its maximum in the middle of the ac voltage cycle, or one-quarter of a cycle later than the voltage. Under the influence of the electric field the generated holes are injected into the space-charge region toward the negative terminal. As the injected holes traverse the drift space, they induce a current $I_e(t)$ in the external circuit as shown in Fig. 6-2-2(d).

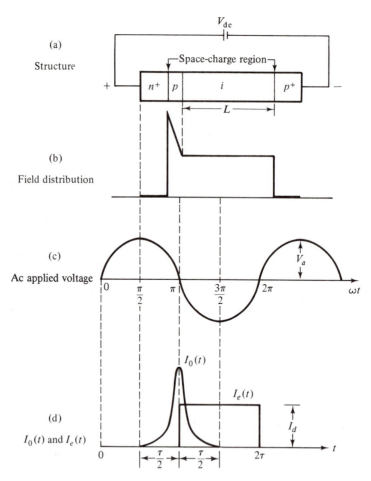

Figure 6-2-2 Field voltage and currents in Read diode (From W. T. Read [1]; reprinted with permission from *The Bell System*, AT&T)

When the holes generated at the n^+-p junction drift through the space-charge region, they cause a decrease of the field in accordance with Poisson's equation.

$$\frac{dE}{dx} = -\frac{\rho}{\epsilon}$$ (6-2-2)

where ρ is the volume charge density and ϵ is the semiconductor permittivity. Because the drift velocity of the holes in the space-charge region is constant, the induced current $I_e(t)$ in the external circuit is simply equal to

$$I_e(t) = \frac{Q}{\tau} = \frac{v_d Q}{L}$$ (6-2-3)

where Q is the total charge of the moving holes
         v_d is the hole drift velocity
         L is the length of the space-charge region

It can be seen that the induced current $I_e(t)$ in the external circuit is equal to the average current in the space-charge region. When the pulse of hole current $I_0(t)$ is suddenly generated at the n^+-p junction, a constant current $I_e(t)$ starts flowing in the external circuit and continues to flow during the time τ in which the holes are moving across the space-charge region. Thus, on the average, the external current $I_e(t)$ due to the moving holes is delayed by $\tau/2$ or 90° relative to the pulsed carrier current $I_0(t)$ generated at the n^+-p junction. Because the carrier $I_0(t)$ is delayed by one-quarter of a cycle or 90° relative to the ac voltage, the external current $I_e(t)$ is then delayed by 180° relative to the voltage as shown in Fig. 6-2-2(d). Therefore the cavity should be tuned to give a resonant frequency, which means $2\pi f = \pi/\tau$. Then

$$f = \frac{1}{2\tau} = \frac{v_d}{2L}$$ (6-2-4)

Because the applied ac voltage and the external current $I_e(t)$ are out of phase by 180°, negative conductance occurs and the Read diode can be used for microwave oscillation and amplification. For example, taking $v_d = 10^7$ cm/sec for silicon, the optimum operating frequency for a Read diode with an i region length of 2.5 μm is 20 GHz.

Example 6-2-3   Characteristics of a Read Diode

Drift region length	$L = 6.25\ \mu$m	
Total hole charge	$Q = 10$ pC	
Hole drift velocity	$v_d = 10^5$ m/s	

(a) Determine the induced current in the external circuit.
(b) Find the resonant frequency.

Solution.   (a) From Eq. (6-2-3) the induced current is

$$I_e(t) = \frac{10^5 \times 1 \times 10^{-11}}{6.25 \times 10^{-6}} = 0.16 \text{ A} = 160 \text{ mA}$$

(b) From Eq. (6-2-4) the resonant frequency is

$$f = \frac{10^5}{2 \times 6.25 \times 10^{-6}} = 8 \text{ GHz}$$

6-2-4 Output Power and Efficiency

The external current $I_e(t)$ approaches a square wave, being very small during the positive half cycle of the ac voltage and almost constant during the negative half cycle. Because the direct current I_d supplied by the dc bias is the average external current or conductive current, it follows that the amplitude of variation of $I_e(t)$ is approximately equal to I_d. If V_a is the amplitude of the ac voltage, the ac power delivered is found to be

$$P = 0.707 V_a I_d \text{ W/unit area} \qquad (6\text{-}2\text{-}5)$$

6-2-5 Quality Factor Q

The quality factor Q of a circuit is defined as

$$Q = \omega \frac{\text{maximum stored energy}}{\text{average dissipated power}} \qquad (6\text{-}2\text{-}6)$$

Because the Read diode supplies ac energy, it has a negative Q in contrast to the positive Q of the cavity. At the stable operating point, the negative Q of the diode is equal to the positive Q of the cavity circuit. If the amplitude of the ac voltage increases, the stored energy, or energy of oscillation, increases faster than the energy delivered per cycle. This is the condition required for a stable oscillation to be possible.

6-3 IMPATT DIODES

In the last section a theoretical Read diode made of a n^+-p-i-p^+ or p^+-n-i-n^+ structure was analyzed. Its basic physical mechanism concerns the interaction of the impact ionization avalanche and the transit time of charge carriers. Consequently, Read-type diodes are called *IMPATT diodes*. These diodes exhibit a differential negative resistance via two effects:

1. The impact ionization avalanche effect, which causes the carrier current $I_0(t)$ and the ac voltage to be out of phase by $90°$.
2. The transit-time effect, which further delays the external current $I_e(t)$ relative to the ac voltage by $90°$.

The first IMPATT operation as reported by Johnston et al. [6] in 1965, however, was obtained from a simple *p-n* junction. The first real Read-type IMPATT diode was reported by Lee et al. [5] as described previously. From the small-signal theory developed by Misawa [7] it has been confirmed that a negative resistance of the IMPATT diode can be obtained from a junction diode with any doping profile.

IMPATT devices are usually operated via continuous-wave (CW) mode and their pulsed power varies from 50 mW to 5 W for a frequency range of from 1 to 60 GHz. These devices can be used as transmitters in low-power Doppler radars, as pumps for parametric amplifiers, or as local oscillators in radars. Their special features are high-power output and high efficiency, but their disadvantages are high noise, narrow band, and nonlinear amplification. In addition, they require a high applied voltage. Some experimental work has demonstrated the IMPATT devices to 150 GHz.

6-3-1 *Physical Structure*

Many IMPATT diodes consist of a high doping avalanche region followed by a drift region in which the field is low enough that the carriers can traverse through it without avalanching. The Read diode is the basic type of the IMPATT diode family. The others are the one-sided abrupt *p-n* junction, the linearly graded *p-n* junction (or double-drift region IMPATT), and the *p-i-n* diode, all of which are shown in Fig. 6-3-1 [6]. The operating principle of these devices is essentially similar to the mechanism described for the Read diode.

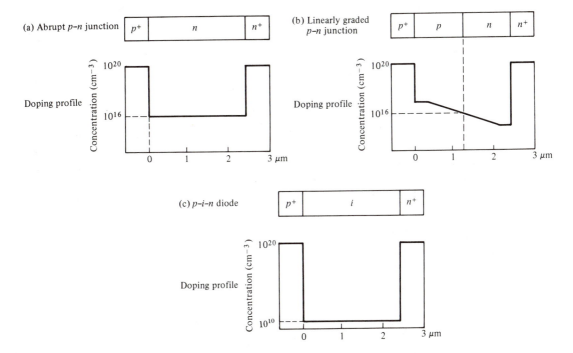

Figure 6-3-1 Three typical silicon IMPATT diodes (From R. L. Johnston et al. [6]; reprinted with permission from *The Bell System*, AT&T)

The double-drift region IMPATTs are particularly useful as oscillators and amplifiers at microwave frequencies. Their breakdown electric field is about 3.5×10^5 V/m, and electron-hole pairs are created by impact ionization in the high-field region close to the p-n junction. Due to the reverse bias the electrons and holes drift through the respective depletion layers on p and n regions, respectively. The double-drift region doubles the negative resistance of the device and so its power output is approximately 2.7 times greater than a corresponding single-drift IM-PATT.

6-3-2 Negative Resistance

Small-signal analysis of a Read diode results in the following expression for the real part of the diode terminal impedance [7]:

$$R = R_s + \frac{2L^2}{v_d \epsilon A} \frac{1}{1 - (\omega^2/\omega_r^2)} \frac{1 - \cos \theta}{\theta} \tag{6-3-1}$$

where R_s is the passive resistance of the inactive region
   v_d is the carrier drift velocity
   L is the length of the drift space-charge region
   A is the diode cross section
   ϵ is the semiconductor dielectric permittivity
   θ is the transit angle, given by

$$\theta = \omega \tau = \omega \frac{L}{v_d} \tag{6-3-2}$$

and ω_r is the avalanche resonant frequency, defined by

$$\omega_r \equiv \left(\frac{2\alpha' v_d I_0}{\epsilon A} \right)^{1/2} \tag{6-3-3}$$

In Eq. (6-3-3) the quantity α' is the derivative of the ionization coefficient with respect to the electric field. This coefficient, the number of ionizations per centimeter produced by a single carrier, is a sharply increasing function of the electric field. The variation of the negative resistance with the transit angle when $\omega > \omega_r$ is plotted in Fig. 6-3-2. The peak value of the negative resistance occurs near $\theta = \pi$. The

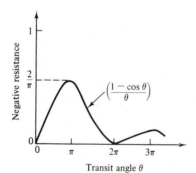

Transit angle θ

Figure 6-3-2   Negative resistance versus transit angle

negative resistance of the diode decreases rapidly for transit angles larger than π and approaching $3\pi/2$. In practice, Read-type IMPATT diodes work well only in a frequency range around the π transit angle—that is,

$$f = \frac{1}{2\tau} = \frac{v_d}{2L} \qquad (6\text{-}3\text{-}4)$$

6-3-3 Output Power and Efficiency

The maximum output power of a single diode at a given frequency is limited by semiconductor materials and the attainable impedance levels in microwave circuitry. The maximum voltage that can be applied across the diode for a uniform avalanche is given by

$$V_m = E_m L \qquad (6\text{-}3\text{-}5)$$

where L is the depletion length and E_m is the maximum electric field. This maximum applied voltage is limited by the breakdown voltage. Furthermore, the maximum current that can be carried by the diode is also limited by the avalanche breakdown process, for the current in the space-charge region causes an increase of the electric field. The maximum current is given by

$$I_m = J_m A = \sigma E_m A = \frac{\epsilon}{\tau} E_m A = \frac{v_d \epsilon E_m A}{L} \qquad (6\text{-}3\text{-}6)$$

Therefore the upper limit of the power input is given by

$$P_m = I_m V_m = E_m^2 \epsilon v_d A \qquad (6\text{-}3\text{-}7)$$

The capacitance across the space-charge region is defined as

$$C = \frac{\epsilon A}{L} \qquad (6\text{-}3\text{-}8)$$

Substitution of Eq. (6-3-8) into (6-3-7) and application of $2\pi f \tau \doteq 1$ yield

$$P_m f^2 = \frac{E_m^2 v_d^2}{4\pi^2 X_c} \qquad (6\text{-}3\text{-}9)$$

It is interesting to note that this equation is identical to Eq. (3-1-47) of the power-frequency limitation for the microwave power transistor. The maximum power that can be given to the mobile carriers decreases as $1/f^2$. This electronic limit for silicon is dominant at high frequencies up to 100 GHz. The efficiency of the IMPATT diodes is usual as

$$\eta = \frac{P_{\text{ac}}}{P_{\text{dc}}} = \left(\frac{V_a}{V_d}\right)\left(\frac{I_a}{I_d}\right) \qquad (6\text{-}3\text{-}10)$$

For an ideal Read-type IMPATT diode the ratio of the ac voltage to the applied voltage is about 0.5 and the ratio of the ac current to the dc current is about $2/\pi$; so the efficiency would be about $1/\pi$ or over 30%. But, for practical IMPATT diodes, the efficiency is usually less than 30% due to the space-charge effect, the reverse-saturation-current effect, the high-frequency skin effect, and the ionization-saturation effect.

At present, IMPATT diodes are the most powerful CW solid-state microwave power sources. Diodes have been fabricated from Ge, Si, and GaAs and can probably be constructed from other semiconductors as well. IMPATT diodes provide potentially reliable, compact, inexpensive, and moderately efficient microwave power sources. The power output data for both the GaAs and Si IMPATTs closely follow the $1/f$ and $1/f^2$ slopes. The transition from the $1/f$ to the $1/f^2$ slope for GaAs falls between 50 and 60 GHz and that for Si IMPATTs falls between 100 and 120 GHz. GaAs IMPATTs show higher power and efficiency in the 40- to 60-GHz region, while Si IMPATTs are produced with higher reliability and yield in the same frequency region. To the contrary, the GaAs IMPATTs have higher powers and efficiencies below 40 GHz than do Si IMPATTs. Si IMPATTs seem to outperform GaAs devices above 60 GHz [22].

Example 6-3-3   CW Output Power of an IMPATT Diode

A certain silicon IMPATT diode has the following parameters:

Drift velocity	$v_d = 10^7$ cm/sec
Drift region length	$L = 5\ \mu$m
Maximum operating voltage	$V_{0\max} = 120$ volts
Maximum operating current	$I_{0\max} = 210$ mA
Efficiency	$\eta = 10\%$

(a) Determine the maximum CW output power.
(b) Find the resonant frequency.

Solution.   (a) Using Eq. (6-3-10), we find that the maximum CW output power is

$$P_{\max} = \eta P_{dc} = 0.1 \times 120 \times 0.21 = 2.52 \text{ W}$$

(b) And from Eq. (6-3-4) the resonant frequency is

$$f = \frac{10^5}{2 \times 5 \times 10^{-6}} = 10 \text{ GHz}$$

6-3-4 Design Examples

An IMPATT diode can be designed as an amplifier or an oscillator, depending on its resonant circuit. In general, the added power in amplifier applications is the same as the output power in oscillator applications. The resonant circuit can be either a waveguide type or a coaxial type. The proper oscillator load impedance is the complex conjugate of the diode impedance, whereas the amplifier load impedance is related to the power gain g of the amplifier by the following equation

$$g = \left(\frac{Z_\ell - Z_d^*}{Z_\ell + Z_d} \right)^2 \tag{6-3-11}$$

where $Z_\ell = R_\ell + jX_\ell$ is the load impedance
$Z_d = R_d + jX_d$ is the diode impedance
$Z_d^* = R_d - jX_d$ is the complex conjugate of the diode impedance

If $X_\ell = -X_d$ is chosen for maximum power transfer, then Eq. (6-3-11) becomes

$$g = \left(\frac{R_\ell - R_d}{R_\ell + R_d}\right)^2 \tag{6-3-12}$$

In addition, amplifier design is very similar to oscillator design. In both cases, the load resistance resonates the diode resistance. In amplifier design the load resistance is larger than the oscillator load resistance.

I. Amplifier Design Example 6-3-1

A certain IMPATT diode has a diode impedance $Z_d = -0.70 + j11$ at 10 GHz. The problem is to design a waveguide-type amplifier for a power gain of 10 dB in TE_{10} mode. The width W of the waveguide is 2.287 cm, and its height H is 1.016 cm for X-band operation.

Design data:

Operating frequency:	$f = 10$ GHz
Cutoff frequency:	$f_c = 6.56$ GHz for TE_{10} mode
Cutoff wavelength:	$\lambda_c = 4.57$ cm
Wavelength in waveguide:	$\lambda_g = 4$ cm
Waveguide impedance:	$Z_g = \dfrac{H}{W} \dfrac{377}{\sqrt{1-(f_c/f)^2}} = 221$ ohms

Coaxial characteristic impedance:

$$Z_0 = \frac{\eta_0}{2\pi\sqrt{\epsilon_r}} \ell n \frac{b}{a}$$

$$= 92 \, \ell og \frac{b}{a} \ \Omega \qquad \text{for } \epsilon_r = 2.25 \tag{6-3-13}$$

Free-space intrinsic impedance:	$\eta_0 = 377 \ \Omega$
Post diameter:	$a = 0.363$ cm

Load impedance:

$$Z_\ell = Z_0 \frac{Z_g + jZ_0 \tan \beta\ell}{Z_0 + jZ_g \tan \beta\ell} \tag{6-3-14}$$

where

$$\beta\ell = \frac{2\pi}{\lambda}\ell = 2\pi\frac{\ell}{\lambda}$$

Figure 6-3-3 shows the graphs for the load resistance and the coaxial characteristic impedance in terms of the coaxial-line length.

The basic amplifier circuit and its equivalent circuit are shown in Fig. 6-3-4.

The coaxial section starts at the center of the waveguide. The waveguide is shorted at one end. The post that supports the diode is located in the center line at $\lambda_g/4$ from the short end. The problem is to design a coaxial transformer matching load impedance Z_ℓ to waveguide impedance Z_g for maximum power transfer.

Design Procedures:

(a) Calculate load resistance R_ℓ and load impedance Z_ℓ for maximum power transfer of 10 dB gain.

(b) From the IMPATT diode data sheet find the electrical length ℓ of the coaxial line that matches the waveguide impedance to the load resistance.

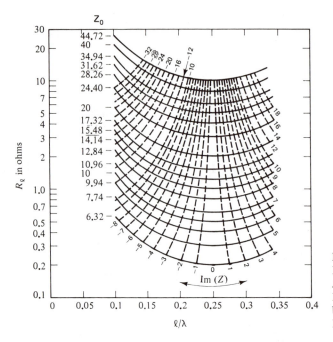

Figure 6-3-3 Data sheet of an IMPATT diode (After HP Application Note 968, IMPATT amplifiers. Reprinted by permission of Hewlett-Packard Company, Inc.)

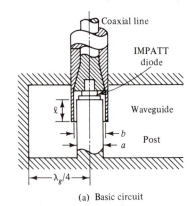

(a) Basic circuit

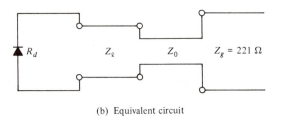

(b) Equivalent circuit

Figure 6-3-4 Circuit diagram and equivalent circuit of an IMPATT amplifier (After HP Application Note 968, IMPATT amplifier. Reprinted by permission of Hewlett-Packard Company, Inc.)

(c) Determine the characteristic impedance Z_0 of the coaxial line which will transfer the load impedance Z_ℓ to the waveguide impedance Z_g.

(d) Compute the values of a and b for the designed coaxial line.

Solution.   (a) From Eq. (6-3-12), the load resistance is found to be

$$R_\ell = 1.35 \ \Omega$$

The load impedance is then

$$Z_\ell = 1.35 - j11 \ \Omega$$

(b) From the diode data sheet in Fig. 6-3-3, it is found

$$\frac{\ell}{\lambda} = 0.132$$

Then phase angle $\beta\ell$ is found to be $47.52°$. The electrical length ℓ is found as 0.40 cm

(c) The coaxial-line characteristic impedance Z_0 can be found from either the diode data graphs in Fig. 6-3-3

$$Z_0 = 12.20 \ \Omega$$

(d) The diameter of the coaxial-line section is given as

$$a = 0.363 \text{ cm}$$

From the coaxial-line Eq. (6-3-13), the ratio of the coaxial outer diameter b to its inner diameter a is obtained as

$$\frac{b}{a} = 1.350$$

Then $b = 0.490$ cm

II. Oscillator Design Example 6-3-2

The real part of an oscillating diode's negative impedance must be exactly equal to the circuit load resistance at the frequency of oscillation, but the diode reactance must cancel the circuit reactance at the resonant frequency. In a waveguide resonant cavity for the TE_{101} mode resonance, the main function of the resonant post is to introduce some shunt capacitance or shunt inductance across the waveguide in order to cancel the circuit reactance for conjugate matching purposes. As a result, resonance will occur and maximum power will be delivered. The height of the resonant post is slightly less than $\lambda_0/4$ and is given by

$$h \simeq \frac{\lambda_0}{4} - \frac{d-r}{2 \ell n(2d/r)} \qquad (6\text{-}3\text{-}15)$$

where λ_0 = free-space wavelength

   r = post radius

   d = distance between the post surface and the inner side of the waveguide

From Example 6-3-1, the following data are found:

$$\lambda_0 = 3 \text{ cm}$$
$$r = 0.181 \text{ cm}$$
$$d = 0.962 \text{ cm}$$

Then the post height h is calculated from Eq. (6-3-15) as

$$h = 0.58 \text{ cm}$$

The post should be completely silver plated to reduce the insertion loss in the waveguide. It is important that the adjustable resonant post be set up as tight as possible so that it will make good contact with the heat sink. Otherwise any loose contact would cause undesirable abrupt mode jumping or would disconnect the heat path to the heat sink. In many cases, the heating problem limits the power output attainable from the CW IMPATT oscillator.

6-4 TRAPATT DIODES

The letter combination TRAPATT is an abbreviation for TRApped Plasma Avalanche Triggered Transit mode. This mode was first reported by Prager et al. [8]. It is a high-efficiency microwave generator capable of operating from several hundred megahertz to several gigahertz.

TRAPATT devices are usually operated in the pulsed mode with a peak power of several hundred watts in the frequency range of 0.5 GHz to 5 GHz. Typical duty cycles are from 0.001 to 0.01. These devices are frequently used as transmitters in radars.

6-4-1 Physical Structure

The basic structure of the oscillator is a semiconductor *p-n* junction diode reverse biased to current densities well in excess of those encountered in normal avalanche operation. High-peak-power diodes are typically silicon n^+-p-p^+ (or p^+-n-n^+) structures with the *n*-type depletion region width varying from 2.5 to 12.5 μm. The doping of the depletion region is generally such that the diodes are well "punched through" at breakdown; that is, the dc electric field in the depletion region just prior to breakdown is well above the saturated drift-velocity level. The device's p^+ region is kept as thin as possible at 2.5 to 7.5 μm. Device diameters range from as small as 50 μm for CW operation to as large as 750 μm at lower frequencies for high-peak-power devices.

6-4-2 Principles of Operation

Approximate analytical solutions for the TRAPATT mode in the p^+-n-n^+ diode have been developed by Clorfeine [9] and DeLoach et al. [10]. These analyses have shown that a high-field avalanche zone propagates through the diode and fills the depletion layer with a dense plasma of electrons and holes which become trapped in the low-field region behind the zone. A typical voltage waveform for the TRAPATT mode of an avalanche p^+-n-n^+ diode operating with an assumed square-wave current drive is shown in Fig. 6-4-1.

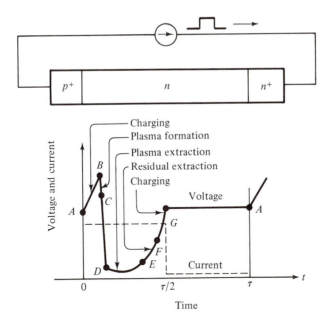

Figure 6-4-1 Voltage and current wave form for the TRAPATT diode (From A. S. Clorfeine [9]; reprinted with permission of RCA Laboratories)

At point A the electric field is uniform throughout the sample and its magnitude is large but less than the value required for avalanche breakdown. Current density is expressed by

$$J = \epsilon_s \frac{dE}{dt} \tag{6-4-1}$$

where ϵ_s is the semiconductor dielectric permittivity of the diode.

The diode current is turned on at the instant of time at point A. Because the only charge carriers present are those due to the thermal generation, the diode initially charges up like a linear capacitor, driving the magnitude of the electric field above the breakdown voltage. When a sufficient number of carriers are generated, the particle current exceeds the external current and the electric field is depressed throughout the depletion region, thereby causing the voltage to decrease. This portion of the cycle is shown by the curve from point B to point C. During this time interval the electric field is sufficiently large for the avalanche to continue and a dense plasma of electrons and holes is created. As some electrons and holes drift out of the ends of the depletion layer, the field is further depressed and "traps" the remaining plasma. The voltage decreases to point D. A long time is required to remove the plasma because the total plasma charge is large compared to the charge per unit time in the external current. The plasma is removed at point E, but a residual charge of electrons remains in one end of the depletion layer and a residual charge of holes in the other end. As the residual charge is removed, the voltage increases from point E to point F. At point F all the charge that was generated internally has been removed. This charge must be greater than or equal to that supplied by the external current; otherwise the voltage will exceed that at point A.

From point F to point G the diode charges up again like a fixed capacitor. At point G the diode current goes to zero for half a period and the voltage remains constant at V_A until the current comes back on and the cycle repeats. The electric field can be expressed as

$$E(x, t) = E_m - \frac{qN}{\epsilon_s} X + \frac{JT}{\epsilon_s} \tag{6-4-2}$$

where N is the doping concentration of the n region and X is the distance.

Thus the value of T at which the electric field reaches E_m at a given distance X into the depletion region is obtained by setting $E(x, t) = E_m$, which yields

$$T = \frac{qN}{J} X \tag{6-4-3}$$

Differentiation of Eq. (6-4-3) with respect to time t results in

$$v_z \equiv \frac{dx}{dt} = \frac{J}{qN} \tag{6-4-4}$$

where v_z is the avalanche-zone velocity.

Example 6-4-2   Avalanche-Zone Velocity of a TRAPATT Diode

A certain TRAPATT diode has the following parameters:

Doping concentration	$N = 10^{15} \text{ cm}^{-3}$
Current density	$J = 10^4 \text{ A/cm}^2$

Calculate the avalanche-zone velocity.

Solution.   Using Eq. (6-4-4), we obtain the avalanche-zone velocity as

$$v_z = \frac{10^4}{1.6 \times 10^{-19} \times 10^{15}} = 6.25 \times 10^7 \text{ cm/sec}$$

It can be seen from Example 6-4-2 that the avalanche-zone velocity is much larger than the scattering-limited velocity. Thus the avalanche zone (or avalanche shock front) will quickly sweep across most of the diode, leaving the diode filled by a highly conducting plasma of holes and electrons whose space charge depresses the voltage to low values. Because of the dependence of the drift velocity on the field at low fields, electrons and holes will drift at velocities determined by low-field mobilities, and transit time of the carriers can become much longer than

$$\tau_s = \frac{L}{v_s} \tag{6-4-5}$$

where v_s is the saturated carrier drift velocity.

Thus the TRAPATT mode can operate at comparatively low frequencies because the discharge time of the plasma—that is, the rate Q/I of its charge to its current—can be considerably greater than the nominal transit time τ_s of the diode at high field. Moreover, the TRAPATT mode is still a transit-time mode in the real sense that the time delay of carriers in transit—that is, the time between injection and collection—is used to obtain a current phase shift favorable for oscillation.

6-4-3 *Output Power and Efficiency*

RF power is delivered by the diode to an external load when the diode is placed in a proper circuit with a load. The main function of this circuit is to match the diode effective negative resistance to the load at the output frequency while reactively terminating (trapping) frequencies above the oscillation frequency in order to ensure TRAPATT operation. To date, the highest pulse power of 1.2 kW has been obtained at 1.1 GHz (5 diodes in series) [11] and the highest efficiency of 75% has been achieved at 0.6 GHz [12].

However, the TRAPATT operation is a rather complicated type of oscillation and it requires good control of both device and circuit properties. In addition, the TRAPATT mode generally exhibits a considerably higher noise figure than the IMPATT mode; and the upper operating frequency appears to be limited practically to below the millimeter wave region.

6-4-4 *Design Example*

The TRAPATT diode is a high-power and high-efficiency device. It often requires a circuit that can support harmonics of the fundamental frequency at high-voltage amplitudes. The design and performance are more complicated for a TRAPATT diode, because of the strong device-circuit interaction that controls the overall device performance. In addition, the high sensitivity to small changes in circuit, operating conditions, or temperature also makes it difficult for the diode to maintain its desired performance for a long time in systems.

Figure 6-4-2 is a schematic diagram of a TRAPATT oscillator. The basic function of the choke is to provide the RF short circuit to the waveguide so as to

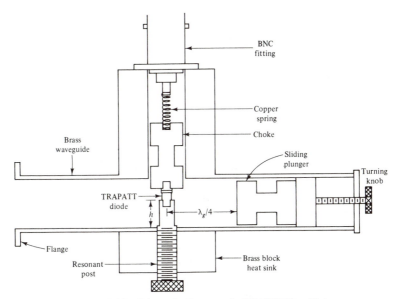

Figure 6-4-2 Schematic diagram of a TRAPATT oscillator

prevent the loss of any RF power through the bias post. The length of each portion of the choke must be a quarter wavelength of the operating frequency in order to achieve a complete short circuit of the element. The diameter of the resonant post for 10 GHz is about 5 mm. The post height h can be found from Eq. (6-3-15).

6-5   BARITT DIODES

BARITT diodes are BARrier Injected Transit-Time diodes as noted and they are the newest addition to the family of active microwave diodes. They have long drift regions similar to those of IMPATT diodes. However, the carriers traversing the drift regions of BARITT diodes are generated by minority-carrier injection from forward-biased junctions rather than being extracted from the plasma of an avalanche region.

6-5-1   Physical Structure

Several different types of device structures have been operated as BARITT diodes, including p-n-p, p-n-ν-p, p-n-metal, and metal-n-metal. For a p-n-ν-p BARITT diode, the forward-biased p-n junction emits holes into the ν region. These holes drift with saturation velocity through the ν region and are collected at the p contact. The diode will exhibit a negative resistance for transit angles between π and 2π. The optimum transit angle is approximately 1.6π [7].

BARITT diodes are much less noisy than IMPATT diodes. Noise figures are as low as 15 dB at C band frequencies with silicon BARITT amplifiers. The major disadvantages of BARITT diodes are relatively narrow bandwidth and power outputs limited to a few milliwatts.

6-5-2   Principles of Operation

A crystal n-type silicon wafer with 11-ohm-cm resistivity and $4 \times 10^{14}/\text{cm}^3$ doping is made of a 10-μm thin slice. Then the n-type silicon wafer is sandwiched between two PtSi Schottky barrier contacts of about 1000 Å thickness. A schematic diagram of a metal-n-metal structure is shown in Fig. 6-5-1(a).

The energy band diagram at thermal equilibrium is shown in Fig. 6-5-1(b), where ϕ_{n1} and ϕ_{n2} are the barrier heights for the metal-semiconductor contacts, respectively. For the PtSi-Si-PtSi structure mentioned previously, $\phi_{n1} = \phi_{n2} = 0.85$ eV. The hole barrier height ϕ_{p2} for the forward-biased contact is about 0.15 eV. Figure 6-5-1(c) shows the energy band diagram when a voltage is applied. The mechanisms responsible for the microwave oscillations are derived from

1. the rapid increase of the carrier injection process caused by the decreasing potential barrier of the forward-biased metal-semiconductor contact, and

2. an apparent $3\pi/2$ transit angle of the injected carrier that traverses the semiconductor-depletion region.

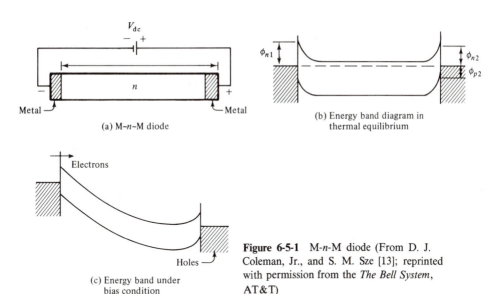

(a) M-*n*-M diode

(b) Energy band diagram in thermal equilibrium

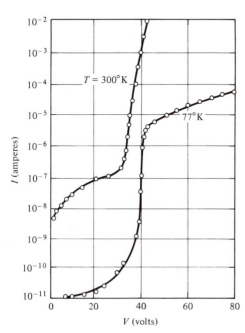

(c) Energy band under bias condition

Figure 6-5-1 M-*n*-M diode (From D. J. Coleman, Jr., and S. M. Sze [13]; reprinted with permission from the *The Bell System*, AT&T)

The rapid increase in terminal current with applied voltage (above 30 V) as shown in Fig. 6-5-2 is caused by thermionic hole injection into the semiconductor as the depletion layer of the reverse-biased contact reaches through the entire device thickness. The critical voltage is approximately given by

$$V_c = \frac{qNL^2}{2\epsilon_s}\tag{6-5-1}$$

Figure 6-5-2 Current versus voltage of a BARITT diode (PtSi-Si-PtSi) (From D. J. Coleman, Jr., and S. M. Sze [13]; reprinted with permission from *The Bell System*, AT&T)

where N is the doping concentration

L is the semiconductor thickness, and

ϵ_s is the semiconductor dielectric permittivity

The current-voltage characteristics of the silicon MSM structure (PtSi-Si-PtSi) were measured at temperatures $77\,°\mathrm{K}$ and $300\,°\mathrm{K}$. The device parameters are $L = 10~\mu$m, $N = 4 \times 10^{14}~\mathrm{cm}^{-3}$, $\phi_{n1} = \phi_{n2} = 0.85$ eV, and area $= 5 \times 10^{-4}~\mathrm{cm}^2$.

The current increase is not due to avalanche multiplication, as is apparent from the magnitude of the critical voltage and its negative temperature coefficient. At $77\,°\mathrm{K}$ the rapid increase is stopped at a current of about 10^{-5} A. This saturated current is to be expected in accordance with the thermionic emission theory of hole injection from the forward-biased contact with a hole barrier height (ϕ_{p2}) of about 0.15 eV.

6-5-3 Microwave Performance

Continuous-wave microwave performance of the M-n-M type BARITT diode was obtained over the entire C band of 4 to 8 GHz. The maximum power observed was 50 mW at 4.9 GHz. The maximum efficiency was about 1.8%. The FM single-side-band noise measure at 1 MHz was found to be 22.8 dB at a 7-mA bias current. This noise measure is substantially lower than that of a silicon IMPATT diode and is comparable to that of a GaAs transferred-electron oscillator. Figure 6-5-3 shows some of the measured microwave power versus current with frequency of operation indicated on each curve for three typical devices tested.

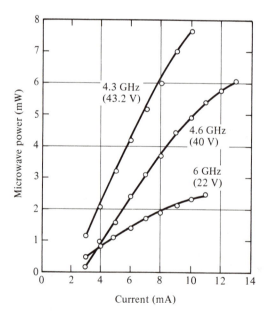

Figure 6-5-3 Power output versus current for three M-n-M devices (From D. J. Coleman, Jr., and S. M. Sze [13]; reprinted with permission from *The Bell System*, AT&T)

The voltage marked in parentheses for each curve indicates the average bias voltage at the diode while the diode is in oscillation. The gain-bandwidth product of a 6-GHz BARITT diode was measured as a 19-dB gain at a 5-mA bias current at 200 MHz. The small-signal noise measure was about 15 dB.

Example 6-5-3   Breakdown Voltage of a BARITT Diode

An M-Si-M-type BARITT diode has the following parameters:

$$
\begin{array}{lll}
\text{Relative dielectric constant of Si} & \epsilon_r = 11.8 \\
\text{Donor concentration} & N = 2.5 \times 10^{21} \text{ m}^{-3} \\
\text{Si length} & L = 5.5 \text{ } \mu\text{m}
\end{array}
$$

Determine the breakdown voltage.

Solution.   Using Eq. (6-5-1), we find that the breakdown voltage is double its critical voltage:

$$
V_{bd} = \frac{qNL^2}{\epsilon_s} = \frac{1.6 \times 10^{-19} \times 2.5 \times 10^{21} \times (5.5 \times 10^{-6})^2}{8.854 \times 10^{-12} \times 11.8}
$$

$$
= 115.81 \text{ V}
$$

and the breakdown electric field is

$$
E_{bd} = \frac{V_{bd}}{L} = \frac{115.81}{5.5 \times 10^{-6}} = 211 \text{ kV/cm}
$$

For safe operation, it is recommended that the operating electric field be less than the breakdown electric field (refer to Appendix 19).

6-5-4   DOVETT Diodes

The *DOVETT* (DOuble VElocity Transit Time) *diode* is the latest addition to the BARITT diode group. Its operational mechanism is similar to the BARITT diode except that the velocity of the carrier near the injection contact is significantly less than that near the collection contact. Because the negative resistance of a DOVETT diode is larger than that of a BARITT diode, it can operate at higher current densities than the latter.

6-6   PARAMETRIC AMPLIFIERS

A *parametric amplifier* is a device that uses a nonlinear reactance (capacitance or inductance) or a time-varying reactance. The word parametric is derived from the term *parametric excitation*, for the capacitance or inductance, which is a reactive parameter, can be utilized to produce capacitive or inductive excitation. Parametric excitation may be subdivided into parametric amplification and oscillation. Many essential properties of nonlinear energy-storage systems were described by Faraday [14] as early as 1831 and by Lord Rayleigh [15] in 1883. The first analysis of the nonlinear capacitance was given by van der Ziel [16] in 1948. In his paper van der

Ziel first suggested that such a device might be useful as a low-noise amplifier because essentially it was a reactive device in which no thermal noise is generated. In 1949 Landon [17] analyzed and presented the experimental results of such circuits when used as amplifiers, converters, and oscillators. In the age of solid-state electronics microwave electronics engineers dreamed of a solid-state microwave device to replace the noisy electron-beam amplifier. Then in 1957 Suhl [18] proposed a microwave solid-state amplifier that used ferrite. The first realization of a microwave parametric amplifier following Suhl's proposal was made by Weiss [19] in 1957. Following the work done by Suhl and Weiss, the parametric amplifier was finally developed.

At present, the *solid-state varactor diode* is the most widely used parametric amplifier. Unlike microwave tubes, transistors, and lasers, the parametric diode is of a reactive nature and thus generates a very small amount of Johnson noise (thermal noise). One of the distinguishing features of a parametric amplifier is that it uses an ac rather than the dc power supply employed by microwave tubes. In this respect, the parametric amplifier is analogous to the quantum amplifier laser or maser, in which an ac power supply is also used.

6-6-1 Nonlinear Reactance

A *reactance* is defined as a circuit element that stores and releases electromagnetic energy as opposed to a *resistance*, which dissipates energy. If the stored energy is predominantly in the electric field, the reactance is said to be *capacitive*; if the stored energy is predominantly in the magnetic field, the reactance is said to be *inductive*. In microwave engineering it is most convenient to speak in terms of voltages and currents rather than electric and magnetic fields. A capacitive reactance may then be considered as a circuit element for which capacitance is the ratio of charge on the capacitor to voltage across the capacitor. Then

$$C = \frac{Q}{V} \tag{6-6-1}$$

If the ratio is not linear, the capacitive reactance is said to be nonlinear. In this case, it is convenient to define a nonlinear capacitance as the partial derivative of charge with respect to voltage. That is,

$$C(v) = \frac{\partial Q}{\partial v} \tag{6-6-2}$$

The analogous definition of a nonlinear inductance is

$$L(i) = \frac{\partial \Phi}{\partial i} \tag{6-6-3}$$

In the operation of parametric devices the mixing effects occur when voltages at two or more frequencies are impressed on a nonlinear reactance.

Small-signal method.   It is assumed that signal voltage v_s is much smaller than pumping voltage v_p and that total voltage across the nonlinear capacitance

$C(t)$ is given by

$$v = v_s + v_p = V_s \cos \omega_s t + V_p \cos \omega_p t \qquad (6\text{-}6\text{-}4)$$

where $V_s \ll V_p$. The charge on the capacitor can be expanded in a Taylor series about point $v_s = 0$, and the first two terms are

$$Q(v) = Q(v_s + v_p) = Q(v_p) + \left. \frac{dQ(v_p)}{dv} \right|_{v_s = 0} v_s \qquad (6\text{-}6\text{-}5)$$

For convenience, it is assumed that

$$C(v_p) = \frac{dQ(v_p)}{dv} = C(t) \qquad (6\text{-}6\text{-}6)$$

where $C(v_p)$ is periodic with a fundamental frequency of ω_p. If the capacitance $C(v_p)$ is expanded in a Fourier series, the resultant is

$$C(v_p) = \sum_{n=0}^{\infty} C_n \cos n \omega_p t \qquad (6\text{-}6\text{-}7)$$

Because v_p is a function of time, the capacitance $C(v_p)$ is also a function of time. Then

$$C(t) = \sum_{n=0}^{\infty} C_n \cos n \omega_p t \qquad (6\text{-}6\text{-}8)$$

Coefficients C_n are the magnitude of each harmonic of the time-varying capacitance. In general, coefficients C_n are not linear functions of the ac pumping voltage v_p. Because the junction capacitance $C(t)$ of a parametric diode is a nonlinear capacitance, the principle of superposition does not hold for arbitrary ac signal amplitudes.

The current through the capacitance $C(t)$ is the derivative of Eq. (6-6-5) with respect to time, and it is

$$i = \frac{dQ}{dt} = \frac{dQ(v_p)}{dt} + \frac{d}{dt} \left[C(t) v_s \right] \qquad (6\text{-}6\text{-}9)$$

Clearly the nonlinear capacitance behaves like a time-varying linear capacitance for signals with amplitudes that are much smaller than the amplitude of the pumping voltage. The first term of Eq. (6-6-9) yields a current at the pump frequency f_p and is not related to the signal frequency f_s.

Large-signal method.   If the signal voltage is not small compared with the pumping voltage, the Taylor series may be expanded about a dc bias voltage V_0 in a junction diode. In a junction diode the capacitance C is proportional to $(\phi_0 - V)^{-1/2} = V_0^{-1/2}$, where ϕ_0 is the junction barrier potential and V is a negative voltage supply. Because

$$\left(V_0 + V_p \cos \omega_p t \right)^{-1/2} \simeq V_0^{-1/2} \left(1 - \frac{V_p}{3 V_0} \cos \omega_p t \right) \qquad \text{for } V_p \ll V_0$$

the capacitance $C(t)$ can be expressed as

$$C(t) = C_0(1 + 2\gamma \cos \omega_p t) \qquad (6\text{-}6\text{-}10)$$

The parameter γ is proportional to the pumping voltage v_p and indicates the coupling effect between the voltages at the signal frequency f_s and the output frequency f_0.

6-6-2 Manley–Rowe Power Relations

Manley and Rowe [20] have derived a set of general energy relations regarding power flowing into and out of an ideal nonlinear reactance. These relations are useful in predicting whether power gain is possible in a parametric amplifier. Figure 6-6-1 shows an equivalent circuit for the Manley–Rowe derivation.

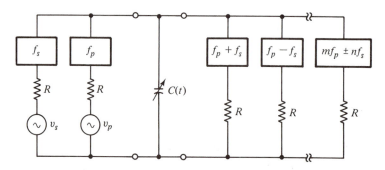

Figure 6-6-1   Equivalent circuit for Manley-Rowe derivation

In Fig. 6-6-1 one signal generator and one pump generator at their respective frequencies f_s and f_p, together with associated series resistances and bandpass filters, are applied to a nonlinear capacitance $C(t)$. These resonating circuits of filters are designed to reject power at all frequencies other than their respective signal frequencies. An infinite number of resonant frequencies of $mf_p \pm nf_s$ are generated in the presence of two applied frequencies of f_s and f_p, where m and n are any integer from zero to infinity.

Each of the resonating circuits is assumed to be ideal. The power loss by the nonlinear susceptances is negligible. That is, the power entering the nonlinear capacitor at the pump frequency is equal to the power leaving the capacitor at the other frequencies through the nonlinear interaction. Manley and Rowe established the power relations between the input power at the frequencies f_s and f_p and the output power at the other frequencies $mf_p \pm nf_s$.

From Eq. (6-6-4) the voltage across the nonlinear capacitor $C(t)$ can be expressed in an exponential form as

$$v = v_p + v_s = \frac{V_p}{2}\left(e^{j\omega_p t} + e^{-j\omega_p t}\right) + \frac{V_s}{2}\left(e^{j\omega_s t} + e^{-j\omega_s t}\right) \qquad (6\text{-}6\text{-}11)$$

The general expression of the charge Q deposited on the capacitor is given by

$$Q = \sum_{m=-\infty}^{\infty} \sum_{n=-\infty}^{\infty} Q_{m,n} e^{j(m\omega_p t + n\omega_s t)} \tag{6-6-12}$$

In order for the charge Q to be real, it is necessary that

$$Q_{m,n} = Q^*_{-m,-n} \tag{6-6-13}$$

The total voltage v can be expressed as a function of the charge Q. A similar Taylor series expansion of $v(Q)$ shows that

$$v = \sum_{m=-\infty}^{\infty} \sum_{n=-\infty}^{\infty} V_{m,n} e^{j(m\omega_p t + n\omega_s t)} \tag{6-6-14}$$

In order for the voltage v to be real, it is required that

$$V_{m,n} = V^*_{-m,-n} \tag{6-6-15}$$

The current flowing through $C(t)$ is the total derivative of Eq. (6-6-12) with respect to time. This is

$$i = \frac{dQ}{dt} = \sum_{m=-\infty}^{\infty} \sum_{n=-\infty}^{\infty} j(m\omega_p + n\omega_s) Q_{m,n} e^{j(m\omega_p t + n\omega_s t)}$$

$$= \sum_{m=-\infty}^{\infty} \sum_{n=-\infty}^{\infty} I_{m,n} e^{j(m\omega_p t + n\omega_s t)} \tag{6-6-16}$$

where

$$I_{m,n} = j(m\omega_p + n\omega_s) Q_{m,n} \quad \text{and} \quad I_{m,n} = I^*_{-m,-n}$$

Because the capacitance $C(t)$ is assumed to be a pure reactance, the average power at the frequencies $mf_p + nf_s$ is

$$P_{m,n} = \left(V_{m,n} I^*_{m,n} + V^*_{m,n} I_{m,n} \right)$$

$$= \left(V^*_{-m,-n} I_{-m,-n} + V_{-m,-n} I^*_{-m,-n} \right) = P_{-m,-n} \tag{6-6-17}$$

Then conservation of power can be expressed as

$$\sum_{m=-\infty}^{\infty} \sum_{n=-\infty}^{\infty} P_{m,n} = 0 \tag{6-6-18}$$

Multiplication of Eq. (6-6-18) by a factor of $(m\omega_p + n\omega_s)/(m\omega_p + n\omega_s)$ and rearrangement of the resultant into two parts yield

$$\omega_p \sum_{m=-\infty}^{\infty} \sum_{n=-\infty}^{\infty} \frac{m P_{m,n}}{m\omega_p + n\omega_s} + \omega_s \sum_{m=-\infty}^{\infty} \sum_{n=-\infty}^{\infty} \frac{n P_{m,n}}{m\omega_p + n\omega_s} = 0 \tag{6-6-19}$$

Because $I_{m,n}/(m\omega_p + n\omega_s) = jQ_{m,n}$, then $P_{m,n}/(m\omega_p + n\omega_s)$ becomes $-jV_{m,n}Q^*_{m,n}$, or $-jV_{-m,-n}Q^*_{-m,-n}$ and is independent of ω_p or ω_s. For any choice of frequencies f_p and f_s, the resonating circuit external to that of the nonlinear capacitance $C(t)$

can be so adjusted that the currents may keep all the voltage amplitudes $V_{m,n}$ unchanged. The charges $Q_{m,n}$ are then also unchanged since they are functions of the voltages $V_{m,n}$. Consequently, frequencies f_p and f_s may be arbitrarily adjusted in order to require

$$\sum_{m=-\infty}^{\infty} \sum_{n=-\infty}^{\infty} \frac{mP_{m,n}}{m\omega_p + n\omega_s} = 0 \tag{6-6-20}$$

$$\sum_{m=-\infty}^{\infty} \sum_{n=-\infty}^{\infty} \frac{nP_{m,n}}{m\omega_p + n\omega_s} = 0 \tag{6-6-21}$$

Equation (6-6-20) can be expressed as two terms:

$$\sum_{m=0}^{\infty} \sum_{n=-\infty}^{\infty} \frac{mP_{m,n}}{m\omega_p + n\omega_s} + \sum_{m=0}^{\infty} \sum_{n=-\infty}^{\infty} \frac{-mP_{m,n}}{-m\omega_p - n\omega_s} = 0 \tag{6-6-22}$$

Because $P_{m,n} = P_{-m,-n}$, we obtain

$$\sum_{m=0}^{\infty} \sum_{n=-\infty}^{\infty} \frac{mP_{m,n}}{mf_p + nf_s} = 0 \tag{6-6-23}$$

Similarly,

$$\sum_{m=-\infty}^{\infty} \sum_{n=0}^{\infty} \frac{nP_{m,n}}{mf_p + nf_s} = 0 \tag{6-6-24}$$

where ω_p and ω_s have been replaced by f_p and f_s, respectively.

Equations (6-6-23) and (6-6-24) are the standard forms for the Manley–Rowe power relations. The term $P_{m,n}$ indicates the real power flowing into or leaving the nonlinear capacitor at a frequency of $mf_p + nf_s$. The frequency f_p represents the fundamental frequency of the pumping voltage oscillator and the frequency f_s the fundamental frequency of the signal voltage generator. The sign convention for the power term $P_{m,n}$ is that power flowing into the nonlinear capacitance or the power coming from the two voltage generators is positive whereas the power leaving from the nonlinear capacitance or the power flowing into the load resistance is negative.

To illustrate, consider the case where the power output flow is allowed at a frequency of $f_p + f_s$ as shown in Fig. 6-6-1. All other harmonics are open circuited. Then currents at the three frequencies f_p, f_s, and $f_p + f_s$ are the only ones in existence. Under these restrictions m and n vary from -1 through 0 to $+1$, respectively. Next, Eqs. (6-6-23) and (6-6-24) reduce to

$$\frac{P_{1,0}}{f_p} + \frac{P_{1,1}}{f_p + f_s} = 0 \tag{6-6-25}$$

$$\frac{P_{0,1}}{f_s} + \frac{P_{1,1}}{f_p + f_s} = 0 \tag{6-6-26}$$

where $P_{1,0}$ and $P_{0,1}$ are the power supplied by the two voltage generators at frequencies f_p and f_s, respectively, and the powers are considered positive. The power $P_{1,1}$ flowing from the reactance into the resistive load at a frequency of $f_p + f_s$ is considered negative. The power gain, which is defined as a ratio of the power delivered by the capacitor at a frequency of $f_p + f_s$ to that absorbed by the capacitor at a frequency of f_s, as shown in Eq. (6-6-25), is given by

$$\text{Gain} = \frac{f_p + f_s}{f_s} = \frac{f_0}{f_s} \quad \text{for modulator} \qquad (6\text{-}6\text{-}27)$$

where $f_p + f_s = f_0$ and $(f_p + f_s) > f_p > f_s$

The maximum power gain is simply the ratio of the output frequency to the input frequency. This type of parametric device is called the *sum-frequency parametric amplifier* or *up-converter*.

If the signal frequency is the sum of the pump frequency and the output frequency, then Eq. (6-6-23) predicts that the parametric device will have a gain of

$$\text{Gain} = \frac{f_s}{f_p + f_s} \quad \text{for demodulator} \qquad (6\text{-}6\text{-}28)$$

where $f_s = f_p + f_0$ and $f_0 = f_s - f_p$

This type of parametric device is called the *parametric down-converter*, and its power gain is actually a loss.

If the signal frequency is at f_s, the pump frequency at f_p, and the output frequency at f_0, where $f_p = f_s + f_0$, then the power $P_{1,1}$ supplied at f_p is positive. Both $P_{1,0}$ and $P_{0,1}$ are negative. In other words, the capacitor delivers power to the signal generator at f_s instead of absorbing it. The power gain may be infinite, which is an unstable condition, and the circuit may be oscillating both at f_s and f_0. This is another type of parametric device, often called a *negative-resistance parametric amplifier*.

6-6-3 Up-Converter, Down-Converter, Negative-Resistance, and Degenerate Amplifiers

In a superheterodyne receiver an RF signal can be mixed with a signal from the local oscillator in a nonlinear circuit (the mixer) to generate the sum and difference frequencies. In a parametric amplifier the local oscillator is replaced by a pumping generator and the nonlinear element by a time-varying capacitor (or inductor) as shown in Fig. 6-6-2.

Here the signal frequency f_s and the pump frequency f_p are mixed in the nonlinear capacitor C. Consequently, a voltage of the fundamental frequencies f_s and f_p, as well as the sum and difference frequencies $mf_p \pm nf_s$, will appear across C. If a resistive load is connected across the terminals of the idler circuit, an output voltage can be generated across the load at the output frequency f_0. The output

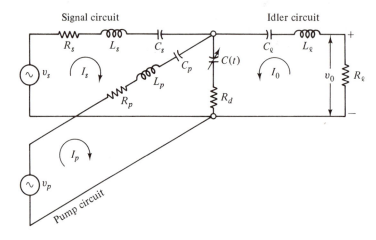

Figure 6-6-2   Equivalent circuit for a parametric amplifier

circuit, which does not require external excitation, is called the *idler circuit*. The output (or idler) frequency f_0 in the idler circuit is expressed as the sum and difference frequencies of the signal frequency f_s and pump frequency f_p. That is,

$$f_0 = mf_p \pm nf_s \qquad (6\text{-}6\text{-}29)$$

where m and n are positive integers from zero to infinite.

If $f_0 > f_s$, the device is called a parametric up-converter. Conversely, if $f_0 < f_s$, the device is known as a parametric down-converter.

Parametric up-converter.   A parametric up-converter has the following properties:

1. The output frequency is equal to the sum of the signal frequency and the pump frequency.
2. There is no power flow in the parametric device at frequencies other than the signal, pump, and output frequencies.

Power Gain.   When these two conditions are satisfied, the maximum power gain of a parametric up-converter [21] is expressed as

$$\text{Gain} = \frac{f_0}{f_s} \frac{x}{\left(1 + \sqrt{1+x}\,\right)^2} \qquad (6\text{-}6\text{-}30)$$

where $f_0 = f_p + f_s$, $x = f_s/f_0 (\gamma Q)^2$, $Q = 1/2\pi f_s C R_d$

R_d is the series resistance of a *p-n* junction diode and γQ is the figure of merit for the nonlinear capacitor. The quantity of $x/(1 + \sqrt{1+x}\,)^2$ may be regarded as a gain-degradation factor. As R_d approaches zero, the figure of merit γQ goes to infinity and the gain-degradation factor becomes equal to unity. As a result, the power gain of a parametric up-converter for a lossless diode is equal to f_0/f_s, which is predicted by the Manley-Rowe relations as shown in Eq. (6-6-27). In a typical microwave diode γQ could be equal to 10. If $f_0/f_s = 15$, the maximum gain, as given by Eq. (6-6-30), is 7.3 dB.

Noise Figure.   One advantage of a parametric amplifier over a transistor amplifier is its low-noise figure, because a pure reactance does not contribute thermal noise to the circuit. The noise figure F for a parametric up-converter [21] is given by

$$F = 1 + \frac{2T_d}{T_0}\left[\frac{1}{\gamma Q} + \frac{1}{(\gamma Q)^2}\right] \qquad (6\text{-}6\text{-}31)$$

where T_d = diode temperature in $°K$
$T_0 = 300°K$ is the ambient temperature in $°K$, and
γQ = figure of merit for the nonlinear capacitor

In a typical microwave diode γQ could be equal to 10. If $f_0/f_s = 10$ and $T_d = 300°K$, the minimum noise figure is 0.86 dB, as calculated by using Eq. (6-6-31).

Bandwidth.   The bandwidth (BW) of a parametric up-converter is related to the factor of merit figure and the ratio of the signal frequency to the output frequency. The bandwidth equation [21] is given by

$$BW = 2\gamma\sqrt{\frac{f_0}{f_s}} \qquad (6\text{-}6\text{-}32)$$

If $f_0/f_s = 10$ and $\gamma = 0.2$, the bandwidth is equal to 1.265.

Example 6-6-3 Up-Converter Parametric Amplifier

An up-converter parametric amplifier has the following parameters:

Output frequency-to-signal frequency ratio $\qquad \dfrac{f_0}{f_s} = 20$

Figure of merit $\qquad\qquad\qquad\qquad\qquad\qquad\quad \gamma Q = 10$
Factor of merit figure $\qquad\qquad\qquad\qquad\quad\;\; \gamma = 0.3$
Diode temperature $\qquad\qquad\qquad\qquad\qquad T_d = 320°K$

(a) Determine the power gain in decibels.
(b) Find the noise figure in decibels.
(c) Calculate the bandwidth.

Solution.   (a) Using Eq. (6-6-30), we obtain

$$\text{Power gain} = 20 \times \frac{100/20}{\left[1 + (1 + 100/20)^{1/2}\right]^2} = 8.4 = 9.24 \text{ dB}$$

(b) Then from Eq. (6-6-31) the noise figure is

$$F = 1 + \frac{2 \times 320}{300}\left(\frac{1}{10} + \frac{1}{100}\right) = 1.23 = 0.90 \text{ dB}$$

(c) From Eq. (6-6-32) the bandwidth is

$$BW = 2 \times 0.3 \times (20)^{1/2} = 2.68$$

Parametric down-converter.   If a mode of down-conversion for a parametric amplifier is desirable, the signal frequency f_s must be equal to the sum of the pump frequency f_p and the output frequency f_0. In other words, the input power must feed into the idler circuit and the output power must move out from the signal circuit as shown in Fig. 6-6-2. The down-conversion gain (actually a loss) is given by [21]

$$\text{Gain} = \frac{f_s}{f_0} \frac{x}{(1 + \sqrt{1 + x})^2} \tag{6-6-33}$$

Negative-resistance parametric amplifier.   If a significant portion of power flows only at signal frequency f_s, pump frequency f_p, and idler frequency f_i, then a regenerative condition with the possibility of oscillation at both the signal frequency and the idler frequency will occur. The idler frequency is defined as the difference between the pump frequency and the signal frequency, $f_i = f_p - f_s$. When the mode operates below the oscillation threshold, the device behaves as a bilateral negative-resistance parametric amplifier.

Power Gain.   Output power is taken from the resistance R_i at a frequency f_i and the conversion gain from f_s to f_i [21] is given by

$$\text{Gain} = \frac{4f_i}{f_s} \cdot \frac{R_g R_i}{R_{Ts} R_{Ti}} \cdot \frac{a}{(1 - a)^2} \tag{6-6-34}$$

where $f_s =$ signal frequency
$f_p =$ pump frequency
$f_i = f_p - f_s$ is the idler frequency
$R_g =$ output resistance of the signal generator
$R_{Ts} =$ total series resistance at f_s
$R_{Ti} =$ total series resistance at f_i
$a = \dfrac{R}{R_{Ts}}$,   and
$R = \dfrac{\gamma^2}{\omega_s \omega_i C^2 R_{Ti}}$ is the equivalent negative resistance.

Noise Figure.   The optimum noise figure of a negative-resistance parametric amplifier [21] is expressed as

$$F = 1 + \frac{2T_d}{T_0} \left[\frac{1}{\gamma Q} + \frac{1}{(\gamma Q)^2} \right] \tag{6-6-35}$$

where $\gamma Q =$ figure of merit for the nonlinear capacitor
$T_0 = 300\,°\text{K}$ is the ambient temperature in degrees Kelvin
$T_d =$ diode temperature in degrees Kelvin.

It is interesting to note that the noise figure given by Eq. (6-6-35) is identical to that for the parametric up-converter in Eq. (6-6-31).

Bandwidth.   The maximum gain-bandwidth of a negative-resistance parametric amplifier [21] is given by

$$\text{BW} = \frac{\gamma}{2} \sqrt{\frac{f_i}{f_s \text{ gain}}} \qquad (6\text{-}6\text{-}36)$$

If gain = 20 dB, $f_i = 4f_s$, and $\gamma = 0.30$, then the maximum possible bandwidth for single-tuned circuits is about 0.03.

Degenerate parametric amplifier.

The degenerate parametric amplifier or oscillator is defined as a negative-resistance amplifier with the signal frequency equal to the idler frequency. Because the idler frequency f_i is the difference between the pump frequency f_p and the signal frequency f_s, the signal frequency is just one-half the pump frequency.

Power Gain and Bandwidth.   The power gain and bandwidth characteristics of a degenerate parametric amplifier are exactly the same as for the parametric up-converter. With $f_s = f_i$ and $f_p = 2f_s$, the power transferred from the pump to the signal frequency is equal to the power transferred from the pump to the idler frequency. At high gain the total power at the signal frequency is almost equal to the total power at the idler frequency. So the total power in the passband will have 3 dB more gain.

Noise Figure.   The noise figures for a single-sideband and a double-sideband degenerate parametric amplifier [21] are given, respectively, by

$$F_{\text{ssb}} = 2 + \frac{2\overline{T}_d R_d}{T_0 R_g} \qquad (6\text{-}6\text{-}37)$$

$$F_{\text{dsb}} = 1 + \frac{\overline{T}_d R_d}{T_0 R_g} \qquad (6\text{-}6\text{-}38)$$

where $\overline{T}$ = average diode temperature in °K,
$T_0 = 300\,°\text{K}$ is the ambient noise temperature in °K,
R_d = diode series resistance in ohms, and
R_g = external output resistance of the signal generator in ohms.

It can be seen that the noise figure for double-sideband operation is 3 dB less than that for single-sideband operation.

6-6-4 Microwave Applications

The choice of which type of parametric amplifier to use will depend on the microwave system requirements. The up-converter is a unilateral stable device with a wide bandwidth and low gain. The negative-resistance amplifier is an inherently bilateral and unstable device with narrow bandwidth and high gain. The degenerate parametric amplifier does not require a separate signal and idler circuit coupled by the diode and is the least complex type of parametric amplifier.

In general, the up-converter has the following advantages over the negative-resistance parametric amplifier:

1. A positive input impedance
2. Unconditional stability and unilateralness
3. A power gain independent of changes in its source impedance
4. No circulator requirement, and
5. A typical bandwidth of 5%.

At higher frequencies where the up-converter is no longer practical, the negative-resistance parametric amplifier operated with a circulator is the right choice. When a low noise figure is required by a system, the degenerate parametric amplifier may be the logical choice because its double-sideband noise figure is less than the optimum noise figure of the up-converter or the nondegenerate negative-resistance parametric amplifier. Furthermore, the degenerate amplifier is a much simpler device to build and uses a relatively low pump frequency. The negative-resistance parametric amplifier may be the better choice in radar systems, for the frequency required by the system may be higher than the X band. But because the parametric amplifier is complicated to fabricate and expensive to produce, there is a tendency in microwave engineering to replace the parametric amplifier by the GaAs metal-semiconductor field-effect transistor (MESFET) amplifier in airborne radar systems.

REFERENCES

[1] Read, W. T. A proposed high-frequency negative-resistance diode. *Bell Syst. Tech. J.*, **37** (1958), 401–446.

[2] Sobol, H., and F. Sterzer. Solid-state microwave power sources. *IEEE Spectrum*, **9** (April 1972), 32.

[3] Coleman, D. J. Jr., and S. M. Sze. A low-noise metal-semiconductor-metal (MSM) microwave oscillator. *Bell Syst. Tech. J.*, **50** (May–June 1971), 1695–1699.

[4] DeLoach, Bernard C., Jr. Modes of avalanche diodes and their associated circuits. *IEEE J. Solid-State Circuits*, **SC-4**, no. 6 (December 1969), 376–384.

[5] Lee, C. A., et al. The Read diode an avalanche, transit-time, negative-resistance oscillator. *Appl. Phys. Lett.*, **6** (1965), 89.

[6] Johnston, R. L., B. C. DeLoach, and G. B. Cohen. A silicon diode microwave oscillator. *Bell Syst. Tech. J.*, **44** (February 1965), 369–372.

[7] Gilden, M. and M. E. Hines. Electronic tuning effects in the Read microwave avalanche diode. *IEEE Trans. on Electron Devices*, **ED-13** (January 1966).

[8] Prager, H. J., et al. High-power, high-efficiency silicon avalanche diodes at ultra high frequencies. *Proc. IEEE (Lett.)*, **55** (April 1967), 586–587.

[9] Clorfeine, A. S., et al. A theory for the high-efficiency mode of oscillation in avalanche diodes. *RCA Review*, **30** (September 1969), 397–421.

[10] DeLoach, B. C., Jr., and D. L. Scharfetter. Device physics of TRAPATT oscillators. *IEEE Trans. on Electron Devices*, **ED-17** (January 1970), 9–21.

[11] Liu, S. G., and J. J. Riska. Fabrication and performance of kilowatt L-band avalanche diodes. *RCA Review*, **31** (March 1970), 3.

[12] Kostichack, D. F. UHF avalanche diode oscillator providing 400 watts peak power and 75 percent efficiency. *Proc. IEEE (Lett.)*, **58** (August 1970), 1282–1283.

[13] Coleman, D. J., Jr., and S. M. Sze. A low-noise metal-semiconductor-metal (MSM) microwave oscillator. *Bell Syst. Tech. J.* (May–June 1971), pp. 1695–1696.

[14] Faraday, M. On a peculiar class of coustical figures; and certain forms assumed by a group of particles upon vibrating elastic surface. *Phil. Trans. Roy. Soc. (London)*, **121** (May 1831), 299–318.

[15] Lord Rayleigh, and J. W. Strutt. On the crispations of fluid resting upon a vibrating support. *Phil. Mag.*, **16** (July 1883), 50–53.

[16] van der Ziel, A. On the mixing properties of nonlinear capacitances. *J. Appl. Phys.*, **19** (November 1948), 999–1006.

[17] Landon, V. D. The use of ferrite cored coils as converters, amplifiers, and oscillators. *RCA Review*, **10**, 387–396.

[18] Suhl, H. Proposal for a ferromagnetic amplifier in the microwave range. *Phys. Rev.*, **106** (April 15, 1957), 384–385.

[19] Weiss, M. T. A solid-state microwave amplifier and oscillator using ferrites. *Phys. Rev.*, **107** (July 1957), p. 317.

[20] Manley, J. M., and H. E. Rowe. Some general properties of nonlinear elements: Part I, general energy relations. *Proc. IRE*, **44** (July 1956), 904–913.

[21] Blackwell, L. A., and K. L. Kotzebue. *Semiconductor-diode Parametric Amplifiers.* Englewood Cliffs, N.J.: Prentice-Hall, Inc. 1961, pp. 41, 42, 45, 53, 57, 62, 70.

[22] Hieslmair, Hans, et al. State of the art of solid-state and tube transmitters. *Microwave J.*, **26**, no. 10 (October 1983).

SUGGESTED READINGS

1. Chang, K. K. N. *Parametric and Tunnel Diodes.* Englewood Cliffs, N.J.: Prentice-Hall, Inc., 1964.

2. DeLoach, B. C., and D. L. Scharfetter. Device physics of TRAPATT oscillators. *IEEE Trans. on Electron Devices*, **ED-17**, no. 1 (January 1970), 9–21.

3. Haddad, G. I., ed. *Avalanche Transit-Time Devices.* Dedham, Mass.: Artech House, 1973.

4. Haddad, G. I., et al. Basic principles and properties of avalanche transit-time devices. *IEEE Trans. on Microwave Theory and Techniques.* **MTT-18**, no. 11 (November 1970), 752–772.

5. Howes, M. J., and D. V. Morgan. *Microwave Devices*, Chapter 3. New York: John Wiley & Sons, 1976.

6. *IEEE Transactions on Electron Devices*. Special issues on microwave solid-state devices.
   Vol. ED-27, No. 2, February 1980
   Vol. ED-27, No. 6, June 1980
   Vol. ED-28, No. 2, February 1981
   Vol. ED-28, No. 8, August 1981

7. *IEEE Transactions on Microwave Theory and Techniques*. Special issues on microwave solid-state devices.
   Vol. MTT-21, No. 11, November 1973
   Vol. MTT-24, No. 11, November 1976
   Vol. MTT-27, No. 5, May 1979
   Vol. MTT-28, No. 12, December 1980
   Vol. MTT-30, No. 4, April 1982
   Vol. MTT-30, No. 10, October 1982

8. Liao, Samuel Y. *Microwave Devices and Circuits*, Chapter 6. Englewood Cliffs, N.J.: Prentice-Hall, Inc., 1980.

9. Milnes, A. G. *Semiconductor Devices and Integrated Electronics*, Chapter 11. New York: Van Nostrand Reinhold Company, 1980.

10. Parker, Don. TRAPATT oscillations in a p-i-n avalanche diode. *IEEE Trans. on Electron Devices*, **ED-18**, no. 5 (May 1971), 281–293.

11. Sze, S. M. Microwave avalanche diodes. *IEEE Proc.*, **59**, no. 8 (August 1971), 1140–1171.

12. Sze, S. M. *Physics of Semiconductor Devices*, Chapter 10, 2nd ed. New York: John Wiley & Sons, 1981.

PROBLEMS

6-2 Read Diode

6-2-1. A Read diode is a theoretical model of avalanche transit-time devices proposed by Read. For example, the drift *i*-region length is 10 μm and the external circuit current I_e is measured as 4 mA.
(a) Calculate the carrier drift time τ in seconds.
(b) Estimate the moving charges in coulombs.
(c) Determine the operating frequency in gigahertz.

6-2-2. A Read diode has the following parameters:

Drift region length	$L = 10 \ \mu$m
Total hole charge	$Q = 20$ pC
Hole drift velocity	$v_d = 2 \times 10^7$ cm/s

(a) Compute the induced current in the external circuit.
(b) Calculate the resonant frequency.

6-3 IMPATT Diodes

6-3-1. The basic operating principle of an IMPATT diode was proposed by Read. Describe this principle of operation for IMPATT diodes.

6-3-2. An IMPATT diode has a drift length L of 4 μm.
   (a) Calculate the operating frequency in gigahertz.
   (b) Compute the transit angle in degrees.

6-3-3. A Ku-band IMPATT diode has a peak output voltage of 150 V and a peak output current of 1 A. The duty cycle is 0.001 at a frequency of 16 GHz and the efficiency is 10%.
   (a) Calculate the peak output power in watts.
   (b) Compute the average output power in watts.
   (c) Determine the input average power in watts.

6-3-4. An IMPATT diode is operating at a frequency of 60 GHz with a CW output power of 15 dBm at a current of 150 mA.
   (a) Determine the CW output power in watts.
   (b) Find the output voltage in volts.

6-3-5. A certain IMPATT diode has the following parameters:

Drift region length	$L = 7\,\mu$m
Maximum operating voltage	$V_{0\,\text{max}} = 180$ V
Maximum operating current	$I_{0\,\text{max}} = 165$ mA
Conversion efficiency	$\eta = 10\%$

   (a) Determine the output power in watts and dBW.
   (b) Find the resonant frequency.

6-3-6. Derive the power-frequency limitation equation of Eq. (3-6-9).

6-3-7. A double-drift region IMPATT diode has the following parameters:

Duty cycle	$DC = 0.25$
Drift velocity	$v_d = 10^7\,\text{cm/s}$
Drift region length	$L = 5\,\mu$m
Pulsed operating voltage	$V = 100$ volts
Pulsed operating current	$I = 900$ mA
Efficiency	$\eta = 11\%$

   (a) Compute the maximum peak output power.
   (b) Calculate the average power.
   (c) Determine the resonant frequency.

6-3-8. A certain IMPATT diode has a diode impedance Z_d of $-0.80 + j12$ at 9 GHz. The problem is to design a waveguide-type amplifier for a power gain of 9 dB in TE_{10} mode. The width of the waveguide is 2.287 cm and its height is 1.016 cm for 9-GHz operation.
   (a) Determine the cutoff frequency in gigahertz.
   (b) Compute the cutoff wavelength in centimeters.
   (c) Find the waveguide wavelength in centimeters.
   (d) Determine the waveguide impedance in ohms.
   (e) Calculate the load resistance R_ℓ and the load impedance Z_ℓ for the maximum power transfer of a 9-dB gain.
   (f) Find the electrical length ℓ of the coaxial line that will match the waveguide impedance to the load resistance from the IMPATT diode data sheet.

(g) Determine the coaxial-line characteristic impedance Z_0 that will transfer the load impedance Z_ℓ to the waveguide impedance Z_g.

(h) Compute the values of a and b for the designed coaxial line.

6-4 TRAPATT Diodes

6-4-1. A TRAPATT diode is operating under the trapped plasma avalanche triggered mode. Describe its principle of operation.

6-4-2. The TRAPATT diode is a high-efficiency microwave device. Its dc voltage and current are 35 V and 453 mA at a frequency of 4 GHz. The output power is 6 W. Calculate its efficiency.

6-4-3. A TRAPATT diode is operating under 60 V at a frequency of 8 GHz. Its output peak power is 50 W at a duty cycle of 0.02.

(a) Calculate the output average power in watts.

(b) Determine the pulse duration in picoseconds.

6-4-4. A certain TRAPATT diode has a doping concentration of 4×10^{15} cm^{-3} and a current density of 3×10^4 A/cm^2. Calculate the avalanche-zone velocity.

6-5 BARITT Diodes

6-5-1. The BARITT diode is a low-noise metal-semiconductor-metal (MSM) microwave device. If the Si semiconductor length L is 10 μm and the doping concentration N is 4×10^{14} cm^{-3},

(a) compute the critical voltage in volts

(b) calculate the current at room temperature.

6-5-2. BARITT diode operates in several different structures. Describe the operational principle for a M-S-M structure.

6-5-3. An M-Si-M BARITT diode has the following parameters:

Relative dielectric constant of Si	$\epsilon_r = 11.8$
Donor concentration	$N = 3 \times 10^{21}$ m^{-3}
Si length	$L = 5$ μm

Determine the breakdown voltage and field.

6-6 Parametric Amplifiers

6-6-1. A parametric amplifier is a nonlinear varactor type of microwave device and its noise figure is extremely low.

(a) Describe the advantages and disadvantages of parametric amplifiers.

(b) Explain their applications.

6-6-2. The figure of merit for a nonlinear diode capacitor in an up-converter parametric amplifier is 10 and the ratio of the output (or idler) frequency f_0 over the signal frequency f_s is 9. The diode temperature is 300° K.

(a) Calculate the maximum power gain in decibels.

(b) Compute the noise figure F in decibels.

(c) Determine the bandwidth (BW) for $\gamma = 0.3$.

6-6-3. A negative-resistance parametric amplifier has the following parameters:

$f_s = 3$ GHz	$R_i = 1$ kΩ	$\gamma = 0.37$
$f_i = 2f_s$	$R_g = 1$ kΩ	$\gamma Q = 12$
$f_p = 10$ GHz	$R_{Ts} = 1$ kΩ	$T_d = 300°$ K
	$R_{Ti} = 1$ kΩ	$C = 0.01$ pF

(a) Determine the power gain in decibels.
(b) Compute the noise figure F in decibels.
(c) Find the bandwidth.

6-6-4. An up-converter parametric amplifier has the following parameters:

Output frequency-to-signal frequency ratio $\dfrac{f_0}{f_s} = 25$

Figure of merit $\gamma Q = 10$
Factor of merit figure $\gamma = 0.4$
Diode temperature $T_d = 330\,°\text{K}$

(a) Compute the power gain in decibels.
(b) Calculate the noise figure in decibels.
(c) Find the bandwidth.

Chapter 7

Optical Fibers and Systems

7-0 INTRODUCTION

The optical fiber is one of the latest additions to signal links in submillimeter-wave technology. Its transmission loss is extremely low and its major applications are in computer links, industrial automation, medical instruments, telecommunications, and military command systems. The operational mechanisms, modes, structures, and characteristics of optical fibers and systems are described in this chapter.

7-1 OPTICAL FIBERS

An optical-fiber cable is an assembly of several optical fibers. The optical fiber is simply a cylindrical waveguide system. The center part consists of a core with a greater refractive index and the outer part of a cladding with a smaller refractive index. If a beam of electromagnetic energy impinges this system through one end-face of the fiber, the light beam will travel through the fiber and emerge from the other end-face.

The basic difference in signal transmissions between a metal transmission line and an optical fiber lies in the carriers. In a metal line the carriers are electrons; in an optical fiber the carriers are photons.

Light wave transmission on an optical fiber has reached a fully commercial stage with carrier wavelengths that are 0.82 to 0.85 μm. The future system is expected to operate with carrier wavelengths in the 1.2- to 1.6-μm range.

The advantages of optical fibers over metal lines in long-distance communication systems are the low-transmission loss and attractive bandwidths. Installed optical-fiber cables have shown losses in the neighborhood of 4 dB/km at wavelengths 0.82 to 0.85 μm. Several experiments have demonstrated that an optical fiber 30 km or longer had a loss below 0.7 dB/km near 1.3 μm. In data communication systems a bit rate of 50 Mbit/sec for a repeater span of at least 10 km was achieved for a multimode fiber.

The key components needed for a light-wave system are optical fibers, light-wave sources [lasers or light-emitting diodes (LEDs)], and light-wave detectors (avalanche photodiodes or *pin* photodiodes). Optical-fiber cables have been used in underwater intercontinental communication systems, intercity and metropolitan telephone systems, data communication systems, computer and switching link systems, automation process systems, and military command and control systems.

7-1-1 *Materials and Fabrications*

The three major materials used in the production of optical fibers are silica, glass, and plastic.

Silica Fibers.   Basically silica fibers are made of silicon dioxide (SiO_2) along with other metal oxides to establish a difference in the refractive index between the core and cladding. Various dopants have been used to increase or decrease the refractive index of silica, such as TiO_2, Al_2O_3, GeO_2, and P_2O_5. The cladding has a lower refractive index than the core, but its coefficient of thermal expansion is higher.

Glass Fibers.   Glass fibers are made of compound glasses with low melting temperatures and long-term chemical stability.

Plastic Fibers.   Plastic fibers consist of plastics having a higher attenuation loss than silica and glass. They are commonly used for short-distance links in computers.

Four fabrication processes for making optical fibers exist.

Outside Vapor-Phase Oxidation (OVPO) Process.   This method was first developed by the Corning Glass Works to deposit high-silica glass from vapor-phase sources. When the mandrel is removed, the "soot" tube is sintered and subsequently drawn to a fiber.

Modified Chemical Vapor Deposition (MCVD) Process.   This process was invented by the Bell Laboratories. For example, the glass of the desired composition is deposited inside a silica tube, layer by layer, via a flame-hydrolysis process, over the length of a mandrel. When deposition is completed, the material is sintered and the mandrel removed. The consolidated tube is then fused and collapsed simultaneously into a perform rod.

Vapor-Phase Axial Deposition (VAD) Process.   This method was introduced by the Nippon Telegraph and Telephone Public Corporation. It is similar to the OVPO process except that deposition occurs on the end of a growing "soot" cylinder.

Plasma Chemical Vapor Deposition (PCVD) Process. The Philip Company pioneered the use of microwave plasma to excite reactants and deposit vitreous material directly inside a silica tube.

All four processes fabricate fibers by using compound SiO_2 (silica) with relatively small amounts of Ge (germanium), P (phosphorus), and sometimes B (boron) as dopants to alter the refractive index and lower the working temperature somewhat below that for pure SiO_2.

7-1-2 Physical Structures

As noted, an optical fiber is simply a cylindrical dielectric waveguide system. It consists of a core at the center and a cladding outside. The core is a cylinder of transparent dielectric rod with a greater refractive index n_1 and the cladding is a second dielectric sheathing or covering, usually glass fused to the core, with a lower refractive index n_2. Generally optical fibers are classified as three types:

1. Monomode step-index fiber—having a core radius of 1 to 16 μm and a cladding radius of 50 to 100 μm

2. Multimode step-index fiber—having a core radius of 25 to 60 μm and a cladding radius of 50 to 150 μm

3. Multimode graded-index fiber—having a core radius of 10 to 35 μm and a cladding radius of 50 to 80 μm

Figure 7-1-1 shows diagrams of these three fiber types with their index profiles.

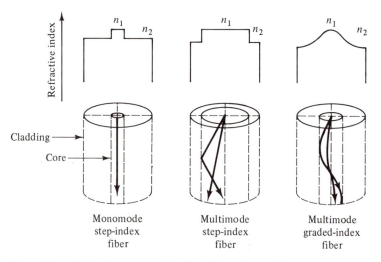

Figure 7-1-1 Physical structures of optical fibers (From Thomas G. Giallorenzi [10]; reprinted by permission of the IEEE, Inc.)

An optical-fiber cable is an assembly of several optical fibers incorporated in a protective material. Many different structures of optical-fiber cables are available. Figure 7-1-2 shows several typical cable configurations [1].

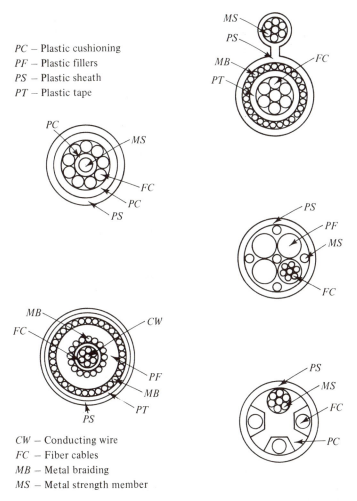

PC – Plastic cushioning
PF – Plastic fillers
PS – Plastic sheath
PT – Plastic tape

CW – Conducting wire
FC – Fiber cables
MB – Metal braiding
MS – Metal strength member

Figure 7-1-2   Diagrams of fiber-cable configurations (Reprinted from G. R. and H. A. Elion [1]; courtesy of Marcel Dekker, Inc.)

One fiber cable may contain a hundred fibers for high-capacity channels. Figure 7-1-3 shows a cable that consists of 12 stacked ribbons, each containing 12 optical fibers [2].

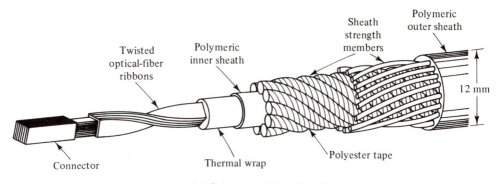

(a) Bell system lightguide cable

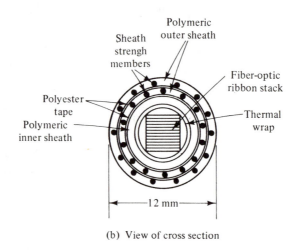

(b) View of cross section

Figure 7-1-3 Optical fiber cable with 144 fibers (From Ira Jacobs and S. E. Miller [2]; reprinted by permission of the IEEE, Inc.)

7-1-3 Losses

In an ideal case, the fiber core is considered as a perfectly transparent material and the cladding as a shield that confines the electromagnetic energy to the core region by total reflection. Otherwise absorption loss, transmission loss, plus some scattering loss and microbending loss, occur.

Absorption Losses.   If the core material is not perfectly transparent due to an impurity, the absorption loss for a length ℓ of the light path through the core glass is given by

$$\text{Absorption loss} = e^{-\alpha\ell} \tag{7-1-1}$$

where α = absorption coefficient per unit length

$\ell = n_1(n_1^2 - \sin^2\theta_m)^{-1/2}$ is the light path length

θ_m = maximum acceptance angle

Scattering Losses.   Fiber waveguide scattering losses are caused mainly by geometric irregularities at the core–cladding interface.

Microbending Losses.   The microbending of an optical fiber causes radiation losses if the fiber is distorted at joints separated 1 mm longitudinally.

In particular, plastic optical fibers have higher absorption losses than silica and glass fibers. In general, they are used for short-distance applications, such as in links in computers.

Example 7-1-3   Absorption Loss of an Optical Fiber

An optical fiber has the following parameters:

Maximum acceptance angle	$\theta_m = 30°$
Core refractive index	$n_1 = 1.40$
Absorption coefficient	$\alpha = 0.47$ Neper/km

Compute: (a) the light path length, and
(b) the absorption loss.

Solution.   (a) The light path length is

$$\ell = \frac{1.40}{\sqrt{(1.40)^2 - \sin^2(30)}} = \frac{1.40}{1.31}$$

$$= 1.07 \text{ km}$$

(b) The absorption loss is

$$\text{Loss} = e^{-\alpha\ell} = e^{-0.47 \times 1.07} = e^{-0.50} = 0.61$$

$$= 10\log(0.61) = -2.20 \text{ db/km}$$

7-1-4 Characteristics

In the twenty-first century microwave electronic devices and circuits will be heavily based on electrooptics. The characteristics of optical fibers are as follows:

Small Size and Light Weight.   An optical-fiber cable is much smaller than a copper wire line. The features of small size and light weight are especially important for underwater cables as well as the area of overcrowded transmission lines.

Low Cost and Low Loss.   The cost of optical-fiber cables and their installation is much lower than for metal cables. The transmission loss of the fibers is very low at 4 dB/km, as with coaxial lines at 150 dB for 0.5 km at 1 GHz or microstrip lines at 150 dB for 0.4 km at 10 GHz.

Immunity of Interference.   An optical-fiber cable does not generate or receive any electrical or electromagnetic noise or interference.

High Reliability and Durability.   Optical-fiber cables are safe to use in any explosive environment and eliminate the hazard of short circuits in wire lines.

High Bandwidth.   An optical-fiber cable can carry more channels than either coaxial or microstrip lines.

High-Security Capability.   Optical-fiber cables are immune to electrical or electromagnetic noise or grounding, such as cross talk or jamming. Therefore they increase the security capability of data transmission. This feature is very important in military communications.

7-2 OPERATIONAL MECHANISMS OF OPTICAL FIBERS

Light wave propagation in an optical fiber with a nondissipative core of radius a and refractive index n_1 embedded in a nondissipative cladding medium of radius b and refractive index n_2 can be analyzed by solving Maxwell's field equations for specific boundary conditions. The resultant waves in an optical fiber are neither transverse electric nor transverse magnetic but hybrid. In other words, there is no nonzero component of both electric E and magnetic H waves in the direction of propagation. A cylindrical coordinate system (r, ϕ, z) for an optical fiber is shown in Fig. 7-2-1.

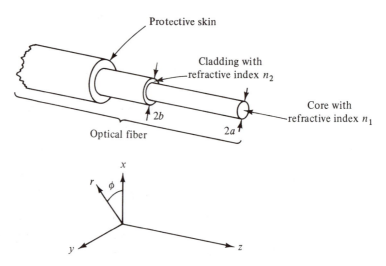

Figure 7-2-1   Schematic diagram of an optical fiber and its cylindrical coordinate system (r, ϕ, z)

The optical properties of a material are usually characterized by a complex refractive index N, which contains two constants—the refractive index n and the extinction index k. That is,

$$N = n - jk \qquad (7\text{-}2\text{-}1)$$

The extinction index k is related to the exponential decay of the wave as it passes through the medium. For nondissipative optical fibers, the extinction index k is equal to zero. The refractive index n is defined as the ratio of the velocity of light in vacuum to the velocity of light in a given medium. For a nonmagnetic medium the refractive index is expressed as

$$n = \frac{v_0}{v_\epsilon} = \frac{\sqrt{\mu_0 \epsilon}}{\sqrt{\mu_0 \epsilon_0}} = \sqrt{\epsilon_r} \qquad (7\text{-}2\text{-}2)$$

where ϵ_r is the relative dielectric constant of the medium. Typical values of n are 1.00 for air, 1.50 for polyethylene, 1.60 for polystyrene, and 8.94 for freshwater.

7-2-1 Wave Equations

As shown in Fig. 7-2-1, the optical-fiber waveguide consists of a core of higher refractive index n_1 and radius a surrounded by a cladding of lower refractive index n_2 and radius b. Both regions are assumed to be perfect insulators with free-space magnetic permeability μ_0. Such a structure can support an infinite number of modes, but for given values of n_1, n_2, a, and b, only a finite number are waveguide modes that have their fields localized in the vicinity of the core. The other unbound modes would correspond, for example, to light striking the core from the side, passing on through the core, and emerging from the other side.

The z axis of the cylindrical coordinate system (r, ϕ, z) is chosen to be along the fiber core axis. A waveguide mode is a coherent distribution of light, which is localized in the core by total internal reflection and which propagates through the guide with a well-defined phase velocity. The z components of electric field E and magnetic field H parallel to the fiber guide are given by the following expressions under the assumption that exponential factor $e^{j(\omega t - \beta_g z)}$ is implied.

In core region
$$E_z = AJ_n(kr)\cos(n\phi) \qquad (7\text{-}2\text{-}3)$$
$$H_z = BJ_n(kr)\sin(n\phi) \qquad (7\text{-}2\text{-}4)$$

In cladding region
$$E_z = CH_n(\chi r)\cos(n\phi) \qquad (7\text{-}2\text{-}5)$$
$$H_z = DH_n(\chi r)\sin(n\phi) \qquad (7\text{-}2\text{-}6)$$

where $J_n(kr) =$ the nth-order Bessel function of the first kind
$H_n(\chi r) =$ the nth-order Hankel function of the first kind
$k =$ the transverse propagation constant in the core region
$\chi =$ the transverse propagation constant in the cladding region
$k^2 = \beta_1^2 - \beta_g^2$ is the separation equation in the core region
$\chi^2 = \beta_g^2 - \beta_2^2$ is the separation equation in the cladding region
$\beta_g = \omega\sqrt{\mu_0 \epsilon_0}$ is the phase constant in free space
$\beta_1 = \omega\sqrt{\mu_1 \epsilon_1}$ is the phase constant in the core region
$\beta_2 = \omega\sqrt{\mu_2 \epsilon_2}$ is the phase constant in the cladding region

The Bessel function J_n exhibits oscillatory behavior for real k, as do the sinusoidal functions. So this function represents a cylindrical standing wave in the core region for $r < a$. The Hankel function H_n represents a traveling wave for real χ in the cladding region for $r > a$, as do the exponential functions. For a nondissipative medium, the Hankel function H_n becomes the modified Bessel function. In order for electric and magnetic fields to be evanescent in the cladding region, χ must be imaginary.

From Eqs. (7-2-3) and (7-2-4) the ϕ components of the field equations in the core region are

$$E_\phi = \left[A \frac{j\beta_g n}{k^2 r} J_n(kr) + B \frac{j\omega\mu_1}{k^2} J_n'(kr) \right] \sin(n\phi) \qquad (7\text{-}2\text{-}7)$$

$$H_\phi = -\left[A \frac{j\omega\epsilon_1}{k^2} J_n'(kr) + B \frac{j\beta_g n}{k^2 r} J_n(kr) \right] \cos(n\phi) \qquad (7\text{-}2\text{-}8)$$

Similarly, from Eqs. (7-2-5) and (7-2-6) the ϕ components for the cladding region are

$$E_\phi = -\left[C \frac{j\beta_g n}{\chi^2 r} H_n(\chi r) + D \frac{j\omega\mu_2}{\chi^2} H_n'(\chi r) \right] \sin(n\phi) \qquad (7\text{-}2\text{-}9)$$

$$H_\phi = \left[C \frac{j\omega\epsilon_2}{\chi^2} H_n'(\chi r) + D \frac{j\beta_g n}{\chi^2 r} H_n(\chi r) \right] \cos(n\phi) \qquad (7\text{-}2\text{-}10)$$

where the primes on J_n and H_n refer to differentiation with respect to their arguments (kr) and (χr), respectively. The tangential field components should be continuous at the interface $r = a$, and therefore

$$AJ_n(ka) = CH_n(\chi a) \qquad (7\text{-}2\text{-}11)$$

$$BJ_n(ka) = DH_n(\chi a) \qquad (7\text{-}2\text{-}12)$$

$$A \frac{j\beta_g n}{k^2 a} J_n(ka) + B \frac{j\omega\mu_1}{k^2} J_n'(ka) = -C \frac{j\beta_g n}{\chi^2 a} H_n(\chi a) - D \frac{j\omega\mu_2}{\chi^2} H_n'(\chi a) \quad (7\text{-}2\text{-}13)$$

$$A \frac{j\omega\epsilon_1}{k^2} J_n'(ka) + B \frac{j\beta_g n}{k^2 a} J_n(ka) = -C \frac{j\omega\epsilon_2}{\chi^2} H_n'(\chi a) - D \frac{j\beta_g n}{\chi^2 a} H_n(\chi a) \quad (7\text{-}2\text{-}14)$$

Once constants A, B, C, and D are determined, the electric and magnetic field equations can be finally derived from Maxwell's equations.

Example 7-2-1   Wave Propagation in an Optical Fiber

A monomode step-index fiber has the following parameters:

Carrier wavelength	$\lambda = 0.82 \ \mu\text{m}$
Carrier frequency	$f = 3.66 \times 10^{14} \ \text{Hz}$
Core radius	$a = 2 \ \mu\text{m}$
Core refractive index	$n_1 = 1.50$
Cladding radius	$b = 80 \ \mu\text{m}$
Cladding refractive index	$n_2 = 1.35$

Compute: (a) the phase constant in the core region,
        (b) the phase constant in the cladding region,
        (c) the phase constant in free space,
        (d) the transverse propagation constant in the core region,
        (e) the transverse propagation constant in the cladding region,
        (f) the value of the Bessel function $J_1(ka)$, and
        (g) the value of the Hankel function $H_1(\chi b)$.

Solution. (a) The phase constant in the core region is

$$\beta_1 = \omega\sqrt{\mu_1\epsilon_1} = \omega\frac{n_1}{c} = 2\pi \times 3.66 \times 10^{14} \times \frac{1.5}{3 \times 10^8}$$

$$= 1.15 \times 10^7 \text{ rads/m}$$

(b) The phase constant in the cladding region is

$$\beta_2 = 2\pi \times 3.66 \times 10^{14} \times \frac{1.35}{3 \times 10^8}$$

$$= 1.03 \times 10^7 \text{ rads/m}$$

(c) The phase constant in free space is

$$\beta_g = \frac{2\pi \times 3.66 \times 10^{14}}{3 \times 10^8} = 7.67 \times 10^6 \text{ rads/m}$$

(d) The transverse propagation constant in the core region is

$$k = \sqrt{(1.15 \times 10^7)^2 - (7.67 \times 10^6)^2} = 8.6 \times 10^6 \text{ rads/m}$$

(e) The transverse propagation constant in the cladding region is

$$\chi = \sqrt{(7.67 \times 10^6)^2 - (1.03 \times 10^7)^2} = j6.9 \times 10^6 \text{ rads/m}$$

(f) The Bessel function $J_1(ka)$ is

$$J_1(ka) = J_1(8.6 \times 10^6 \times 2 \times 10^{-6}) = J_1(17.20)$$

$$= -0.12814$$

(g) The Hankel function $H_1(\chi b)$ is

$$H_1(\chi b) = H_1(j6.9 \times 10^6 \times 80 \times 10^{-6}) = H_1(j552)$$

The Hankel function vanishes in the cladding region because its argument is imaginary.

7-2-2   Total Internal Reflection and Numerical Aperture (NA)

Total internal reflection. If the light wave in medium 1 propagates into medium 2, then Snell's law states that

$$\frac{\sin\phi_2}{\sin\phi_1} = \frac{\beta_1}{\beta_2} = \frac{v_2}{v_1} = \sqrt{\frac{\epsilon_1}{\epsilon_2}} = \frac{n_1}{n_2} \qquad \text{for } \mu_1 = \mu_2 = \mu_0 \qquad (7\text{-}2\text{-}15)$$

where ϕ_1 = incident angle in medium 1
    ϕ_2 = transmission angle in medium 2
    β_1 = incident phase constant in medium 1
    β_2 = transmission phase constant in medium 2
    v_1 = wave velocity in medium 1
    v_2 = wave velocity in medium 2
    ϵ_1 = dielectric permittivity of medium 1
    ϵ_2 = dielectric permittivity of medium 2
    n_1 = refractive index of medium 1
    n_2 = refractive index of medium 2

The total reflection occurs at $\phi_2 = 90°$ and then the incident angle in medium 1 is given by

$$\phi_1 = \phi_c = \arcsin \frac{n_2}{n_1} \qquad (7\text{-}2\text{-}16)$$

The angle specified by Eq. (7-2-16) is called the *critical incident angle* for total reflection. A wave incident on the interface of the core and cladding in an optical fiber at an angle equal to or greater than the critical angle will be totally reflected. There is a real critical angle only if $n_1 > n_2$. Thus the total internal reflection occurs only if the wave propagates from the core toward the cladding because the value of $\sin \phi_c$ must be equal to or less than unity.

Numerical aperture (NA).   In an optical-fiber waveguide, the light wave is incident on one end-face of the fiber. Figure 7-2-2 shows a diagram for the light wave impinging on a fiber.

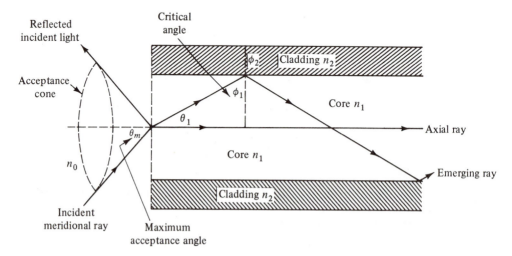

Figure 7-2-2   Wave propagation in an optical fiber

Several terminologies used in optical-fiber cables and systems are defined as follows:

Acceptance Angle.   The acceptance angle is defined as any angle measured from the longitudinal center line up to the maximum acceptance angle of an incident ray that will be accepted for transmission along a fiber.

Acceptance Cone.   The acceptance cone is a cone whose angle is equal to twice the acceptance angle.

Critical Angle.   The critical angle is the smallest angle made by a meridional ray in an optical fiber that can be totally reflected from the innermost interface. Thus it determines the maximum acceptance angle at which a meridional ray can be accepted for transmission along a fiber.

Maximum Acceptance Angle.   The maximum acceptance angle depends on the refractive indices of the two media that determine the critical angle.

Numerical Aperture (NA).   A numerical aperture is defined as a number that expresses the light-gathering power of a fiber.

For an optical fiber the most important parameter is the numerical aperture, which is expressed as

$$NA = \sin\theta \tag{7-2-17}$$

where θ is the acceptance angle.

According to Snell's law,

$$n_0\sin\theta = n_1\sin\theta_1 = n_1\cos\phi_1 \tag{7-2-18}$$

and

$$n_1\sin\phi_1 = n_2\sin\phi_2 \tag{7-2-19}$$

For $n_1 > n_2$, the total reflection occurs at

$$\sin\phi_1 > \frac{n_2}{n_1} \tag{7-2-20}$$

Then

$$\cos\phi_1 < \left(1 - \frac{n_2^2}{n_1^2}\right)^{1/2} \tag{7-2-21}$$

Substitution of Eq. (7-2-21) into (7-2-20) by using the trigonometrical identity yields

$$n_0\sin\theta < \left(n_1^2 - n_2^2\right)^{1/2} \tag{7-2-22}$$

or

$$\sin\theta < \frac{1}{n_0}\left(n_1^2 - n_2^2\right)^{1/2} \tag{7-2-23}$$

Equation (7-2-23) determines the incident acceptance angle of a meridional ray for total internal reflection in an optical fiber. Then the maximum acceptance angle is related to the following expressions.

$$\sin \theta_m = \frac{1}{n_0}\left(n_1^2 - n_2^2\right)^{1/2} \qquad \text{for } n_1^2 < \left(n_2^2 + n_0^2\right) \qquad (7\text{-}2\text{-}24)$$

$$\sin \theta_m = 1.0 \qquad \text{for } n_1^2 > \left(n_2^2 + n_0^2\right) \qquad (7\text{-}2\text{-}25)$$

Normally the optical fiber is immersed in free space or in a vacuum, $n_0 = 1.0$. Then Eqs. (7-2-24) and (7-2-25) become, respectively,

$$\sin \theta_m = \left(n_1^2 - n_2^2\right)^{1/2} \qquad \text{for } n_1^2 < \left(n_2^2 + 1\right) \qquad (7\text{-}2\text{-}26)$$

$$\sin \theta_m = 1.0 \qquad \text{for } n_1^2 > \left(n_2^2 + 1\right) \qquad (7\text{-}2\text{-}27)$$

Equation (7-2-26) is a measure of the numerical aperture of an optical fiber.

$$\text{NA} = \left(n_1^2 - n_2^2\right)^{1/2} \qquad (7\text{-}2\text{-}28)$$

The value of the numerical aperture also indicates the acceptance of impinging light, the degree of openness, the light-gathering ability, and the acceptance cone. The light-gathering power for a meridional ray in an optical fiber is given by

$$P = \left(\text{NA}\right)^2 = n_1^2 - n_2^2 \qquad (7\text{-}2\text{-}29)$$

Example 7-2-2 Characteristics of a Multimode Step-Index Fiber

A certain multimode step-index fiber has the following parameters:

Core refractive index	$n_1 = 1.54$
Cladding refractive index	$n_2 = 1.49$
Air refractive index	$n_0 = 1.00$

(a) Determine the critical angle for total reflection in the core region.
(b) Find the maximum incident angle at the interface between air and the fiber end.
(c) Compute the light-gathering power.

Solution. (a) Using Eq. (7-2-16), we find that the critical angle is

$$\phi_c = \arcsin\left(\frac{1.49}{1.54}\right) = 75.36°$$

(b) Then from Eq. (7-2-24) the maximum incident angle is

$$\theta_m = \arcsin\left[\frac{(1.54)^2 - (1.49)^2}{1^2}\right]^{1/2}$$

$$= \arcsin(0.39)$$

$$= 22.79°$$

(c) And from Eq. (7-2-29) the light-gathering power is

$$P = (1.54)^2 - (1.49)^2 = 0.15$$

7-2-3 Light-Gathering Power

In an optical-fiber system the light source is considered a Lambertian source—that is, a source whose intensity is given by

$$I = I_0 \cos \theta \tag{7-2-30}$$

where I_0 = light intensity at the normal direction
θ = angle between the normal to the source and the direction of measurement

Figure 7-2-3 shows a diagram for the light output from a Lambertian source through a spherical surface of radius r. The amount of light collected by a fiber that is normal to a Lambertian source is defined by the dielectric boundary of the fiber or the maximum acceptance angle θ_m and it is given by

$$\text{Light} = \int_0^{\theta_m} I_0 \cos \theta \, dA = \int_0^{\theta_m} I_0 \cos \theta \, 2\pi \sin \theta \, d\theta \tag{7-2-31}$$

$$= \pi I_0 \sin^2 \theta_m$$

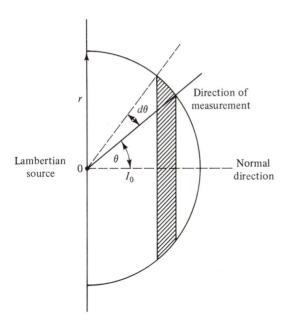

Figure 7-2-3 Light from a Lambertian source

The total light emitted by the Lambertian source is πI_0 and the fraction of light collected by the fiber is the light-collecting efficiency, which is

$$\begin{aligned} \eta_m &= \sin^2 \theta_m = (\text{NA})^2 & \text{for NA} < 1.0 \\ \eta_m &= 1.0 & \text{for NA} = 1.0 \end{aligned} \tag{7-2-32}$$

In practice, the light-collecting efficiency is much less than unity due to many factors, such as the angled end-face effect, the fiber-curvature effect, and the varying diameter effect.

The light-gathering power of an optical fiber is given by [3]

$$P = n_0^2 - \frac{2}{\pi} \left\{ \left(n_1^2 - n_2^2\right)^{1/2} \left[n_0^2 - \left(n_1^2 - n_2^2\right)\right]^{1/2} \right.$$

$$\left. + \left[n_0^2 - 2\left(n_1^2 - n_2^2\right)\right] \cos^{-1} \left(\frac{n_1^2 - n_2^2}{n_0^2}\right)^{1/2} \right\} \qquad (7\text{-}2\text{-}33)$$

or $\quad P = 1 - \frac{2}{\pi} \left\{ (\text{NA})\left[1 - (\text{NA})^2\right]^{1/2} + \left[1 - 2(\text{NA})^2\right] \cos^{-1}(\text{NA}) \right\} \quad$ for $n_0 = 1.0$

$$(7\text{-}2\text{-}34)$$

where $(\text{NA})^2 = n_1^2 - n_2^2$ is the meridional light-gathering power.

7-2-4   Wave Modes and Cutoff Wavelengths

Wave Modes.   For metallic circular waveguides the wave modes are usually designated by TE_{np} and TM_{np}, meaning Transverse Electric and Transverse Magnetic waves, respectively. In dielectric cylindrical waveguides only the cylindrically symmetric ($n = 0$) modes are either transverse electric (TE_{0p}) or transverse magnetic (TM_{0p}). The other modes are all hybrid; in other words, both electric field E_z and magnetic field H_z coexist.

The designation of the hybrid modes is based on the relative contributions of waves E_z and H_z to a transverse component of the field at some reference point. If electric wave E_z makes the larger contribution, the mode is considered E like and is designated EH_{np}. Conversely, if magnetic wave H_z makes the dominant contribution, the mode is designated HE_{np}. The method of designation is arbitrary, for it does not depend on any particular component of the chosen field, the reference point, and how far the wavelength is from the cutoff. The use of two letters, such as EH and HE, however, merely implies the hybrid nature of these modes.

The subscripts on EH_{np} and HE_{np} modes refer to the nth order of the Bessel function and the pth rank, where the rank gives the successive solutions of the boundary condition equation involving a Bessel function of the first kind J_n.

Cutoff Wavelengths.   For the cutoff condition, $J_n(ka) = 0$, which means that the limit of the argument of the Hankel function (χa) approaches zero [4]. Then the separation equation in the cladding region as shown in Eq. (7-2-6) becomes

$$\beta_g^2 = \omega^2 \mu_2 \epsilon_2 \qquad (7\text{-}2\text{-}35)$$

and the separation equation in the core region is

$$k^2 = \omega^2 \mu_1 \epsilon_1 - \omega^2 \mu_2 \epsilon_2 \qquad (7\text{-}2\text{-}36)$$

Let the argument $(k_{np}a)$ of the Bessel function of the first kind be X_{np}. Then

$$k_{np} = \frac{X_{np}}{a} \tag{7-2-37}$$

The cutoff condition is given by

$$X_{np} = \frac{2\pi a}{\lambda_0}\left(n_1^2 - n_2^2\right)^{1/2} \tag{7-2-38}$$

and the free-space cutoff wavelength is expressed as

$$\lambda_0 = \frac{2\pi a}{X_{np}}\left(n_1^2 - n_2^2\right)^{1/2} \tag{7-2-39}$$

For a given core refractive index n_1, a cladding refractive index n_2, and a core radius a, the free-space cutoff wavelength for a monomode EH_{01} operation is

$$\lambda_{0c} = \frac{2\pi a}{2.405}\left(n_1^2 - n_2^2\right)^{1/2} \tag{7-2-40}$$

The cutoff parameter X_{np} (or $k_{np}a$) is usually called the *V number* of the fiber. The number of propagating modes in the step-index fiber is proportional to its V number.

Example 7-2-4   Cutoff Wavelength of Monomode Optical Fibers

A monomode optical fiber has the following parameters:

Core radius	$a = 5\ \mu m$
Core refractive index	$n_1 = 1.40$
Cladding refractive index	$n_2 = 1.05$

Calculate the cutoff wavelength for the monomode EH_{01} operation.

Solution.   The argument of the Bessel function of the first kind for the EH_{01} mode is 2.405. Then the cutoff wavelength is

$$\lambda_{0c} = \frac{2\pi \times 5 \times 10^{-6}}{2.405}\left[(1.40)^2 - (1.05)^2\right]^{1/2}$$

$$= 12.11\ \mu m$$

This means that the signal with a wavelength larger than 12.11 μm will be cut off.

The wave modes capable of propagating in an optical fiber are those for which X_{np} are less than the values determined by Eq. (7-2-38). Because X_{np} forms an increasing sequence for fixed n and increasing p or for a fixed p and increasing n, the number of allowed modes increases with the square of the radius a. That is, the total number of modes for an optical fiber is given by [5]

$$\text{Modes} = \frac{16}{\lambda_0^2}\left(n_1^2 - n_2^2\right)a^2 \tag{7-2-41}$$

The electromagnetic field is guided only partially within the core region whereas outside the core the electromagnetic field is evanescent in a direction normal to propagation. Among the electromagnetic wave modes there is one—namely, the HE_{11} mode—that has no cutoff wavelength. Only this HE_{11} mode can propagate in an optical fiber for a wavelength greater than the highest cutoff wavelength of the other modes.

The modes of operation for optical fibers are commonly classified into single mode, such as monomode step-index fiber, and multimode, such as multimode step-index and multimode graded-index fiber. Figure 7-2-4 shows a plot of Bessel functions used for determining the cutoff conditions of optical-fiber modes.

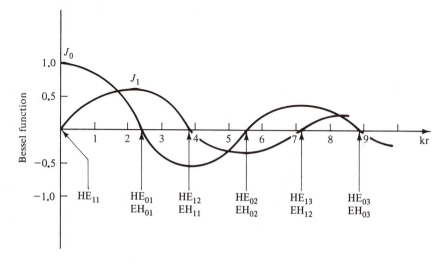

Figure 7-2-4 Cutoff modes of optical fibers

The cut-off V numbers of the step- and parabolic-index optical fibers for a few lower order modes are tabulated in Table 7-2-1.

TABLE 7-2-1 CUT-OFF V NUMBERS OF THE STEP- AND PARABOLIC-INDEX FIBERS FOR A FEW LOWER ORDER MODES

	Cut-off V numbers			Cut-off V numbers	
Hybrid modes	Step-index	Parabolic-index	Hybrid modes	Step-index	Parabolic-index
HE_{11}	0	0	HE_{13}	7.016	9.158
$HE_{01}, EH_{01}, HE_{21}$	2.405	3.518	EH_{12}, HE_{32}	7.016	9.645
HE_{12}	3.832	5.068	EH_{41}, HE_{61}	7.588	11.938
EH_{11}, HE_{31}	3.832	5.744	EH_{22}, HE_{42}	8.417	11.760
EH_{21}, HE_{41}	5.136	7.848	$HE_{03}, EH_{03}, HE_{23}$	8.654	11.424
$HE_{02}, EH_{02}, HE_{22}$	5.520	7.451	EH_{51}, HE_{71}	8.771	13.959
EH_{31}, HE_{51}	6.380	9.158	EH_{32}, HE_{52}	9.761	13.833

7-3 STEP-INDEX FIBERS

A step-index fiber is the one that has an abrupt change in refractive indices between the core and cladding materials. Two modes can exist in a step-index fiber: monomode fiber and multimode fiber. If the core radius is very small in comparison with the wavelength of the light source, say 1 to 16 μm, only a single mode is propagated. If the core radius is large enough, say 30 μm, multimodes coexist.

7-3-1 Monomode Step-Index Fibers

A monomode step-index fiber is a low-loss optical fiber with a very small core. The fiber requires a laser source for the input signals because of the very small acceptance aperture (or acceptance cone). When the small core radius approaches the wavelength of the light source, only a single mode is propagated.

    The core radius is from 1 to 16 μm and the differential in refractive indices between the core and cladding is about 0.6%. The bit rate is from 20 Mbit/s-km to 19 Gbit/s-km. This fiber is ideally suitable for long-haul and high-bandwidth applications, such as telecommunication systems.

    According to Eq. (7-2-38), the value of X_{np} required for a monomode operation in a step-index fiber must be less than 2.405. That is,

$$X_{np} = \frac{2\pi a}{\lambda_0}\left(n_1^2 - n_2^2\right)^{1/2} < 2.405 \qquad (7\text{-}3\text{-}1)$$

Figure 7-3-1 shows the diagrams for a monomode step-index fiber, light path, and signal waveform profile.

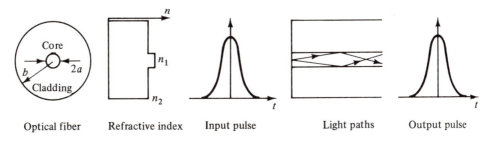

Optical fiber   Refractive index   Input pulse   Light paths   Output pulse

Figure 7-3-1 Diagram for monomode step-index fiber (From Thomas G. Giallorenzi [10]; reprinted by permission of the IEEE, Inc.)

Example 7-3-1 **Determination of Core Radius for a Monomode Optical Fiber**

Given: Core refractive index		$n_1 = 1.51$
Cladding refractive index		$n_2 = 1.49$
Light source—Nd^{3+}:YAG laser		$\lambda_0 = 1.064\ \mu$m

Determine the core radius that will support only a single mode operation.

Solution. From Eq. (7-3-1) the maximum core radius is

$$a \leq \frac{2.405 \times 1.064 \times 10^{-6}}{2\pi\sqrt{(1.51)^2 - (1.49)^2}}$$

$$a \leq 1.70 \ \mu\text{m}$$

7-3-2 Multimode Step-Index Fibers

A multimode step-index fiber has a large core and large numerical aperture (NA); so it can couple efficiently to a light source LED. The core radius is from 25 to 60 μm and the differential in refractive indices is from 1 to 10%. The bit rate is under 100 Mbit/s-km. It is suitable for low-bandwidth, short-haul and low-cost applications, such as links in computers. Figure 7-3-2 shows the diagrams for a multimode step-index fiber, light path and signal waveform profile.

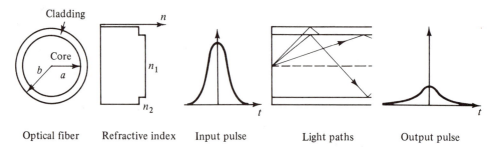

Optical fiber   Refractive index   Input pulse   Light paths   Output pulse

Figure 7-3-2 Diagrams for multimode step-index fiber (From Thomas G. Giallorenzi [10]; reprinted by permission of the IEEE, Inc.)

Example 7-3-2 Computations of a Multimode Step-Index Fiber
    A multimode step-index fiber has the following parameters:

Core refractive index	$n_1 = 1.45$
Numerical aperture	NA $= 0.35$
V number	$X_{np} = 100$
Wavelength of LED source	$\lambda_0 = 0.87 \ \mu$m

(a) Calculate the cladding refractive index n_2.
(b) Compute the core radius a.
(c) Determine the cladding radius b.

Solution. (a) The cladding refractive index is

$$n_2 = \left(n_1^2 - \text{NA}^2\right)^{1/2} = \sqrt{(1.45)^2 - (0.35)^2} = 1.41$$

(b) The core radius is obtained from Eq. (7-2-38), as

$$a = \frac{\lambda_0 X_{np}}{2\pi(\text{NA})} = \frac{0.87 \times 100}{6.283 \times 0.35} = 39.56 \ \mu\text{m}$$

(c) In order for electric and magnetic fields to be evanescent in the cladding region, the propagation constant χ of the Hankel function must be imaginary. That is,

$$\chi = j\sqrt{\beta_2^2 - \beta_g^2} = j\sqrt{\omega^2 \mu_2 \epsilon_2 - \omega^2 \mu_0 \epsilon_0}$$

$$= j\frac{2\pi}{\lambda_0}\left(n_2^2 - 1\right)^{1/2} = j\frac{6.2832}{0.87}\sqrt{(1.41)^2 - 1}$$

$$= j7.15$$

We see from Appendix 13 that the argument of the Hankel function could be $4X_{np} = 400$. Then the cladding radius is

$$b = \frac{400}{7.15} = 55.94 \ \mu m$$

7-4 GRADED-INDEX FIBER

A graded-index fiber is the one in which the refractive index of the core region is decreased monotonically from the center and converged into a flat at the cladding region. Much of the longer path lengths are through the lower index of the refractive material and so the increased velocity over this part of the path compensates somewhat for the longer path lengths and the spread in group velocity between modes is not as great as for the step-index fiber. In order to meet bandwidth requirements for telecommunication applications, multimode fibers are usually fabricated with graded-index profiles to reduce the intermodal dispersion. The graded-index fiber has a relatively large core; so it can support multimode operation. This type of fiber is suitable for high-bandwidth and medium-haul applications. Table 7-4-1 lists some typical parameters for graded-index fibers.

TABLE 7-4-1 PARAMETERS OF GRADED-INDEX FIBER

Parameters	Values
Bit rate	140–1000 Mbit/s-km
Core radius	10–35 μm
Cladding radius	50–80 μm
Losses	2–5 dB/km
Numerical aperture (NA)	0.15–0.25
Number of modes at 0.9 μm	140–900
Pulse dispersion/mode	0.1–4.0 ns/km
Refractive index n_1	1.47–1.50
Deviation index Δ	0.7–30%

Figure 7-4-1 shows the diagrams for a multimode graded-index fiber, light path, and signal waveform profiles.

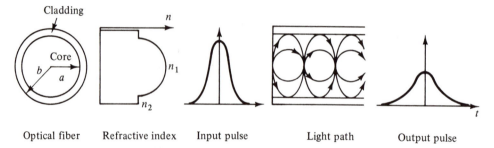

Optical fiber   Refractive index   Input pulse   Light path   Output pulse

Figure 7-4-1 Diagram for graded-index fiber (From Thomas G. Giallorenzi [10]; reprinted by permission of the IEEE, Inc.)

7-4-1 Refractive-Index Profiles

In a multimode graded-index fiber the refractive index is expressed by [6]

$$n(r) = n_1\left[1 - 2\Delta\left(\frac{r}{a}\right)^{\alpha}\right]^{1/2} \quad \text{for } r \leq a \qquad (7\text{-}4\text{-}1)$$

and

$$n(r) = n_1(1 - 2\Delta)^{1/2} \quad \text{for } r > a \qquad (7\text{-}4\text{-}2)$$

where n_1 = the refractive index at the center of the core region

$$\Delta = \frac{n_1 - n_2}{n_2} \simeq \frac{n_1 - n_2}{n_1} \simeq \frac{n_1^2 - n_2^2}{2n_1^2} \quad \text{is the deviation index}$$

α = a parameter that is also called the power-law coefficient
   between 1 and ∞

Figure 7-4-2 shows a cross-sectional diagram of a circular symmetric index profile in a multimode fiber.

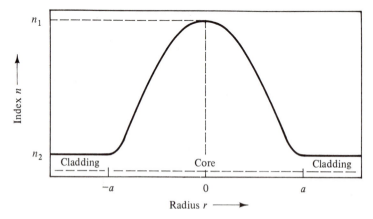

Figure 7-4-2 Cross-sectional diagram for graded refractive index (From John E. Midwinter [5–7]; reprinted by permission of *The Bell System*, AT&T.)

All profiles reach a constant cladding index n_2 at $r = a$ whereas the index is n_1 for $r = 0$ at the core center. The core profile has a cone shape or nearly parabolic figure for $\alpha = 1$ and becomes convergent to the form of the step profile for $\alpha = \infty$. Figure 7-4-3 illustrates a few of the index profiles as defined by Eq. (7-4-1) for the small deviation index Δ.

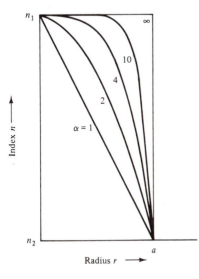

Figure 7-4-3   Graded index profiles (From John E. Midwinter [5–7]; reprinted by permission of *The Bell System*, AT&T.)

The total number of modes in a graded-index fiber are given by

$$N = \left(\frac{2\pi a}{\lambda_0} n_1 \right)^2 \Delta \left(\frac{\alpha}{\alpha + 2} \right) \qquad (7\text{-}4\text{-}3)$$

For a parabolic profile at $\alpha = 2$, the number of modes in a graded-index fiber is equal to half the number that exists in a step-index fiber ($\alpha = \infty$). That is,

$$N \text{ (parabolically graded)} = \frac{N(\text{step})}{2} \qquad (7\text{-}4\text{-}4)$$

Then the total number of modes in a graded-index fiber can be expressed as

$$N = \frac{V^2}{2} \left(\frac{\alpha}{\alpha + 2} \right) \qquad (7\text{-}4\text{-}5)$$

where $V = X_{np}$ is defined by Eq. (7-2-38).

Example 7-4-1   Number of Modes in a Graded-Index Fiber

A graded-index fiber has the following parameters:

Core radius	$a = 20 \ \mu\text{m}$
Power law coefficient	$\alpha = 2$
Deviation index	$\Delta = 0.03$
Core refractive index	$n_1 = 1.52$
Wavelength of light source LED	$\lambda_0 = 0.87 \ \mu\text{m}$

(a) Compute the total number of modes for the graded-index fiber.
(b) Calculate the number of modes for a step-index profile.
(c) Determine the graded refractive index.

Solution.   (a) From Eq. (7-4-3) the number of modes for the graded-index fiber is

$$N = \left(\frac{6.283 \times 20 \times 10^{-6}}{0.87 \times 10^{-6}} \times 1.52 \right)^2 \times 0.03 \left(\frac{2}{2+2} \right) = 723 \text{ modes}$$

(b) From Eq. (7-4-4) the number of modes for a step-index profile is

$$N = 2 \times 723 = 1446 \text{ modes}$$

(c) From Eq. (7-4-2) the graded refractive index is

$$n_2 = 1.52(1 - 2 \times 0.03)^{1/2} = 1.47$$

7-4-2 Wave Patterns

Graded-index fibers offer multimode propagation in a relatively large-core fiber coupled with low-mode dispersion. A typical bit rate is 1400 Mbit/sec over a link of 8 to 10 km in conjunction with a GaAs light source, whereas a step-index fiber provides an upper limit of perhaps 10 to 20 Mbit/s-km. There are three wave patterns in a graded-index fiber [7] as shown in Fig. 7-4-4.

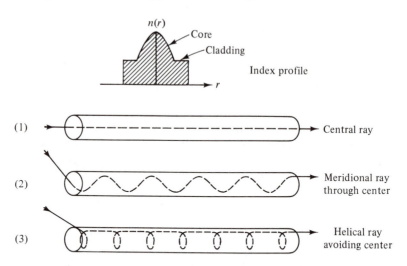

Figure 7-4-4 Wave patterns in graded-index fibers (From John E. Midwinter [5–7]; reprinted by permission of John Wiley & Sons, Inc.)

Center Ray.   Because the index distribution is graded, with a high index in the center surrounded by a steadily decreased index, the medium behaves similarly to a series of lenses. A ray traversing the center of the medium follows the axis, traveling in high index material all the way but traversing the shortest possible physical path from the end to end.

Meridional Ray.   A ray leaving the center of the fiber at an angle to the axis is bent by the index profile (lenslike structure) and curves around after some characteristic distance and crosses the axis once again. It continues in a sinusoidal path to the end. It thus travels a greater distance than the center ray, but much of its path is in the low-index material. This ray is called the meridional ray—that is, the ray passes through the axis of the fiber while being internally reflected.

Skew Ray (Helical Ray).   The third ray is a helical wave, formed when light is launched at a skew angle along the surface of a constant radius cylinder at some intermediate radius and index. Once again the physical path length is longer than that taken by the center ray, but it traverses in lower index material. This ray is called the skew ray or helical ray, which never intersects the axis of the fiber while being internally reflected.

7-5   OPTICAL-FIBER COMMUNICATION SYSTEMS

An optical-fiber communication system is essentially a dielectric-filled cylindrical waveguide with some electronic components and circuits (see Fig. 7-5-1).

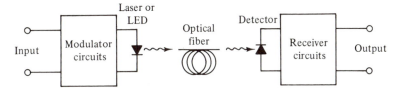

Figure 7-5-1   Optical-fiber communication system

An incoming binary-digital signal switches the light source laser or LED ON and OFF. When the light is ON, the light impinges on one end of the optical fiber, passes through it by total internal reflection at the core-cladding interface, and emerges at the far end. Then the emerged light impinges on the photodetector, which converts the photon light to electron-hole pairs in the detector and a current pulse is generated. Consequently, the receiver circuits will reproduce the information signal in a binary-digital form of ZERO and ONE.

The design of an optical-fiber communication system is based on four basic requirements: the desired data rate (bandwidth), the signal-to-noise ratio (SNR), the distance between terminals, and the type of source information (digital or analog).

7-5-1   Light Sources

Light sources for optical-fiber communication systems must have certain characteristics, such as long lifetime, high efficiency, low cost, high capacity, and sufficient output power. There are two light sources—lasers and LEDs—and they are described in Chapters 8 and 10, respectively. The power needed for various fiber communications is usually no more than a few milliwatts. At this power level the fibers are linear. When the power reaches the level of 100 mW, however, single-mode fibers begin to exhibit nonlinearity. A laser power of about 2.5 W marks the onset of nonlinearity for multimode fibers.

LED Light Sources.   Because the LED emits light randomly in all directions from its junction, the light is incoherent. Therefore the transmission of LED-generated signals inherently involves multimodes. The GaAs LED is an adequate source for data links in the wavelength range of 0.82 to 0.85 μm.

Laser Light Sources.   The laser beam is coherent and an ideal light source for monomode fibers. Because the light is very narrow, it can be coupled efficiently into the fibers and it reduces the effect of the intrinsic chromatic dispersion.

Generally gas lasers like CO_2 and helium–neon are too large, too expensive, and too inefficient for optical-fiber communications. At the wavelength of 1.06 μm the N^{3+}:YAG (Neodymium: Yttrium Aluminum Garnet) laser is one of the most useful light sources for single-mode fibers. Ternary AlGaAs and InGaAs lasers are commonly used at the 0.82- to 0.85-μm range. Figure 7-5-2 is a diagram of a light-source AlGaAs laser coupled to a glass fiber [8].

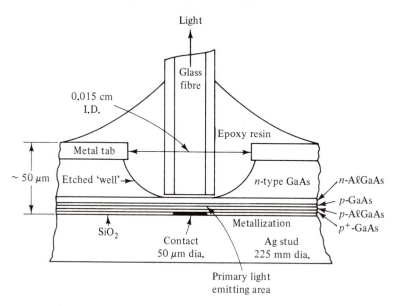

Figure 7-5-2   Diagram of an AlGaAs laser coupled to a fiber (After C. A. Burrus and B. I. Miller [8]; reprinted by permission of *Optics Communications*.)

7-5-2 Light Detectors

The function of light detectors in an optical-fiber communication system is to detect the optical signal from the fiber and convert it into an electrical signal. The GaAs photodetector is the suitable choice at the wavelength region of 0.82 to 0.85 μm. The *pin* photodetectors, such as InGaAs and InGaAsP, are commonly used at the 1.3- to 1.5-μm region. At the receiver side, a low-noise GaAs MESFET preamplifier is usually used immediately after the photodetector for a better signal-to-noise ratio at the output terminals.

7-5-3 Applications

Several assumptions could be made about communication applications with multimode optical fibers:

1. The index profile of the fibers is circularly symmetric about the core axis.
2. The core diameter is large enough, say 100 wavelengths, so that many modes can be expected to propagate.
3. The index difference is small enough that modes can be considered transverse electromagnetic (TEM).
4. The index variation is very small over distances of a wavelength and thus the local plane wave approximation can be made.

Optical-fiber cables are primarily used in computers, industrial automation, medical instruments, military systems, and telecommunications.

Computer Applications.   Computer terminals and their internal links require a very high data rate up to several Gbit/s-km. Optical fibers can easily meet this requirement. Because the fibers offer freedom from electromagnetic interference and from grounding problems, the signal quality is greatly improved.

Industrial Automation.   Optical-fiber uses in industrial automation are in the areas of process control, discrete manufacturing automation, and transportation. Transportation would include airways, shipping, highways, railways, and so on.

Medical Instruments.   Historically the medical profession was the first to use the optical fiber for the diagnosis and treatment of disease. In the late 1930s a guided light was used to provide illumination for simple medical inspection instruments. Medical instruments that could be made of optical fibers are the cardioscope, colonoscope, endoscope, gastroscope, and opthalmoscope.

Military Applications.   The major advantage of optical-fiber cables is the absence of cross talk, RF interference, grounding problems, and outside jamming influences. Such fibers greatly increase the security level. Consequently, optical fibers are commonly used as communication links in aircraft, missiles, submarines, ocean-surface vessels, and military command and control systems.

Telecommunications.   The primary use of optical fibers is in telephone and telegraph. Optical-fiber telecommunications systems can solve many inherent problems that occur in wire systems, such as ringing, echoes, and cross talk. Such fibers are commonly used in the areas of voice telephones, video phones, telegraph services, and various news broadcast systems.

Table 7-5-1 lists the applications and characteristics of the three major optical-fiber systems [9].

TABLE 7-5-1   APPLICATIONS AND CHARACTERISTICS OF THREE MAJOR OPTICAL FIBERS

	Monomode step-index fiber	Multimode graded-index fiber	Multimode step-index fiber
Source	Requires laser	Laser or LED	Laser or LED
Bandwidth	Very very large > 3 GHz-km	Very large 200 MHz- to 3 GHz-km	Large < 200 MHz-km
Splicing	Very difficult due to small core	Difficult but doable	Difficult but doable
Example of application	Submarine cable system	Telephone trunk between central offices	Data links
Cost	Less expensive	Most expensive	Least expensive

(After A. H. Cherin [9]; reprinted by permission of McGraw-Hill Book Company.)

7-5-4   Design Example

The choice of a light source for a specific optical-fiber system is an essential part of the design process. The output power of a light source is usually expressed in terms of its input current. In practice, the light source is terminated in a fiber connector. For a LED light source and a 50-μm core fiber, the coupling loss may be as high as 20 dB whereas for a laser to the same fiber, the coupling loss can be as low as 3 dB.

An optical-fiber system is to be designed to meet the following specifications:

LED source-to-fiber coupling loss	20 dB
Laser source-to-fiber coupling loss	3 dB
Fiber length	4 km
Fiber loss	4 dB/km
Number of splices	1 splice/0.5 km
Splice loss	0.5 dB/splice
Fiber-to-detector coupling loss	0.1 dB
Minimum power required for receiver	−40 dBm

Solution.

1. Selection of LED source
   (a) Choose the LED 1A83 (ASEA-HAFO shown in Appendix 12) as the light source that has an output power of 10 mW from a drive current 0.1 A.
   $$P_{in} = 10 \text{ mW} = 10 \text{ dBm}$$
   (b) Then the total link loss is
   $$\text{Loss} = 20 \text{ dB} + 4 \times 4 \text{ dB} + 8 \times 0.5 \text{ dB} + 0.1 \text{ dB} = 40.1 \text{ dB}$$
   (c) The power available at the receiver is
   $$P_{av} = 10 \text{ dBm} - 40.1 \text{ dB} = -30.1 \text{ dBm}$$
   (d) Finally, the power margin at the receiver is
   $$P = -30.1 \text{ dBm} - (-40 \text{ dBm}) = 9.9 \text{ dBm}$$
   $$= 9.772 \text{ mW}$$

2. Selection of laser source
   (a) A CW-laser diode has an output power 10 mW from a drive current 0.2 A.
   $$P_{in} = 10 \text{ mW} = 10 \text{ dBm}$$
   (b) The total link loss is
   $$\text{Loss} = 3 \text{ dB} + 4 \times 4 \text{ dB} + 8 \times 0.5 \text{ dB} + 0.1 \text{ dB} = 23.1 \text{ dB}$$
   (c) The power available at the receiver is
   $$P_{av} = 10 \text{ dBm} - 23.1 \text{ dB} = -13.1 \text{ dBm}$$
   (d) The power margin at the receiver is
   $$P = -13.1 \text{ dBm} - (-40 \text{ dBm}) = 26.9 \text{ dBm}$$
   $$= 489.8 \text{ mW}$$

REFERENCES

[1] Elion, G. R., and Herbert A. Elion. *Fiber Optics in Communications Systems*, p. 35. New York: Marcel Dekker, Inc., 1978.

[2] Jacobs, Ira, and S. E. Miller. Optical transmission of voice and data. *IEEE Spectrum*, **14**, no. 2 (February 1977), 39.

[3] Potter, R. J., et al. Light-collecting properties of a perfect circular optical fiber. *J. Opt. Soc. Am.*, **53** (1963), 256.

[4] Schelkunoff, S. A. *Electromagnetic Waves*, p. 247. New York: D. Van Nostrand Company, 1964.

[5–7] Midwinter, John E. *Optical Fibers for Transmission*, pp. 83, 82, 108. New York: John Wiley & Sons, 1979.

[8] Burrus, C. A., and B. I. Miller. Small area, double heterostructure AlGaAs electro-lunescent diode source for optical-fiber transmission lines. *Opt. Commun.*, **4** (1971), 307.

[9]   Cherin, Allen H. *An Introduction to Optical Fibers*, p. 3. New York: McGraw-Hill Book Company, 1983.

[10] Giallorenzi, Thomas G. Optical communications research and technology: Fiber optics. *Proc. IEEE*, **66**, no. 7 (July 1978).

SUGGESTED READINGS

1. Allan, W. B. *Fiber Optics: Theory and Practice*. London: Plenum Press, 1973.

2. Cherin, Allen H. *An Introduction to Optical Fibers*. New York: McGraw-Hill Book Company, 1983.

3. Elion, Glenn R., and H. A. Elion. *Fiber Optics in Communications Systems*. New York: Marcel Dekker, Inc., 1978.

4. Howes, M. J., and D. V. Morgan, eds. *Optical Fibre Communications*. New York: John Wiley & Sons, 1980.

5. *IEEE Journal of Quantum Electronics*, vol. QE-18, no. 4 (April 1982). Special issue on optical guided wave technology.

6. *IEEE Proceedings*, vol. 68, no. 6 (June 1980). Special issue on optical communications.

7. *IEEE Proceedings*, vol. 61, no. 12. (December 1973). Special issue on optical fibers.

8. *IEEE Proceedings*, vol. 68, no. 10 (October 1980). Special issue on optical-fiber communications.

9. *IEEE Proceedings*, vol. 66, no. 7 (July 1978). Special issue on optical-fiber light paths.

10. *IEEE Transactions on Electron Devices*, vol. ED-29, no. 9 (September 1982). Special issue on optical fibers.

11. *IEEE Transactions on Microwave Theory and Techniques*, vol. MTT-30, no. 4 (April 1982). Special issue on optical guided wave technology.

12. Kao, Charles K. *Optical Fiber Systems: Technology, Design, and Applications*. New York: McGraw-Hill Book Company, 1982.

13. Midwinter, John E. *Optical Fibers for Transmission*. New York: John Wiley & Sons, 1979.

PROBLEMS

7-1   Optical Fibers

7-1-1. The absorption loss of an optical fiber is described by Eq. (7-1-1), where ℓ is the light path length. Verify that

$$\ell = \frac{n_1}{\left(n_1^2 - \sin^2\theta_{\max}\right)^{1/2}} .$$

7-2   Operational Mechanisms

7-2-1. The numerical aperture of an optical fiber is defined as $\text{NA} = \sin\theta$, where θ is the acceptance angle. Verify Eq. (7-2-28).

7-2-2. A certain optical fiber has $n_1 = 1.30$ and $n_2 = 1.02$. Calculate the meridional light-gathering power.

7-2-3. The wave equations in the core region ($r < a$) of an optical fiber are represented by the Bessel functions as shown in Eqs. (7-2-3) and (7-2-4). Using these two equations and Maxwell's equations, derive Eqs. (7-2-7) and (7-2-8).

7-2-4. The wave equations in the cladding region ($r > a$) of an optical fiber are represented by the Hankel functions as shown in Eqs. (7-2-5) and (7-2-6). Using these two equations and Maxwell's equations, derive Eqs. (7-2-9) and (7-2-10).

7-2-5. For cutoff wavelengths, the Hankel function H_n must vanish at the interface between the core and cladding ($r = a$). Derive the free-space cutoff wavelength Eq. (7-2-39).

7-2-6. The cladding refractive index n_2 of an optical fiber is 1.10 and the core radius a is 30λ. Determine the wave component $K_v(u)$ in the cladding region for a distance r of 31λ.

7-2-7. Using Fig. 7-2-2, derive the equation of maximum acceptance angle as shown in Eq. (7-2-24).

7-2-8. A multimode step-index fiber has the following parameters:

Carrier wavelength	$\lambda = 0.85\ \mu m$
Carrier frequency	$f = 3.53 \times 10^{14}\ Hz$
Core radius	$a = 30\ \mu m$
Core refractive index	$n_1 = 1.48$
Cladding radius	$b = 60\ \mu m$
Cladding refractive index	$n_2 = 1.20$

Compute: **(a)** the phase constant in the core region,
(b) the phase constant in the cladding region,
(c) the phase constant in free space,
(d) the transverse propagation constant in the core region,
(e) the transverse propagation constant in the cladding region,
(f) the value of the Bessel function $J_1(ka)$, and
(g) the value of the Hankel function $H_1(\chi b)$.

7-3   Step-Index Fibers

7-3-1. A step-index fiber operates in a single mode EH_{01}. Its radius a is 5 μm and its refractive indexes are $n_1 = 1.15$ and $n_2 = 1.08$. Determine the cutoff wavelength and frequency.

7-3-2. A multimode step-index fiber has a core radius a of 40 μm and a cladding radius b of 50 μm. The refractive index n_1 of the core is 1.40 and the refractive differential between n_1 and n_2 is 10%.
(a) Calculate the light-gathering power.
(b) Compute the cutoff wavelength.
(c) Determine the number of modes.

7-3-3. Describe the characteristics and applications of monomode and multimode step-index fibers.

7-3-4. A multimode step-index fiber has the following parameters:

Core radius	$a = 40\ \mu m$
Core refractive index	$n_1 = 1.52$
Cladding refractive index	$n_2 = 1.48$
Wavelength of light source LED	$\lambda_0 = 0.87\ \mu m$

(a) Calculate the numerical aperture of the fiber.
(b) Compute the maximum acceptance angle of the fiber.
(c) Estimate the maximum number of modes.
(d) Determine the light-gathering power.

7-3-5. A multimode step-index fiber has the following parameters:

Core refractive index	$n_1 = 1.50$
Numerical aperture	$NA = 0.37$
V number	$X_{np} = 120$
Wavelength of light source Nd^{3+}:YAG	$\lambda_0 = 1.06 \ \mu m$

(a) Calculate the cladding refractive index n_2.
(b) Compute the core radius a.
(c) Determine the cladding radius b.

7-4 Graded-Index Fibers

7-4-1. A graded-index fiber has $n_1 = 1.50$ and $n_2 = 1.05$. Determine the refractive index at $r = a/2$ for $\alpha = 4$.

7-4-2. The graded-index fiber can support multiple modes. Its core radius a is 20 μm and its refractive indexes are $n_1 = 1.50$ and $\Delta = 0.08$. Determine the number of modes for $\alpha = 2$ at $\lambda_0 = 1.064 \ \mu$m.

7-4-3. Describe the characteristics and applications of the multimode graded-index fibers.

7-4-4. Using Eq. (7-4-3), derive the equation of the total number of modes for a graded-index fiber as shown in Eq. (7-4-5).

7-4-5. A graded-index fiber has the following parameters:

Core radius	$a = 30 \ \mu m$
Core refractive index	$n_1 = 1.50$
Deviation index	$\Delta = 0.02$
Power law coefficient	$\alpha = 2$
Wavelength of light source	$\lambda_0 = 1.06 \ \mu m$

(a) Calculate the number of modes for the graded-index fiber.
(b) Compute the graded refractive index.
(c) Determine the number of modes for a step-index fiber.

Chapter 8

Solid-State Lasers and Laser Modulators

8-0 INTRODUCTION

The word *Maser* is an acronym for Microwave Amplification by Stimulated Emission of Radiation; the word *Laser* stands for Light Amplification by Stimulated Emission of Radiation. Unlike the operation of ordinary microwave tubes, such as klystron and traveling-wave tubes, and microwave solid-state devices, such as Gunn diodes and avalanche diodes, which can be analyzed in terms of classic mechanics, the electrical behavior of masers and lasers can be described only in terms of quantum electronics and statistical mechanics. That is, the interaction of electromagnetic fields in masers and lasers results in bound aggregates of charges rather than in free charges as in electron beams in microwave tubes. Masers and lasers are highly directional coherent power-source devices used for the generation and amplification of radiation. Their noise figure is extremely low and many applications have been found in medicine, communications, space exploration, military, and metals technology. Since 1954, many types of masers and lasers have been invented and built. This chapter, however, deals with the ruby laser, the solid-state lasers, and the laser modulators.

The concept of masers and lasers is based on Einstein's discovery of stimulated emission. This discovery led to a search for methods of establishing population inversion in a suitable material and hence to the construction of a radiation amplifier. In 1954 Townes and his students at Columbia University built the first microwave amplifier and oscillator based on the stimulated emission principle [1]. In 1956 Bloembergen proposed a three-level solid-state maser [2] and the first successful such maser was achieved in 1957 by Scovil, Feher, and Seidel with the paramagnetic

Gd^{3+} ion at 9 GHz [3]. The first ruby laser was constructed in 1960 by Maiman [4]. It was quickly followed by the four-level lasers of Sorokin and Stevenson [5] and the He-Ne laser of Javan [6].

The common feature of masers and lasers is the conversion of atomic or molecular energy to electromagnetic radiation by means of the process known as *stimulated emission of radiation*. When the wavelength of the emitted radiation is in the vicinity of 1 cm, the devices are called *masers*. When the radiated frequency approaches the visible or nearly visible range at 10^5 GHz, the devices are called *optical masers* or *lasers*. There is no sharp limit separating the regions in which the terms maser and laser are used. Both optical and microwave frequencies are shown in Table 8-0-1 and Fig. 8-0-1.

In 1962 Dumke showed that laser action was indeed possible in direct bandgap semiconductors, such as gallium arsenide (GaAs) [7]. The pulsed radiation at 8400 Å was obtained from a liquid-nitrogen-cooled, forward-biased GaAs *p-n* junction. Many new laser materials have been found since then. The wavelength of coherent radiation has been extended through the visible into the ultraviolet and out to the midinfrared [8–11].

TABLE 8-0-1 ELECTROMAGNETIC SPECTRUM

I. Electromagnetic waves

EM Waves	Frequency	Wavelength
Very long waves	3–30 kHz	100–10 km
Long waves	30–300 kHz	10–1 km
Medium waves	300–3000 kHz	1–0.1 km
Short waves	3–30 MHz	100–10 m
Ultrashort waves (meter waves)	30–300 MHz	10–1 m
Microwaves (centimeter waves)	0.3–30 GHz	100–1 cm
Ultramicrowaves (millimeter waves)	30–3000 GHz	10–0.1 mm
Infrared ray	0.003–0.429 MGHz	100–0.7 μm
Visible light	0.429–0.698 MGHz	0.7–0.43 μm
Ultraviolet ray	0.698–100 MGHz	0.43–0.003 μm
x ray	0.01–1000 GGHz	300–0.003 Å
Gamma ray	1 $\approx$ 1M GGHz	3–1μ Å

II. Radio Frequencies

Bands	Frequency	Wavelength
VLF (very low frequency)	3–30 kHz	100–10 km
LF (low frequency)	30–300 kHz	10–1 km
MF (medium frequency)	300–3000 kHz	1–0.1 km
HF (high frequency)	3–30 MHz	100–10 m
VHF (very high frequency)	30–300 MHz	10–1 m
UHF (ultrahigh frequency)	300–3000 MHz	100–10 cm
SHF (superhigh frequency)	3–30 GHz	10–1 cm
EHF (extreme high frequency)	30–300 GHz	10–1 mm

TABLE 8-0-1   ELECTROMAGNETIC SPECTRUM (Continued)

III. Light frequencies

Colors	Frequency	Wavelength
Infrared	$3 \times 10^{12} - 4.29 \times 10^{14}$ Hz	100–0.7 μm
Red	$4.29 \times 10^{14} - 4.92 \times 10^{14}$ Hz	0.7–0.61 μm
Orange	$4.92 \times 10^{14} - 5.08 \times 10^{14}$ Hz	0.61–0.59 μm
Yellow	$5.08 \times 10^{14} - 5.26 \times 10^{14}$ Hz	0.59–0.57 μm
Green	$5.26 \times 10^{14} - 6.00 \times 10^{14}$ Hz	0.57–0.50 μm
Blue	$6.00 \times 10^{14} - 6.67 \times 10^{14}$ Hz	0.50–0.45 μm
Violet	$6.67 \times 10^{14} - 6.98 \times 10^{14}$ Hz	0.45–0.43 μm
Ultraviolet	$6.98 \times 10^{14} - 10^{17}$ Hz	0.43–0.003 μm

IV. Band designation

Frequency (GHz)	0.1	0.2	0.3	0.5	1	2	3	4	6	8	10	20	30	40	60	100
Old designation	VHF		UHF		L		S		C		X	Ku	K	Ka	millimeter	
New designation	A		B		C	D	E	F	G	H	I	J		K	L	M

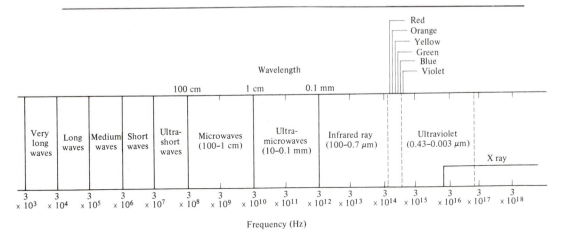

Figure 8-0-1   Electromagnetic spectrum

8-1   *TRANSITION PROCESSES*

Three basic transition processes are involved in the operation of masers and lasers: absorption, spontaneous emission, and stimulated emission. Figure 8-1-1 shows the three transition processes between two energy levels E_1 and E_2. The black dots indicate the state of the electrons. The initial state is at the left and the final state, after the process has occurred, is at the right. The energy level E_1 is the ground state and E_2 is the excited state. Any transition between these states involves, according to Planck's law, the emission or absorption of a photon of energy with frequency ν_{12}, given by

$$h\nu_{12} = E_2 - E_1 \qquad \text{eV} \qquad (8\text{-}1\text{-}1)$$

where h = Planck's constant, which is 6.625×10^{-34} J-s
  ν = the frequency in hertz

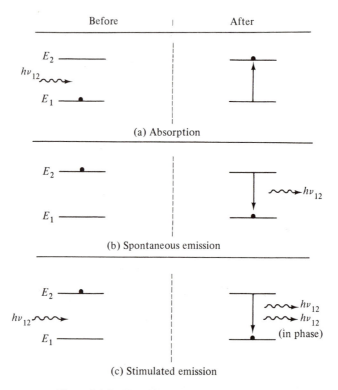

Figure 8-1-1 Transition processes of electrons

When an electron recombines from the minimum conduction band to the maximum valence band, the released photon energy is expressed as

$$h\nu_{12} = E_c - E_v = E_g \qquad \text{eV} \qquad (8\text{-}1\text{-}1a)$$

where E_c = minimum conduction band in electronvolts
  E_v = maximum valence band in electronvolts
  E_g = gap energy in electronvolts

The wavelength in free space is

$$\lambda_0 = \frac{1.242}{E_g} \qquad \mu m \qquad (8\text{-}1\text{-}1b)$$

Most electrons are in the ground state at room temperature. When a photon of energy exactly equal to $h\nu_{12}$ impinges the atoms, an electron in the state E_1 will absorb the photon and thereby move to the excited state E_2. This is the *absorption process* as shown in Fig. 8-1-1(a). The excited state of the electron is unstable and after a short time and without any external stimulus the excited electron will transit to the ground state by giving off a photon of energy $h\nu_{12}$. This process is called *spontaneous emission* as shown in Fig. 8-1-1(b). The lifetime for spontaneous emission—that is, the average time of the excited state—varies from 10^{-9} to 10^{-3} sec, depending on various semiconductor parameters, such as bandgap and density of recombination centers. When a photon of energy $h\nu_{12}$ impinges on an electron while it is still in the excited state E_2, the electron will be immediately stimulated to make its transition back to the ground state E_1 by giving off two photons of energy $h\nu_{12}$. This process is called *stimulated emission* as shown in Fig. 8-1-1(c). The waves of the stimulated photons are in phase with the radiation field. If this process continues and other electrons are stimulated to emit photons in the same fashion, a large radiation can build up. This radiation will be monochromatic since each photon will have an energy of exactly one $h\nu_{12}$ and it will be coherent because all the photons emitted are in phase and reinforcing each other. This process of stimulated emission can be explained quantum mechanically to relate the probability of emission to the intensity of the radiation field.

Classical gas atoms follow a Boltzmann distribution for two discrete energy levels E_2 and E_1 with $E_2 > E_1$. The instantaneous population n_2 of atoms in state E_2 is related to the instantaneous population n_1 of atoms in state E_1 at thermal equilibrium by

$$\frac{n_2}{n_1} = \frac{\exp[-E_2/(kT)]}{\exp[-E_1/(kT)]} = \exp\left[-\frac{(E_2 - E_1)}{kT}\right]$$

$$= \exp\left(\frac{-h\nu_{12}}{kT}\right) \qquad (8\text{-}1\text{-}2)$$

In Eq. (8-1-2) it is assumed that the two levels have an equal number of states. The exponential term $\exp[-(E_2 - E_1)/(kT)]$ is conventionally called the *Boltzmann factor*. Because $E_2 > E_1$, it follows that $n_1 \gg n_2$. That is, most electrons are in the lower energy level as expected. The total transition rate from state E_2 to state E_1 in the presence of the field is given by Fig. 8-1-1(b) and (c).

$$\alpha_{21} = B_{21} n_2 \rho(\nu) + A_{21} n_2 \qquad (8\text{-}1\text{-}3)$$

Total transition = stimulated + spontaneous
emission   emission

where $\rho(\nu) = \dfrac{2\pi h \nu^5}{c^3} \dfrac{1}{e^{h\nu/(kT)} - 1}$ is the energy density per unit frequency

$c = 3 \times 10^8$ cm/sec   is the velocity of light in vacuum

$k = 1.38 \times 10^{-23}$ W-s/°K   is Boltzmann's constant

$T =$ the absolute temperature in degrees Kelvin, to be determined

$B_{21} =$ a coefficient to be determined

$A_{21} =$ a coefficient to be determined

At steady state, the transition rate from level 2 to level 1 is equal to that from level 1 to level 2; thus

$$B_{12} n_1 \rho(\nu) = B_{21} n_2 \rho(\nu) + A_{21} n_2 \qquad (8\text{-}1\text{-}4)$$

Absorption = stimulated + spontaneous
              emission   emission

This relation was described by Einstein and so coefficients B_{12}, A_{21}, and B_{21} are called *Einstein coefficients*. Two special cases are of interest.

CASE 1:   The ratio of the stimulated-to-spontaneous emission rate at thermal equilibrium is

$$\frac{\text{Stimulated emission}}{\text{Spontaneous emission}} = \frac{B_{21}}{A_{21}} \rho(\nu) \qquad (8\text{-}1\text{-}5)$$

In order to have a large ratio of stimulated emission over spontaneous emission, the photon field energy density $\rho(\nu)$ must be large. In the laser this condition is achieved by providing an optical resonant cavity in which the photon energy density can build up to a large value.

CASE 2:   The ratio of the stimulated emission to absorption is given by

$$\frac{\text{Stimulated emission}}{\text{Absorption}} = \frac{B_{21}}{B_{12}} \frac{n_2}{n_1} \qquad (8\text{-}1\text{-}6)$$

This equation indicates that in order to have a large ratio of the stimulated emission rate over the absorption rate, n_2 must be much larger than n_1 ($n_2 \gg n_1$). This condition is called *population inversion*, and is discussed in the next section.

Example 8-1-1   Relative Population of Two Energy Levels in the GaAsP Laser

A red-light junction laser GaAsP has a wavelength of 0.69 μm and operates at room temperature.
(a) Determine the energy difference of the two levels in electronvolts.
(b) Calculate the population ratio of n_1 over n_2.

Solution.   (a) Using Eq. (8-1-1a), we obtain the energy difference as

$$E_2 - E_1 = h\nu = \frac{hc}{\lambda} = \frac{6.625 \times 10^{-34} \times 3 \times 10^{8}}{0.69 \times 10^{-6}}$$

$$= 2.88 \times 10^{-19} \text{ J}$$

$$= 1.80 \text{ eV}$$

(b) Then from Eq. (8-1-2) the population ratio is

$$\frac{n_1}{n_2} = \exp\left(\frac{h\nu}{kT}\right) = \exp\left(\frac{2.88 \times 10^{-19}}{1.38 \times 10^{-23} \times 300}\right) = \exp(69.57)$$

$$= 1.64 \times 10^{30}$$

8-2 POPULATION INVERSION

As noted previously, most electrons are in the low energy level at the normal condition—that is, $n_1 \gg n_2$. However, the necessary condition for the stimulated emission rate over the absorption rate is $n_2 \gg n_1$. That is, the population is said to be inverted when there are more electrons in the excited state E_2 than in the ground state E_1. If photons of energy $h\nu_{12}$ are incident on a device where the population level E_2 is inverted with respect to level E_1, stimulated emission will exceed absorption and more photons of energy $h\nu_{12}$ will leave the device than enter it. Such a phenomenon is called *quantum amplification*.

The inversion condition for a solid-state laser can be analyzed by means of Fig. 8-2-1. Figure 8-2-1(a) shows the equilibrium condition at $T = 0°\text{K}$ for an intrinsic

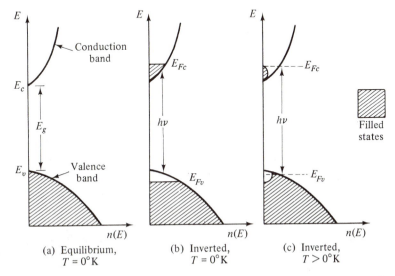

(a) Equilibrium,   (b) Inverted,   (c) Inverted,
    $T = 0°\text{K}$ $T = 0°\text{K}$ $T > 0°\text{K}$

Figure 8-2-1 Energy versus density of states in a semiconductor (From M. I. Nathan [13]; reprinted by permission of the IEEE, Inc.)

semiconductor in which the shaded area represents the filled states. Figure 8-2-1(b) shows the situation for an inverted population at 0°K. This inversion can be achieved by a pumping force with a photon energy greater than the bandgap E_g. The valence band is devoid of electrons down to an energy level E_{Fv} (Fermi level in the valence band) and the conduction band is filled up to the level E_{Fc} (Fermi level in the conduction band). Then photons with energy $h\nu$ such that $E_g < h\nu < (E_{Fc} - E_{Fv})$ will cause downward transition and hence stimulated emission. At some finite temperatures the carrier distributions will be distorted in energy as shown in Fig. 8-2-1(c). For conduction-to-valence band transitions in an intrinsic semiconductor, the necessary condition for stimulated emission to be dominant over absorption is

$$h\nu < (E_{Fc} - E_{Fv}) \qquad (8\text{-}2\text{-}1)$$

After the completion of the first ammonia maser by Townes in 1954, several new methods were proposed for obtaining population inversion in various materials. The best-known method was the one that used three levels of energy bands. A high-frequency signal incident on the material excites the atomic system from the ground state E_1 to the top excited state E_3. In this manner, a population inversion can be created between the excited state E_3 and the metastable state E_2 or possibly between the state E_2 and the ground state E_1. This principle was first proposed by Basov and Prokhorov in 1955 [14]. Shortly thereafter Bloembergen discovered the same principle independently and suggested its application to paramagnetic ions in crystals [2]. A three-level pumping system is shown in Fig. 8-2-2.

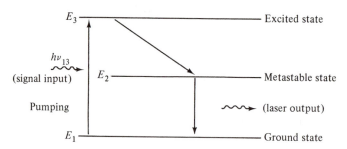

Figure 8-2-2   Three-level pumping diagram

In addition to the p-n junction method of excitation described, other means of pumping, such as optical pumping, electron beam pumping, and avalanche break-down pumping, are also in common use. Guillaume and Debever first used an electron beam with an energy level of 50 keV to bombard the semiconductor crystal [1]. A typical setup is shown in Fig. 8-2-3. Electrons in the valence band are "pumped" into the conduction band by the impinging electrons. A hole is created for each electron thus excited, so that even in a pure semiconductor the result is a degenerate electron-hole population. As the electron beam penetrates the material, the electron-hole pairs form degenerate populations. Consequently, spontaneous emission builds up until a threshold condition is reached and lasing occurs.

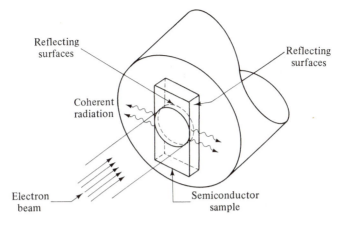

Figure 8-2-3 Electron beam pumping scheme (From S. M. Sze [12]; reprinted by permission of John Wiley & Sons, Ltd.)

Example 8-1-2   Population Inversion

A junction laser GaAs has the following parameters:

Gap energy	$E_g = 1.43$ eV
Operating temperature	$T = 300°$K
Intrinsic carrier concentration	$n_i = 10^{13}$ m^{-3}
Equal electron and hole densities	$n_n = n_p$

Compute the minimum carrier concentration for population inversion.

Solution.   For equal electron and hole concentrations, $n_n = n_p$, we have

$$E_c - E_i = E_i - E_v = \tfrac{1}{2}E_g = 0.715 \text{ eV}$$

Then from Eq. (3-4-3) the population inversion is

$$n_n = n_p = n_i \exp\left(\frac{E_c - E_i}{kT}\right)$$

$$= 10^{13} \exp\left(\frac{0.715}{26 \times 10^{-3}}\right)$$

$$= 10^{13} \exp(27.5)$$

$$= 8.77 \times 10^{24} \text{ m}^{-3}$$

8-3 RESONANT CAVITY

One required structure in order for laser action to occur is the resonant cavity. Multiple reflections of light intensity are produced in the cavity to achieve the laser output. There are several types of resonant cavities, such as the Fabry–Perot cavity, cylindrical cavity, triangular cavity, and rectangular or nondirectional cavity (see Fig. 8-3-1).

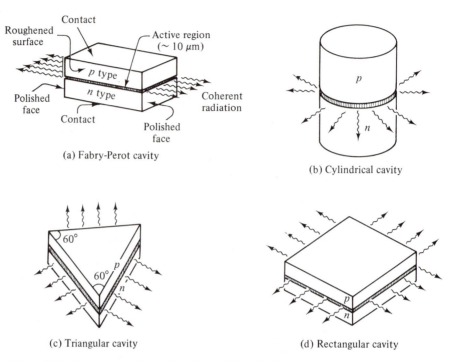

(a) Fabry-Perot cavity

(b) Cylindrical cavity

(c) Triangular cavity

(d) Rectangular cavity

Figure 8-3-1   Resonant cavities for laser beams (From S. M. Sze [12]; reprinted by permission of John Wiley & Sons, Ltd.)

The most common resonant cavity for producing a laser beam is the Fabry–Perot cavity illustrated in Fig. 8-3-1(a). A pair of parallel faces that are perpendicular to the plane of the p-n junction are cleaved or polished with a highly reflected material, such as aluminum (Al) or silver (Ag). The top and the bottom sides, which are parallel to the junction, are metalized with a conducting material. The remaining two faces are roughened or sawed to suppress the radiation. The two parallel-polished planes are used to produce multiple reflections in the cavity in order to achieve a very high intensity of laser beam. Figure 8-3-2 shows the multiple reflections within a resonant cavity.

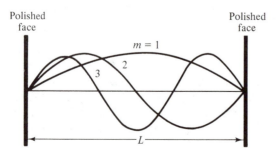

Figure 8-3-2   Resonance within a laser cavity

8-3-1 Axial Modes

In order for resonance to occur in a laser cavity, the distance between the two parallel-polished faces must be equal to an integral number of half wavelengths in the cavity. That is,

$$L = m\frac{\lambda}{2} \qquad (8\text{-}3\text{-}1)$$

where m = an integer
$\lambda = \dfrac{v}{v}$ is the photon wavelength within the laser cavity material
$v = c/n$ is the light velocity in the laser cavity material
$c = 3 \times 10^8$ m/sec is the velocity of light in free space
$n = \sqrt{\epsilon_r}$ is the refraction index of the laser cavity material
$v =$ (photon energy)$/h$ is the frequency of photon
$\epsilon_r =$ relative dielectric constant of the laser cavity material, and
$h = 6.625 \times 10^{-34}$ J-sec is Planck's constant.

If wavelength λ_0 in free space is used, the index of refraction n of the laser material must be considered. That is,

$$\lambda = \frac{\lambda_0}{n} \qquad (8\text{-}3\text{-}2)$$

In practice, it is not necessary to adjust mechanically the cavity length L to an integral number of half wavelengths for resonance and Eq. (8-3-1) is automatically satisfied over some portion of the cavity because the wavelength of laser beam is very short and many values of the integral number m will satisfy the resonant condition.

Equation (8-3-1) may also be expressed as

$$v = \frac{mc}{2nL} \qquad (8\text{-}3\text{-}3)$$

When $\lambda \ll L$, a large number of modes have occurred in the cavity. Each value of m that satisfies Eq. (8-3-1) defines an axial (or longitudinal) mode of the cavity. The frequency separation between adjacent modes ($\Delta m = 1$) is given by

$$\Delta v = \frac{c}{2nL} \qquad (8\text{-}3\text{-}4)$$

The modes of oscillation of the laser cavity will consist of a large number of frequencies, each given by Eq. (8-3-3) and separated by Δv as shown in Eq. (8-3-4).

Example 8-3-1 Axial Modes of a GaAsP Laser Cavity

A laser cavity has a length L of 40 cm and the wavelength of the laser light is 0.6943 μm in free space.
(a) Determine the number of axial modes.
(b) Calculate the number of oscillation frequencies.
(c) Find the frequency separation.

Solution.   (a) The number of axial modes is obtained by using Eq. (8-3-1).

$$m = \frac{2nL}{\lambda_0} = \frac{2 \times 3.32 \times 40 \times 10^{-2}}{0.6943 \times 10^{-6}} = 3.83 \times 10^6 \text{ modes}$$

(b) Then from Eq. (8-3-3) the number of oscillating frequencies is

$$\nu = \frac{mc}{2nL} = \frac{3.83 \times 10^6 \times 3 \times 10^8}{2 \times 3.32 \times 40 \times 10^{-2}} = 4.33 \times 10^{14} \text{ Hz}$$

(c) Finally, from Eq. (8-3-4) the frequency separation is

$$\Delta\nu = \frac{c}{2nL} = \frac{3 \times 10^8}{2 \times 3.32 \times 40 \times 10^{-2}} = 0.113 \text{ GHz}$$

8-3-2 Transverse Modes

A laser cavity probably has waves that propagate in the off-axis direction and replicate themselves after traveling some distance. This type of wave is said to be a *transverse mode*.

8-4 SOLID-STATE RUBY LASER

During 1957 and early 1958 many masers using Cr^{3+} in the now-famous ruby crystals were constructed in several laboratories. The first published result was made by Makhov [15] at the University of Michigan. It reported a ruby maser pumped at 24 GHz and operating at 9.3 GHz. Rubies were used in masers of greatly varying construction and performance.

8-4-1 Physical Description

The first ruby laser built at the Hughes Research Laboratory in 1960 by Maiman [4] was made of a circular ruby rod having mirrored ends. A xenon flash tube surrounded the ruby rod. A bank of capacitors discharged through the xenon tube, causing it to emit a very intense flash of light with a lifetime of several milliseconds. The original ruby developed by Maiman and its simplified cavity diagram are shown in Fig. 8-4-1.

8-4-2 Principles of Operation

The ruby belongs to a family of gems consisting of corundum (Al_2O_3) with various types of impurities. A pink ruby used by Maiman, for instance, contains about 0.05% chromium ion (Cr^{3+}). This ion absorbs radiation in two wide bands located in the blue and green portions of the spectrum and the resulting excitation is followed by a very fast radiationless process. Three basic requirements for achieving

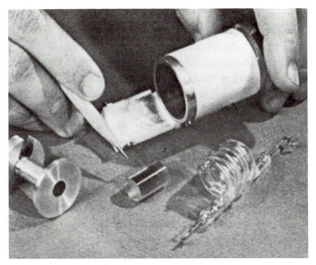

(a) Maiman's original ruby laser

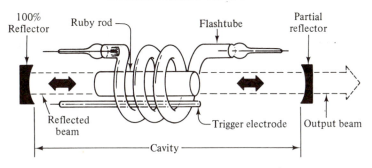

(b) Simplified diagram of Maiman's ruby laser

Figure 8-4-1   Maiman's ruby laser and its simplified cavity diagram

laser operation are

1. A method of excitation or pumping of atoms from the ground energy level to higher energy levels
2. A sufficient large population inversion
3. A resonant cavity to produce positive feedback of the radiation

The basic principle of operation for the ruby laser can be explained by means of the three-level model theory described earlier. The flash light has a center frequency at a wavelength of about 5500 Å. When the pumping flash light heats the ruby rod, chromium atoms (Cr^{3+}) in corundum are elevated from the ground state E_1 to the excited state E_3. The lifetime of an excited state E_3 is less than 1 microsecond, so the excited ions Cr^{3+} quickly lose some of their excitation energy through nonradiative transitions to the metastable state E_2. The lifetime of the metastable state E_2 is about 3 milliseconds. This state then slowly decays by spontaneously emitting a sharp doublet, the components of which are at 6943 Å and 6929 Å at 300°K. Under very intense excitation the population of this metastable state E_2 can become greater than that of the ground state E_1.

The circular ruby rod was of 1-cm dimension and coated on two faces with silver. Some photons travel longitudinally along the axis of the rod and are reflected at the ends, thus causing further stimulated emission. Population inversion is sustained by the pumping light source and continuous photon amplification results in the generation of a very intense beam, which emerges through the partly transmissive mirror at one end of the ruby rod.

Once the stimulated emission begins, however, the population in the metastable state quickly decreases. Thus the laser output consists of an intense spike lasting from a few nanoseconds to microseconds. After the stimulated emission spike, population inversion builds up again and a second spike results. Consequently, the metastable state never reaches a highly inverted population of electrons. Such a population can be achieved by keeping the coherent photon field in the ruby rod from building up and thus preventing stimulated emission until after a larger population inversion is gained. This process is called *Q-switching* or *Q-spoiling*, where Q is the quality of the resonant structure. A *Q*-switched ruby laser is shown in Fig. 8-4-2. Here one face of the ruby rod is not silver coated but instead is provided by a fast-rotating external mirror.

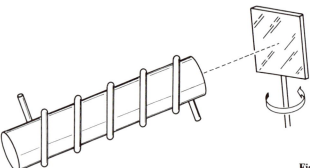

Figure 8-4-2   Q-switched ruby laser

When the mirror plane is aligned exactly perpendicular to the laser axis, a resonant structure exists; but as the mirror rotates away from this position, there is no laser action because of no buildup of photons by multiple reflections. So a very large inverted population builds up as the mirror rotates off-axis. When the mirror again returns to the on-axis position, stimulated emission with a highly intense laser pulse again occurs. This structure is called a *Q-switched laser*. By saving the electron population for a single pulse, a large amount of energy is emitted in a very short time. If the total energy in the pulse is 10 J and the pulse width is 100 nsec, for example, then the pulse power is 100 MW (megawatts).

Essentially there are two basic methods of Q-switching: passive and active. In passive Q-switching a material is used with an optical transmission at the laser frequency that varies with intensity of the light. Such materials include liquid cells, Uranyl glass, exploding films, and semiconductor reflectors. Active Q-switching can be accomplished mechanically by Porro prisms, electrooptically by Kerr or Pockels cells, or by acoustic deflection. The obvious advantage of active Q-switching is the precise control that it offers, with the Kerr cell giving optimum control. The laser modulators are described in Section 8-8.

8-4-3 Microwave Characteristics

All classical sources of light have the unfortunate characteristic of being polychromatic. It is possible to obtain quasi-monochromaticity by using filters, but the radiation intensity is greatly reduced. Because radiation in the oscillating modes of laser cavities is very directional, the ruby laser output beams are extremely directional and highly monochromatic. They are because the round-trip distance between the two silver-coated parallel faces is an integral number of wavelengths. It is this monochromaticity that gives a laser its coherence, which, in turn, contributes to the high-intensity (brightness) output. These unique features exhibited by the ruby laser source—collimation (directionality), coherence, and high intensity—are found in no other radiation source. Table 8-4-1 shows the characteristics of commercially available ruby lasers [10].

TABLE 8-4-1 COMMERCIALLY AVAILABLE RUBY LASERS

Laser	Wavelength (μm)	Mode of operation	Power output (watts) Average	Peak	Repetition rate (pps)	Pulse length	Processing uses
Ruby	0.6943	Pulsed	1–20		1	0.3–6 msec	Welding, material removal
Ruby	0.6943	Q-switched		10^5	1	0.3–2 msec	Welding, material removal
Ruby	0.6943	Q-switched		10^8	1	5–50 nsec	Vaporation

(After M. Eleccion [10]; reprinted by permission of the IEEE, Inc.)

8-5 SOLID-STATE JUNCTION LASERS

Since Townes and his students invented the first maser in 1954, many lasers—solid-state lasers, neutral gas laser, ion gas laser, molecular gas laser, and liquid laser—have been built. During the years 1957 to 1966 Nishizawa and Watanabe [16], Basov et al. [17], and Aigrain [18] independently proposed the p-n junction semiconductor laser. In 1962 Dumke demonstrated that laser action was possible in direct bandgap semiconductors [7]. When only one junction exists in a single type of laser material, such as GaAs, it is referred to as a *homojunction laser*. If there are multiple layers in the laser structure, such as $Al_xGa_{1-x}As$, it is called a *heterojunction laser*.

There are three types of lasers:

1. Far- and middle-infrared lasers, such as the PbSe laser for military and spectroscopy applications.
2. Near-infrared lasers, such as the GaAs laser for low-loss transmission of optical fibers.
3. Visible light lasers, such as the GaAsP laser for signal display.

The important features of semiconductor lasers are extreme monochromaticity (time coherence) and high directionality (space coherence), features that are similar to those of other lasers, such as the solid-state ruby laser and the He-Ne gas laser. Solid-state junction lasers, however, differ from other lasers in several basic respects.

1. The quantum transition of a solid-state junction laser occurs between energy bands (conduction and valence bands) rather than between discrete energy levels as in the solid-state ruby laser.
2. The size of a solid-state junction laser is very small, typically 50 μm $\times$ 25 μm $\times$ 50 μm.
3. The characteristics of a solid-state junction laser are strongly influenced by properties of the junction material, such as doping and band tailing.
4. Population inversion of a solid-state junction laser takes place in the very narrow junction region and the pumping is supplied by a forward bias across the junction.

8-5-1 p-n Junction Laser Materials

The most common type of semiconductor used for *p-n* junction lasers consists of the III-V compounds listed in Table 8-5-1. It can be seen from Table 8-5-1 that the solid-state junction laser materials are made up of binary and ternary semiconductor compounds. The notation for the ternary compounds usually consists of forms like $A_xB_{1-x}C$ or DE_xF_{1-x}, where A and B (or D) are the elements of group III, C (or E and F) the element of group V, and x represents the percentage of the element atoms concerned.

According to solid-state electronic theory, semiconductors can be classified into direct and indirect bandgap types. Direct bandgap semiconductors, such as GaAs, GaSb, and InP, have the lowest conduction band minimum and the highest valence band maximum at the same wave vector in the Brillouin zone. Indirect bandgap semiconductors, such as Ge, Si, AlGa, and GaP, have the extrema at

TABLE 8-5-1   MATERIALS OF SOLID-STATE p-n JUNCTION LASERS.

Material Scientific name	Symbol	Bandgap E_g (eV)	Wavelength λ_0 (μm)	Light wave Infrared	Light wave Visible	Light wave Ultra-violet
Lead-tin selenide	$Pb_x Sn_{1-x} Se$	0.10	12.42	√		
Lead selenide	PbSe	0.14	8.86	√		
Lead-tin telluride	$Pb_x Sn_{1-x} Te$	0.20	6.20	√		
Lead telluride	PbTe	0.19	6.53	√		
Indium antimonide	InSb	0.23	5.39	√		
Lead sulphide	PbS	0.29	4.26	√		
Indium arsenide-antimonide	$InAs_x Sb_{1-x}$	0.39	3.18	√		
Indium arsenide	InAs	0.36	3.44	√		
Gallium-indium arsenide	$Ga_x In_{1-x} As$	0.70	1.80	√		
Gallium antimonide	GaSb	0.70	1.77	√		
Indium arsenide-phosphide	$InAs_x P_{1-x}$	0.80	1.55	√		
Indium phosphide	InP	1.26	0.98	√		
Gallium arsenide	GaAs	1.43	0.87	√		
Gallium-indium phosphide	$Ga_x In_{1-x} P$	1.60	0.78	√		
Aluminum-gallium arsenide	$Al_x Ga_{1-x} As$ ($x = 0.37$)	1.80	0.69		Red	
Gallium arsenide-phosphide	$GaAs_x P_{1-x}$ ($x = 0.40$)	1.80	0.69		Red	
Cadmium sulphide-selenide	$CdS_x Se_{1-x}$	2.50	0.50		Blue	

different wave vectors. In direct III-V compounds radiative transitions take place faster than nonradiative and impurity transitions. In indirect semiconductors radiative recombination is improbable and nonradiative recombinations occur faster than radiative recombination. As a result, the radiation efficiency of the direct-type semiconductors is much higher than that of the indirect type. Therefore most p-n junction lasers are made of direct III-V compounds.

The laser cutoff wavelength is determined from Eq. (8-1-1b) as

$$\lambda_{0c} > \frac{1.242}{E_g} \qquad \mu m \qquad (8\text{-}5\text{-}1)$$

where E_g is the bandgap energy in electronvolts.

Figure 8-5-1 is a diagram showing the bandgap energy changes from direct GaAs to indirect AlAs as x increases from 0 to 1 [19] and Table 8-5-2 lists the characteristics of the most commonly used laser and LED materials.

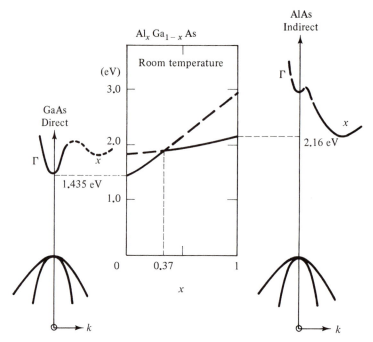

Figure 8-5-1 Diagram of direct and indirect semiconductors (After H. C. Casey and M. B. Parish [19]; reprinted by permission of the American Institute of Physics.)

TABLE 8-5-2   CHARACTERISTICS OF SELECTED LASER AND LED MATERIALS.

Material	Direct	Indirect	Recombination Constant B (m^3/sec)
Si		√	1.79×10^{-21}
Ge		√	5.25×10^{-20}
GaP		√	5.37×10^{-20}
GaAs	√		7.21×10^{-16}
InP	√		1.26×10^{-15}
InAs	√		8.50×10^{-17}
InSb	√		4.58×10^{-17}

Example 8-5-1   Direct and Indirect Semiconductors

Aluminum-gallium arsenide, $Al_xGa_{1-x}As$, can be changed from direct GaAs to indirect AlAs as x increases from 0 to 1 (see Fig. 8-5-1).

(a) Name the materials when x is 0 and 1.
(b) Find the bandgap energies of the named materials.
(c) Calculate the wavelengths of the emitted light.
(d) Identify the types of light.

Solution.   (a) The materials are GaAs for $x = 0$ and AlAs for $x = 1$.

(b) Their bandgap energies are 1.43 eV for GaAs and 2.16 eV for AlAs.

(c) The wavelengths of the emitted light are

$$\lambda_0 = \frac{1.24}{1.43} = 0.87\,\mu\text{m} \qquad \text{for GaAs}$$

$$\lambda_0 = \frac{1.24}{2.16} = 0.57\,\mu\text{m} \qquad \text{for AlAs}$$

(d) The types of light are infrared for GaAs and yellow for AlAs.

8-5-2 Principles of Operation

Laser action processes.   The basic principles underlying laser action can be developed in four steps.

1. Atoms (and also ions and molecules) of the material exhibit internal resonances at certain discrete characteristic frequencies. These internal atomic resonances occur at frequencies ranging from the audio up to and beyond the optical regions.

2. A signal applied to an atom at or near one of its internal resonances will cause a measurable response in the atom. Depending on the circumstances, the atom may absorb energy from the signal or it may emit energy into the signal.

3. The strength and the sign of the total response that will be obtained from a collection of many atoms of the same kind depend on the population difference—that is, the difference in population between the lower and upper quantum energy levels responsible for that particular transition. Under normal positive-temperature thermal equilibrium conditions this net response will always be absorptive.

4. If this population difference can somehow be inverted so that there are more atoms in the upper energy level than in the lower, then the total response of the collection of atoms will also invert. That is, the total response will change from a net absorptive to a net emissive condition. The atoms will then give net energy to the signal and thus amplify it.

In examining the basic principle for laser action, the question naturally arises as to whether all semiconductors can have laser action. The answer is no. Only the III-V compounds can, for III-V compound semiconductors have direct and indirect bandgap energy levels as described previously.

Population inversion.   Population inversion in the *p-n* junction is achieved by applying a high forward-bias voltage across the junction. In order for stimulated emission to occur, the applied voltage must be at least as great as the bandgap voltage V_g, so that the electrons that have filled the conduction band on the *n* region have energy of at least $E_v + E_g$, where E_v is the energy level of the valence band in

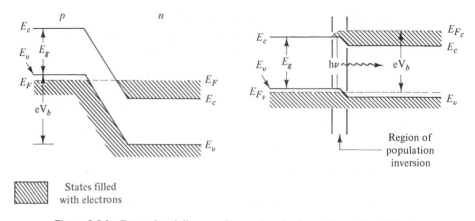

States filled
with electrons

Figure 8-5-2 Energy-band diagram of a *p-n* junction laser (From Marcel Eleccion [10]; reprinted by permission of the IEEE, Inc.)

the *n* region and E_g is the bandgap energy, which is equal to eV_g. The energy band diagram for the *p-n* junction laser is shown in Fig. 8-5-2.

It can be seen from the figure that the height of the potential barrier energy eV_b at the *p-n* junction must be greater than the bandgap energy E_g. Otherwise if the applied voltage is equal to or larger than the barrier potential V_b, the potential barrier will be reduced to zero and very excessive currents will destroy the device. In order for the bandgap energy E_g to be less than the potential barrier energy eV_b, the Fermi level must be below the valence band energy E_v in the *p* region and above the conduction band energy E_c in the *n* region [see Fig. 8-5-2(a)]. This is why the *p-n* junction should be heavily doped in order to have high-enough currents for population inversion.

When a sufficiently high voltage is applied to the *p-n* junction in the forward direction, electrons and holes are injected into and across the transition layer in considerable concentration. As a result, the layer at the junction is far from being depleted of carriers. Consequently, this layer contains a high concentration of electrons in the conduction band and a high concentration of holes in the valence band. This change of concentrations in the two bands is called *population inversion* and the layer at the junction over which population inversion takes place is called the *inversion region* [see Fig. 8-5-2(b)]. Furthermore, population inversion at the junction can be well explained by using the concept of Fermi energy levels, which was shown in Fig. 8-2-1.

When electrons make transitions from the *n* region to empty states in the valence band at the *p* region, they emit photons with energy approximately equal to the bandgap energy E_g. In addition, holes may flow from the *p* region to the *n* region and emit photons by recombining with electrons. In any case, for high-enough forward-bias voltage, this active region exists in the vicinity of the *p-n* junction and a maximum gain over a reasonable distance occurs in the plane of the junction. It is for this reason that the Fabry–Perot resonant cavity is extensively used; here a feedback path is provided directly in the plane of maximum gain.

8-5-3 GaAs Laser

The solid-state junction laser is also called an *injection laser* because its pumping method consists of electron-hole injection in a *p-n* junction. The semiconductor that has been extensively used for junction lasers is the GaAs. The GaAs injection laser is basically a planar *p-n* junction in a single crystal of GaAs. A typical structure of a GaAs *p-n* junction laser is shown in Fig. 8-5-3 [10]. Here the two parallel, semireflective end faces form a Fabry–Perot resonant cavity that enhances the optical Q of the system. The other set of faces is roughened or sawed to suppress all but the modes propagating between the end faces.

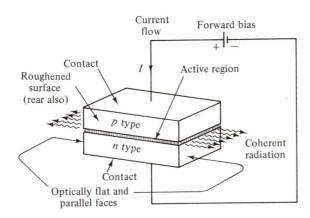

Figure 8-5-3 Structure of a GaAs *p-n* junction laser (From Marcel Eleccion [10]; reprinted by permission of the IEEE, Inc.)

   The junction laser is made of gallium arsenide (GaAs) by the standard diffusion method. Both the *p*-type material and the *n*-type material are very highly doped up to 10^{18} cm^{-3} and so the Fermi levels are in the valence band and the conduction band for the *p* and *n* regions, respectively. The doping is not quite as heavy as in the tunnel diode; thus the current voltage characteristic does not have a negative-resistance region.

   The techniques for fabricating a *p-n* junction laser can be described as follows [20]. The starting material is an *n*-type GaAs wafer doped with silicon in the range of 2 to 4×10^{18} cm^{-3}. A *p*-type layer is grown on the wafer by the liquid-phase epitaxial process. The wafer is lapped to a thickness of 75 μm and the top and bottom surfaces are metalized. The wafer is then cleaved into slivers 300 to 500 μm wide. The junction depths range from 2 to 100 μm. The next step is to evaporate a reflecting coating onto one of the cleaved facets of the sliver so that the laser can emit from only one facet.

Example 8-5-3   GaAs Laser

An *n*-type GaAs laser has the following parameters:

Electron density	$N_d = 3 \times 10^{18} \text{ cm}^{-3}$
Intrinsic concentration	$n_i = 10^6 \text{ cm}^{-3}$
Bandgap energy	$E_g = 1.43 \text{ eV}$
Temperature	$T = 300°\text{K}$

(a) Calculate the energy difference between the Fermi and intrinsic Fermi levels.
(b) Determine the Fermi energy level.

Solution.   (a) According to the basic carrier density equation, the energy difference between the Fermi and intrinsic energy levels is

$$E_F - E_i = kT \ell n \left(\frac{N_d}{n_i} \right) = 26 \times 10^{-3} \ell n \left(\frac{3 \times 10^{18}}{10^6} \right)$$

$$= 0.747 \text{ eV}$$

(b) The Fermi energy level is

$$E_F = E_i + 0.747 = \frac{E_g}{2} + 0.747 = \frac{1.43}{2} + 0.747$$

$$= 1.462 \text{ eV}$$

The Fermi energy level is 0.032 eV above the minimum conduction energy band.

8-6 SEMICONDUCTOR COMPOUND LASERS

As noted, laser action may take place in a slab of semiconductor in the presence of population inversion and in a resonant cavity.

8-6-1 Compound Laser Materials

Table 8-6-1 lists the materials for the semiconductor compound lasers. Their laser action may occur from electron excitation by either electron beam or the optical pumping method.

8-6-2 Nd³⁺:YAG Laser

The Nd^{3+}:YAG laser is one of the most important laser systems in existence. The laser uses Nd^{3+} (trivalent neodymium) as impurities in YAG (yttrium-aluminum-garnet, $Y_3Al_5O_{12}$). Laser emission occurs at a wavelength of 1.064 μm at room

TABLE 8-6-1   MATERIALS OF COMPOUND SEMICONDUCTOR LASERS.

| Material | | Bandgap | Wavelength | Light wave | | |
Scientific name	Symbol	E_g (eV)	λ_0 (μm)	Infrared	Visible	Ultra-violet
Indium antimonide	InSb	0.18	6.90	√		
Mercury-cadmium telluride	$Hg_xCd_{1-x}Te$	0.30	4.12	√		
Tellurium	Te	0.33	3.76	√		
Indium arsenide	InAs	0.36	3.44	√		
Cadmium phosphide	Cd_3P_2	0.60	2.07	√		
Gallium antimonide	GaSb	0.83	1.50	√		
Trivalent neodymium: yttrium-aluminum-garnet ($Y_3Al_5O_{12}$)	Nd^{3+}:YAG	1.167	1.064	√		
Trivalent-neodymium: glass	Nd^{3+}:Glass	1.17	1.062	√		
Cadmium tin phosphide	$CdSnP_2$	1.24	1.00	√		
Gallium arsenide	GaAs	1.43	0.866	√		
Cadmium telluride	CdTe	1.50	0.83	√		
Cadmium silicon arsenide	$CdSiAs_2$	1.60	0.77	√		
Cadmium selenide	CdSe	1.70	0.73	√		
Gallium selenide	GaSe	2.10	0.59		Yellow	
Zinc telluride	ZnTe	2.30	0.54		Green	
Cadmium sulphide	CdS	2.43	0.51		Green	
Zinc selenide	ZnSe	2.67	0.47		Blue	
Zinc oxide	ZnO	3.30	0.38			√
Gallium nitride	GaN	3.45	0.36			√
Zinc sulphide	ZnS	3.60	0.34			√

temperature. It is a four-level laser. The width of the gain line width at room temperature is about 6 cm^{-1}. The spontaneous lifetime for the laser transition is 5.5×10^{-4} sec. The room-temperature cross section at the center of the laser transition is 9×10^{-19} cm^2. In comparison with the ruby laser, which has a cross section of 1.22×10^{-20} cm^2, the optical gain of the Nd^{3+}:YAG laser is about 75 times greater.

When an Nd^{3+} ion is placed in a host crystal lattice, it is subject to the electrostatic field (or crystal field) of the surrounding ions. The crystal field of the host material interacts with the electron energy levels and splits some of the energy levels. The ground and first excited state energy levels of the Nd^{3+} ion split into several groups of levels. As a result, the crystal field modifies the transition probabilities between the various energy levels of the Nd^{3+} ion so that some transition, which is forbidden in the free ion, occurs.

The Nd^{3+}:YAG laser is a four-level system that differs from the three-level structure of the ruby laser.

Example 8-6-2   Nd^{3+}:YAG Laser

A Nd^{3+}:YAG laser has a bandgap energy E_g of 1.167 eV. Determine its emitting wavelength and the type of light color.

Solution.   The wavelength is

$$\lambda_0 = \frac{1.242}{1.167} = 1.064 \ \mu m$$

It is an infrared.

8-6-3 Nd^{3+}:Glass Laser

Instead of using YAG as the host material, the Nd^{3+} ion may be present as an impurity atom in glass. The glass base may be SiBaRb, Ba(PO$_3$)$_2$, LaBBa, SiPbK, or LaAlSi. For example, the Nd^{3+}:SiBaRb laser has an emission wavelength at 1.059 μm.

Glass with a high optical homogeneity can serve as an excellent host material for the Nd^{3+} ion. Local electric fields within the glass modify the energy levels of the neodymium ion in much the same way as the crystal field in YAG.

Nd^{3+}:glass lasers are operated in the pulsed mode and their output spectral line width is greater than in Nd^{3+}:YAG. Their advantages are high energy output and low cost compared to the Nd^{3+}:YAG.

Example 8-6-3   Nd^{3+}:Glass Laser

A Nd^{3+}:glass laser has a bandgap energy E_g of 1.17 eV. Determine its emitting wavelength and the type of light color.

Solution.   The wavelength is

$$\lambda_0 = \frac{1.242}{1.17} = 1.062 \ \mu m$$

It is an infrared.

8-7 LASER APPLICATIONS

As noted, the advantages of the laser beam are its time coherence (monochromaticity) and space coherence (directionality). Time coherence makes the laser beam ideal for communication purposes and space coherence makes it suitable for space communications.

Solid-state junction laser diodes are available with emitting junction widths of 75 μm to 1.4 mm, and produce power outputs of 2 to 70 W, respectively, at driving currents ranging from 10 to 250 A at 300°K. The power output of the laser diodes must be restricted to 1.0 to 1.5 W per 25 μm of emitting facet to prevent degradation of the diode.

At room temperature the solid-state junction laser diode can sustain pulses of 200 nsec and repetition rates of 10 to 20 kHz, depending on the effectiveness of the heat sinking. Because heat generation is the factor that limits operation, a direct tradeoff can be made between pulse width and repetition rate. Thus at narrow pulse widths (less than 10 nsec), repetition rates in excess of 50 kHz have been obtained with laser diodes. Average power output in the milliwatt range can be achieved by pulse operation at 300°K.

The basic concept of holography, which is more than 20 years old, is simply that the diffraction pattern of light from an object is a transform or coded record of the object. If such a diffraction pattern could be stored, it should be possible to reconstruct an image of the object. The original problem with holography was that it is easy to record the magnitude of the diffraction pattern, but it is not easy to record the phase. Laser beams can be used to photograph three-dimensional objects by storing both magnitude and phase information on a film (hologram). Then the hologram can be illuminated by a laser beam to reproduce the object in three dimensions [21].

A large class of laser applications depend not so much on coherence or monochromaticity as on the unprecedented brightness or energy per unit area that can be obtained by focusing a laser beam with a lens. This brightness, a byproduct of the beam's coherence, is a unique feature of laser light and can be many orders of magnitude greater than the brightest light produced by conventional light sources.

The energy density of the image formed by a lens in a laser beam can be used to heat, melt, or even vaporize small areas of any material. This capability promises to find wide use in the field of microelectronics. Several examples are described next. A microcapacitor can be made by using a laser beam to cut a meander path through a 0.3-μm thick gold-conducting film that is vapor-deposited on a sapphire substrate of a microwave integrated circuit. The cut is 6 μm wide. The main advantages of laser light for microelectronics are the very small size of the focused image, the absence of contamination of the very pure materials required in such circuits, and the precise control of the energy used. In fabricating microwave integrated circuits, the laser beam can also be used to weld connections between parts of the circuit. Precision resistors can be adjusted in a completely automatic machine that measures the resistance and pulses the laser to vaporize conductive material until the desired resistance is obtained.

A pulse laser beam has been used to pierce holes in diamond chips or in a alumina ceramic. The thickness of the chip is about 1.5 mm. The diameter of the holes varies between 38 to 76 μm. Twenty-four pulses each with an energy of a tenth of a joule are required to drill a hole all the way through the ceramic.

The high intensity and directionality of a laser beam can also be used to align jigs for mechanical tooling by photoelectric centering of a target and to make surveys over wooded land by determining the laser beam path through the objects in order to lay power transmission lines or commercial telephone cables.

A small laser alarm system can be used to detect intruders when its invisible beam of light is interrupted. When operated, the laser transmitter generates a pulse beam that, in combination with a detector, can seal a path, an area, or a space against undetected intrusion. Other applications for GaAs laser diodes are bomb fuses, secure communications, range-finding radar, data transmission, target designation, collision avoidance, and direct fire simulation. Finally, laser beams are also used in the medical field, such as in restoring detached retinas.

8-8 *MICROWAVE LASER MODULATORS*

The microwave modulation of laser light is increasingly useful in both military and commercial applications [22–38]. In general, there are three types of laser modulators: electrooptic, magnetooptic, and traveling-wave electrooptic. Most crystals are dielectric in nature—that is, they do not carry electric current. Some crystals, however, have a sizable electrooptic effect and do carry current when a certain level of electric voltage is applied. All laser modulators use some type of solid or liquid crystals to produce either an electrooptic or a magnetooptic effect. The selection of a particular crystal for any application depends on such items as the wavelength of light source, the beam diameter, the degree of retardation, and the frequency of modulation.

8-8-1 *Electrooptic Modulator (EOM)*

The electrooptic modulator uses the electrooptic effect. The application of an electric field or voltage across a crystal or liquid medium causes a small change in its refractive index. The velocity of light in a medium is inversely proportional to the refractive index of that medium. If the index is decreased, the light travels faster through the medium; if it is increased, the light travels slower. In other words, the medium becomes birefringent under the influence of an electric field; that is, the light beams of two different linear polarizations travel through the medium with different velocities. Consequently, the phases of the two waves will not be in time phase as they travel through the medium. Two types of electrooptic-modulator configurations exist and are often referred to as the *longitudinal mode* and the *transverse mode*, depending on whether the applied electric field or voltage is parallel or perpendicular to the direction of light propagation.

Longitudinal mode. When a modulating voltage is applied parallel to the optic axis of a crystal, the crystal will exhibit the Pockels effect. This type of device is called the *longitudinal electrooptic modulator* (LEOM).

The *Pockels effect*, which was named after F. Pockels, who studied the effect in 1893, is a linear electrooptic effect exhibited in certain crystals that have the property of advancing or retarding the phase of the polarized light waves when a voltage is applied parallel to the optic axis of the crystal. The effect is linearly proportional to the first power of the applied voltage.

A *Pockels cell* is a device that uses the Pockels effect as a fast light modulator or shutter for laser pulsing. A Pockels cell is usually built with a crystal placed between two crossed polarizers, aligned on its optic axis to the direction of light propagation, and a modulating voltage applied parallel to the optic axis of the crystal (see Fig. 8-8-1).

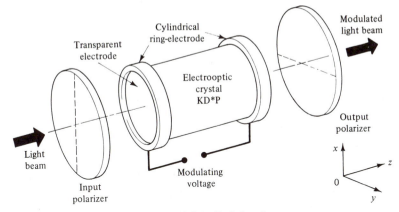

Figure 8-8-1 Pockels cell

Because the crystals are anisotropic, their properties vary in different directions and they must be described by a group of terms referred to collectively as the *second-rank electrooptic tensor* r_{ij}. Fortunately, for an uniaxial crystal, only the component r_{63} is needed. Table 8-8-1 lists the electrooptic constants and refractive indexes of several commonly used crystals for longitudinal electrooptic modulators.

TABLE 8-8-1 CHARACTERISTICS OF CRYSTALS FOR LONGITUDINAL ELECTROOPTIC MODULATORS

Crystal	Electrooptic constant r_{63} ($\times 10^{-12}$ m/V)	Half-wave Voltage in kV $\lambda_0 = 0.5461\ \mu m$	$\lambda_0 = 1.064\ \mu m$	Refractive index n_0
ADA ($NH_4H_2AsO_4$) Ammonium dihydrogen arsenate	5.5	12.59	24.52	1.58
ADP ($NH_4H_2PO_4$) Ammonium dihydrogen phosphate	8.5	9.15	17.82	1.52
KDP (KH_2PO_4) Potassium dihydrogen phosphate.	10.6	7.48	14.59	1.51
KD*P (KD_2PO_4) Potassium dideuterium phosphate	26.4	2.95	5.74	1.52
KDA (KH_2AsO_4) Potassium dihydrogen arsenate	13.0	5.43	10.57	1.57
RDP (RbH_2AsO_4) Rubidium dihydrogen arsenate	11.0	7.35	14.33	1.50

The crystals used for longitudinal electrooptic modulators are normally uniaxial in the absence of an applied electric voltage. That is, the refractive index has only one value in the direction of light propagation through the optic axis. When an electric voltage is applied parallel to the crystal optic axis and the light propagation is assumed to be in the z direction, as shown in Fig. 8-8-1, however, the refractive indexes in the xy plane will change under the influence of the crystal birefringence and will be oriented in the shape of an ellipsoid. Light waves propagated through the optic axis will be polarized in the directions of the two new refractive indexes and will travel with different velocities. The two indexes are orthogonal at zero voltage and define the radius of a circle projected from the index ellipsoid. As the applied voltage is increased, the index circle is elongated in a direction parallel to one of the induced axes. When a collimated laser light is sent through the input polarizer, the light is resolved linearly into two orthogonal components in the plane defined by the axis of the polarization of the polarizer and the light transmittance at the output polarizer is zero with zero-applied voltage. As the applied voltage is increased, the two components travel at different speeds through the crystal and gradually approach to be in time phase, and the light transmittance is increased accordingly. (See Section 2-2.)

The retardation or the phase change of the two polarized waves at the output polarizer is given by

$$\theta = \frac{2\pi n_0^3 r_{63} V}{\lambda_0} \qquad \text{in radians} \qquad (8\text{-}8\text{-}1)$$

where n_0 = refractive index of crystal for the ordinary wave
λ_0 = wavelength in vacuum in meters
V = applied voltage in volts
r_{63} = electrooptic constant of crystal in meters per volt

If a maximum light transmittance through the output polarizer is required, the phase change of the two linearly polarized light waves through the crystal must be one π or a half wavelength. Therefore the half-wave voltage is expressed by

$$V_{\lambda/2} = \frac{\lambda_0}{2n_0^3 r_{63}} \quad \text{volts} \qquad (8\text{-}8\text{-}2)$$

The output light intensity can be written

$$I_0 = I_i \sin^2\left(\frac{\pi V}{2V_{\lambda/2}}\right) \qquad (8\text{-}8\text{-}3)$$

where I_i is the input light intensity to the input polarizer.

Early Pockels cells were made of ammonium dihydrogen phosphate ($NH_4H_2PO_4$): that is, ADP—or of potassium dihydrogen phosphate (KH_2PO_4): that is, KDP. Both crystals are still in use. Because potassium dideuterium phosphate (KD_2PO_4), known as KD*P, has the highest electrooptic coefficient of any known isomorph and yields the same retardation with less than half the voltage needed by KDP and ADP, the crystal KD*P is widely used in Pockels cells for laser modulation at present. In addition, these three major crystals have quite different relative dielectric constants, such as 50 for KD*P, 42 for KDP, and 12 for ADP [39].

Example 8-8-1   Half-Wave Voltage of a KD*P Crystal

The dielectric crystal KD*P is widely used in Pockels cells for laser modulation because its half-wave voltage is much less than that of the other crystals. Determine its half-wave voltage for the Nd^{3+}:YAG laser source.

Solution.   From Eq. (8-8-2) the half-wave voltage of the KD*P crystal is

$$V_{\lambda/2} = \frac{1.064 \times 10^{-6}}{2 \times (1.52)^3 \times 26.4 \times 10^{-12}} = 5.74 \text{ kV}$$

This means that the KD*P crystal can yield the same output light intensity with a voltage less than half of that needed by the other crystals, such as KDP and ADP.

If maximum retardation is required, then the power needed to obtain the required maximum phase difference must be maximum. Consequently, from Eq. (8-8-1) the phase angle is

$$\theta_m = \frac{2\pi n_0^3 r_{63} V_m}{\lambda_0} \qquad (8\text{-}8\text{-}3a)$$

When the load resistance R_ℓ of a resonant circuit is chosen as much larger than the circuit series resistance R_s, the finite modulation bandwidth is given by

$$\Delta f = \frac{1}{2\pi R_\ell C} \qquad (8\text{-}8\text{-}3b)$$

Then the required maximum power may be expressed as

$$P_m = \frac{V_m^2}{2R_\ell} = \frac{\theta_m^2 \lambda_0^2 A \epsilon \Delta f}{4\pi n_0^6 r_{63}^2 L} \qquad (8\text{-}8\text{-}3c)$$

where $C = \dfrac{\epsilon A}{L}$ is the crystal capacitance

$A =$ crystal cross section

$L =$ crystal length

$\Delta f =$ finite modulation bandwidth in the region of 10^8 to 10^9 Hz

Example 8-8-1A   Power Requirement for Maximum Retardation

A KD*P Pockels cell has the following parameters:

Relative dielectric constant	$\epsilon_r = 50$
Electrooptical constant	$r_{63} = 26.4 \times 10^{-12}$ m/V
Crystal refractive index	$n_0 = 1.52$
Wavelength of light source	$\lambda_0 = 0.54$ μm
Crystal cross section	$A = 3 \times 10^{-4}$ m^2
Crystal length	$L = 40$ mm
Maximum retardation angle	$\theta_m = 10°$
Finite modulation bandwidth	$\Delta f = 0.1$ GHz

Calculate the required maximum power.

Solution. Using Eq. (8-8-3c), we obtain the required maximum power as

$$P_m = \frac{(\pi/18)^2 (0.54 \times 10^{-6})^2 (3 \times 10^{-4})(8.854 \times 10^{-12} \times 50)(0.1 \times 10^9)}{4\pi (1.52)^6 (26.4 \times 10^{-12})^2 (40 \times 10^{-3})}$$

$$= 27.33 \text{ W}$$

Transverse mode. When a modulating voltage is applied normal to the optic axis of a crystal, the crystal will exhibit the Kerr effect. This type of device is called the *transverse electrooptic modulator* (TEOM).

The *Kerr effect*, which was named after John Kerr, who discovered the effect in 1875 while studying light refraction in a glass by an electric field, is a quadratic electrooptic effect exhibited in certain crystals or liquids when a voltage is applied normal to the optic axis of the crystal or the direction of the light propagation. The effect is not linear but varies as the square of the applied voltage.

A *Kerr cell* is a transverse electrooptic modulator that is generally made with certain liquids, such as nitrobenzene, or crystals, such as barium titanate. Table 8-8-2 lists the Kerr constants for several commonly used liquids at $\lambda_0 = 5893$ Å and $T = 20°C$.

TABLE 8-8-2   VALUES OF KERR CONSTANTS AT $\lambda_0 = 5893$ Å AND $T = 20°C$

Material	Kerr constant K in m/V^2
Benzene (C_6H_6)	0.67×10^{-14}
Carbon disulfide (CS_2)	3.56×10^{-14}
Nitrotoluene ($C_5H_7NO_2$)	1.37×10^{-12}
Nitrobenzene ($C_6H_5NO_2$)	2.44×10^{-12}
Water (H_2O)	5.10×10^{-14}

When a Kerr-effect medium is inserted between two crossed polarizers and a modulating voltage is applied normal to the direction of light propagation, the device works as a Kerr cell electrooptic modulator. A Kerr cell is illustrated in Fig. 8-8-2.

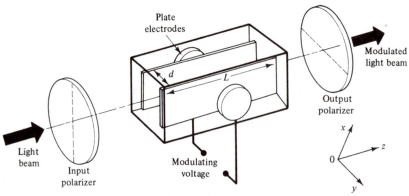

Figure 8-8-2   Kerr cell

When the modulating voltage is off, no light is transmitted by the output polarizer. When the modulating voltage is on, the liquid becomes doubly refracting and light is transmitted. The change in phase of the two polarized waves in the Kerr cell is given by

$$\phi = \frac{2\pi K V^2 L}{d^2} \qquad \text{in radians} \qquad (8\text{-}8\text{-}4)$$

where K = Kerr constant of medium in meters per volt squared
V = applied voltage in volts
L = length of electrodes in the optic axis in meters
d = separation of electrodes in meters

A Kerr cell requires between five to ten times the voltage that a Pockels cell needs to obtain the same optical effect. For this reason, as well as the fact that liquids used in Kerr cells are poisonous and explosive, Pockels cells have replaced Kerr cells in most laser modulators.

Example 8-8-1B   Kerr-Effect Modulation

A Kerr cell is constructed with carbon disulfide (CS_2) and has the following parameters:

Wavelength of LED (GaAsP) light	$\lambda_0 = 0.589\ \mu m$
Kerr constant of carbon disulfide	$K = 3.56 \times 10^{-14}\ m/V^2$
Length of electrodes	$L = 1\ cm$
Separation of electrodes	$d = 2\ mm$
Applied voltage	$V = 37.5\ kV$

Determine the changing phase in radians and degrees.

Solution.   From Eq. (8-8-4) the changing phase is

$$\phi = \frac{2\pi K V^2 L}{d^2} = \frac{6.2832 \times 3.56 \times 10^{-14} \times (37.5 \times 10^3)^2 \times 10^{-2}}{(2 \times 10^{-3})^2}$$

$$= 0.7863\ \text{rad} = 45°$$

8-8-2 Magnetooptic Modulator (MOM)

When a modulating magnetic field is applied parallel to the optic axis of a magnetic crystal and a constant magnetic field is normal to the axis, the crystal will exhibit the Faraday effect. This type of device is called a *magnetooptic modulator.*

The *Faraday effect*, which was named after Michael Faraday, who discovered the effect in 1845 when he studied the light refraction of glass in a strong magnetic field, is a magnetooptic rotation exhibited in a crystal by a magnetic field. When a magnetic crystal is subjected to a magnetic field, the light passing through the crystal is rotated. By experiment the amount of rotation is found to be proportional to the magnetic flux density B, the distance L for the light to travel through the crystal or medium, and the Verdet constant V of the crystal or medium. Therefore the rotation is given by

$$\psi = VBL \qquad \text{in minutes of arc} \qquad (8\text{-}8\text{-}5)$$

where V = Verdet constant, which is defined as the rotation in minutes of arc per meter per tesla

B = magnetic flux density in webers per square meter (tesla)

L = length of the medium in meters

Various gases, liquids, and solids are used for Faraday-effect modulators. Their Verdet constants V at the wavelength of 5893 Å are tabulated in Table 8-8-3.

TABLE 8-8-3 VERDET CONSTANTS AT $\lambda_0 = 5893$ Å

Material	T in °C	Verdet constant V ($\times 10^4$)
Acetone	15	1.109
Carbon disulfide (CS_2)	20	4.230
Ethyl alcohol	25	1.112
Glass (phosphate crown)	18	1.610
Glass (light flint)	18	3.170
Phosphorus (P)	33	13.260
Quartz (perpendicular to axis)	20	1.660
Salt (NaCl)	16	3.585
Water (H_2O)	20	1.310
YIG ($Y_3Fe_5O_{12}$)		

Faraday-effect modulator. A magnetooptic modulator is a device that uses the Faraday rotation as shown in Fig. 8-8-3. The magnetic crystal yttrium-iron-garnet (YIG) is placed between two crossed polarizers. The constant magnetic field is applied normal to the crystal; the modulating magnetic field is parallel to the crystal. The resultant component of the two magnetic fields along the crystal axis causes the Faraday rotation. As current flowing through the coil is changed, the magnetic field in the optic axis is also changed. As a result, the Faraday rotation along the optic

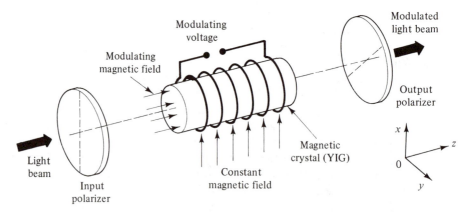

Figure 8-8-3 Magnetooptic modulator

axis is altered. Consequently, a linear modulation is obtained at the output polarizer if the Faraday effect causes a rotation of certain degrees with respect to the input polarizer. The magnetooptic modulator is normally used for large-bandwidth frequency modulation of light.

Example 8-8-2   Faraday-Effect Modulator

The Verdet constant V for the magnetic crystal YIG is not available. The product of VB measured at room temperature is 1750 degrees per centimeter at a wavelength of 0.588 μm under a saturation magnetic intensity of 0.0199 weber/m^2 or 199 gausses. Determine the length L of the magnetic crystal YIG for a Faraday rotation of 45°.

Solution.   The crystal length is

$$L = \frac{5 \times 360 \times 45}{1750} = 1.05 \text{ cm}$$

8-8-3 Traveling-Wave Electrooptic Modulator (TWEOM)

When an electrooptic material slab is inserted between a parallel-plate transmission line, the device will exhibit the optic phase modulation. This type of device is called the *traveling-wave electrooptic modulator* as shown in Fig. 8-8-4.

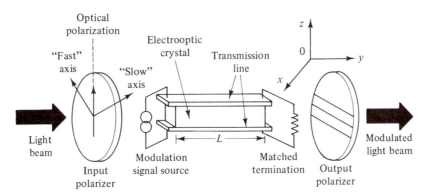

Figure 8-8-4   Traveling-wave electrooptic modulator (After A. Yariv [36]; reprinted by permission of Holt, Rinehart and Winston, CBS College Publishing.)

The electrooptic crystal is placed between the parallel-plate transmission line. The modulating voltage is applied to the input end of the line. Two linearly polarized light waves from the input polarizer travel through the crystal slab. The time for the light to travel the crystal is

$$\tau = \frac{L}{v} = \frac{Ln_0}{c} = \frac{L\sqrt{\epsilon_r}}{c} \qquad \text{in seconds} \qquad (8\text{-}8\text{-}6)$$

where v = velocity of light in crystal in meters per second
   L = length of crystal slab in meters
   n_0 = refractive index of crystal
   c = velocity of light in vacuum, and
   ϵ_r = relative dielectric constant of crystal.

The phase of the output light from the output polarizer is related to the input light at the input polarizer by

$$\Phi = \omega\tau = \frac{\omega L}{v} \qquad \text{in radians} \qquad (8\text{-}8\text{-}7)$$

The phase modulation can be accomplished by varying the velocity of light in the crystal. The change or modulation of the phase angle $\Delta\Phi$ is related to the changing velocity of the light wave in the crystal by

$$\Delta\Phi = \frac{\Phi}{v}\,\Delta v = \frac{\omega L}{v^2}\,\Delta v \qquad (8\text{-}8\text{-}8)$$

The velocity of light in an electrooptic crystal is a function of the applied electric field. The changes in the refractive index in the major axes x, y, and z of the index ellipsoid, as given by Eqs. (2-2-20) through (2-2-22), are repeated here.

$$n_x = n_0 - \frac{n_0^3}{2} r_{63} E_z \qquad (8\text{-}8\text{-}9)$$

$$n_y = n_0 + \frac{n_0^3}{2} r_{63} E_z \qquad (8\text{-}8\text{-}10)$$

and
$$n_z = n_e \qquad (8\text{-}8\text{-}11)$$

where n_0 = refractive index of the ordinary wave
         r_{63} = electrooptic constant of crystal in meters per volt
         n_e = refractive index of the extraordinary wave, and
         E_z = applied electric field in the z direction.

Changes in the velocity of light propagating parallel to the x and y axes of the crystal are given by

$$\Delta v_x = -\tfrac{1}{2} c n_0^2 r_{63} E_z \qquad (8\text{-}8\text{-}12)$$

and
$$\Delta v_y = +\tfrac{1}{2} c n_0^2 r_{63} E_z \qquad (8\text{-}8\text{-}13)$$

Substituting Eq. (8-8-12) or (8-8-13) into Eq. (8-8-8) and replacing velocity v by c/n_0 yield the changing phase angle $\Delta\Phi$ as

$$\Delta\Phi = \frac{n_0^4 \omega r_{63} L E_z}{2c} = \frac{\pi n_0^4 r_{63} L E_z}{\lambda_0} \qquad (8\text{-}8\text{-}14)$$

where λ_0 is the wavelength in vacuum in meters.

This equation shows that the phase change $\Delta\Phi$ is directly proportional to the applied electric field E_z and the length L of the path traveled by the light. When the electric field can be adjusted in such a way, a maximum light intensity is transmitted. The traveling-wave electrooptic modulator is often used for large-bandwidth and low-power modulation of light.

Example 8-8-3   Traveling-Wave Electrooptic Modulator (TWEOM)

A traveling-wave electrooptic modulator is constructed with a KDP crystal and has the following parameters:

Refractive index of KDP	$n_0 = 1.51$
Electrooptic constant of KDP	$r_{63} = 10.60 \times 10^{-12}$ m/V
Length of KDP	$L = 1$ cm
Wavelength of Nd^{3+} : YAG laser	$\lambda_0 = 1.064 \ \mu m$
Electric field	$E_z = 482.7$ kV/m

Determine the changing phase in radians and degrees.

Solution.   Using Eq. (8-8-14), we see that the changing phase angle is

$$\Delta\Phi = \frac{3.1416 \times (1.51)^4 \times 10.6 \times 10^{-12} \times 10^{-2} \times 482.7 \times 10^3}{1.064 \times 10^{-6}}$$

$$= 0.7855 \text{ rad}$$

$$= 45°$$

REFERENCES

[1] Gordon, J. P., H. J. Zeiger, and C. H. Townes. Molecular microwave oscillator and new hyperfine structure in the microwave spectrum of NH_3. *Phys. Rev.*, **95** (1954), 282–284.

[2] Bloembergen, N. Proposal for a new type solid-state maser. *Phys. Rev.*, **104**, no. 2 (October 15, 1956), 324–327.

[3] Scovil, H. E. D., G. Feher, and H. Seidel. Operation of a solid-state maser. *Phys. Rev.*, **105** (1957), 762–763.

[4] Maiman,-T. H. Stimulated optical radiation in ruby masers. *Nature*, **187** (August 6, 1960), 493–494.

[5] Sorokin, P. P., and M. J. Stevenson. Stimulated infrared emission from trivalent uranium. *Phys. Rev. Lett.*, **5** (1960), 577.

[6] Javan, A., et al. Population inversion and continuous optical maser oscillation in a gas discharge, containing a He-Ne mixture. *Phys. Rev. Lett.*, **6** (1961), 106.

[7] Dumke, W. P. Interband transitions and maser action. *Phys. Rev.*, **127** (1962), 1559.

[8] Burns, G., and M. I. Nathan. *p-n* junction lasers. *Proc. IEEE*, **52** (July 1964), 770–794.

[9] Bergh, A. A., and P. J. Dean. Light-emitting diodes. *Proc. IEEE*, **60**, no. 2 (February 1972), 156–224.

[10] Eleccion, Marcel. The family of lasers: a survey. *IEEE Spectrum*, **9**, no. 4 (March 1972), 23–40.

[11] Levine, A. K. Lasers. *Amer. Scient.*, **51** (1963), 14.

[12] Sze, S. M. *Physics of Semiconductor Devices*, 2nd ed., New York: John Wiley & Sons, Inc., 1981, pp. 20, 708.

[13] Nathan, M. I. Semiconductor lasers. *Proc. IEEE*, **54** (1966), 1276.

[14] Basov, N. G., and A. M. Prokhorov. *Zh. Eksperim. i Tear. Fiz.* **28** (1955), 249. [English translation: *Soviet Phys. JETP*, **1** (1955), 184.]

[15] Makhov, G., et al. Maser action in ruby. *Phys. Rev.*, **109** (1958), 1399–1400.

[16] Nishizawa, J. I., and Y. Watanabe. *Electronics* (December 11, 1967), p. 117.

[17] Basov, N. G., et al. *Soviet Phys. Uspekhi*, **3** (1961), 7.

[18] Aigrain, P. Unpublished lecture at the International Conference on Solid-State Physics in Electronics and Telecommunications. Brussels, 1958.

[19] Casey, H. C., and M. B. Parish. Composition dependence of the $Ga_{1-x}Al_xAs$ direct and indirect energygaps. *J. Appl. Phys.*, **40** (1969), 4970.

[20] Glicksman, R. Technology and design of GaAs laser and noncoherent IR-emitting diodes.
Part I.   Solid state technology, vol. 13, no. 9, pp. 29–35, September 1970.
Part II. Solid state technology, vol. 13, no. 10, pp. 39–44, October 1970.

[21] Herriott, D. R. Applications of laser light. *Scientific American*, **219**, no. 3 (September 1968), 141–156.

[22] Liao, Samuel Y. Pockels cell subsystem for laser modulation. El Segundo, Calif.: Hughes Aircraft Company, August 1980.

[23] Goldstein, Robert. Pockels cell primer. *Laser Focus Magazine*, February 1968.

[24] Johnson, B. C., et al. Optical rise-time measurements on KD*P transmission-line Pockels cells. *J. Appl. Phys.*, **49**, no. 1 (January 1978), 75–80.

[25] White, G., and G. M. Chan. Traveling-wave electrooptic modulators. *Optics Communications*, **5**, no. 5 (August 1972), 374–379.

[26] Peters, C. J. Gigacycle bandwidth coherent light traveling-wave phase modulator. *IEEE Proc.*, **53**, no. 1 (January 1963), 147–153.

[27] Rigrod, W. W., and I. P. Kaminow. Wide-band microwave light modulation. *IEEE Proc.*, **53**, no. 1 (January 1963), 137–140.

[28] Kaminow, I. P., and J. Liu. Propagation characteristics of partially loaded two-conductor transmission line for broadband light modulator. *IEEE Proc.*, **53**, no. 1 (January 1963), 132–136.

[29] West, Edward A. Extending the field of view of KD*P electrooptic modulators. *Applied Optics*, **17**, no. 18 (September 15, 1978), 3010–3013.

[30] Nelson, Donald. The modulation of laser light. *Scientific American*, **218**, no. 6 (June 1968), 17–32.

[31] Ploss, Richard S. A review of electrooptic materials, methods and uses. *Optical Spectra* (January–February 1969), pp. 63–67.

[32] Steinmetz, L. L., et al. Cylindrical, ring-electrode KD*P electrooptic modulator. *Applied Optics*, **12**, no. 7 (July 1973), 1468–1471.

[33] Letellier, J. P. Parallel-plate transmission-line Pockels cell. Naval Research Laboratory, Washington, D.C., October 1972.

[34] McLellan, E. J., and J. F. Figueira. Ultrafast Pockels cells for the infrared. *Rev. Sci. Instrum.*, **50**, no. 10 (October 1979).

[35] Goldberg, G. K. Nonnormal incidence in field induced biaxial KD*P. *Applied Optics*, **8**, no. 5 (May 1969), 1037–1039.

[36] Yariv, Ammon. *Introduction to Optical Electronics*. 2nd ed. New York: Holt, Rinehart and Winston, 1976.

[37] Liao, Samuel Y. *Microwave Devices and Circuits*. Englewood Cliffs, N.J.: Prentice-Hall, Inc., 1980.

[38] Howland, M. M., et al. Very fast, high peak-power, planar triode amplifiers for driving optical gates. Livermore, Calif.: Lawrence Livermore Laboratory, UCRL 82538, June 12, 1979.

[39] Wilson, J., and J. F. B. Hawkes. *Optoelectronics*: *An Introduction*. Englewood Cliffs, N.J.: Prentice-Hall, Inc., 1983, p. 104.

SUGGESTED READINGS

1. *IEEE Proceedings*. vol. 70, no. 6 (June 1982). Special issue on laser applications.

2. Kaminskii, A. A. *Laser Crystals*. Berlin: Springer-Verlag, 1981.

3. Levine, A. K., and A. J. De Maria, ed. *Lasers*. New York: Marcel Dekker, Inc., 1971.

4. Liao, Samuel Y. *Microwave Devices and Circuits*, Chapter 6. Englewood Cliffs, N.J.: Prentice-Hall, Inc., 1980.

5. Milnes, A. G. *Semiconductor Devices and Integrated Electronics*, Chapter 12. New York: Van Nostrand Reinhold Company, 1980.

6. Ross, Monte, ed. *Laser Applications*. New York: Academic Press, 1977.

7. Smith, W. V. *Laser Applications*. Dedham, Mass.: Artech House, 1972.

8. Sze, S. M. *Physics of Semiconductor Devices*, Chapter 12. New York: John Wiley & Sons, 1981.

9. Verdeyen, J. T. *Laser Electronics*. Englewood Cliffs, N.J.: Prentice-Hall, Inc., 1981.

10. Wilson, J., and J. F. B. Hawkes, *Optoelectronics*: *An Introduction*. Englewood Cliffs, N.J.: Prentice-Hall, Inc., 1983.

11. Yariv, Ammon. *Introduction to Optical Electronics*. 2nd ed. New York: Holt, Rinehart and Winston, 1976.

PROBLEMS

8-1   Transition Processes

8-1-1. There are three transition processes for laser actions. Describe each in detail.

8-1-2. The photon energy must be equal to or larger than the bandgap energy of the material in order for laser activity to occur. The AlGaAs laser has a bandgap energy of 1.80 eV.
  (a) Determine the frequency of the photon energy that will stimulate a laser action.
  (b) Identify the type of light.

8-1-3. Two green-light lasers, ZnTe (Zinc Telluride) and CdS (Cadmium sulphide), have wavelengths of 0.54 and 0.51 μm, respectively.
  (a) Determine their energy differences between the two levels in electronvolts at room temperature.
  (b) Compute their population ratios of n_1 over n_2.

8-2   Population Inversion

8-2-1. The electron and hole concentrations are equal at 10^{13} m^{-3} for a band-to-band transition in an intrinsic GaAs semiconductor at room temperature. Calculate the minimum carrier concentration for population inversion.

8-2-2. A junction laser GaAsP has the following parameters:

Gap energy	$E_g = 1.80$ eV
Operating temperature	$T = 300°$K
Intrinsic carrier concentration	$n_i = 10^7$ cm^{-3}
Equal electron and hole concentrations	$n_n = n_p$

Calculate the minimum carrier concentration for population inversion.

8-3   Resonant Cavity

8-3-1. A laser cavity has a length L of 30 cm. The laser wavelength is 0.87 μm.
   (a) Find the number of axial modes.
   (b) Compute the number of oscillating frequencies.
   (c) Calculate the frequency separation.

8-4   Ruby Laser

8-4-1. The operating principle for a ruby laser is based on the three-level model theory. Describe the theory in detail.

8-4-2. It is known that the ruby laser emits red light at a wavelength of 0.6943 μm. Estimate the bandgap energy in electronvolts for the ruby gem (Al_2O_3).

8-5   Junction Lasers

8-5-1. The GaAs and GaAsP junction lasers have bandgap energies of 1.43 eV and 1.80 eV, respectively.
   (a) Compute the wavelengths in micrometers for the emitted light.
   (b) Identify the types of light.

8-5-2. From the laser-action equation we have $h\nu = E_g$ in electronvolts. Verify that $\lambda = 1.242/E_g$ in micrometers.

8-5-3. Aluminum-gallium arsenide laser material ($Al_xGa_{1-x}As$) can be changed from direct GaAs to indirect AlGa when x is varied from 0 to 1 as shown in Fig. 8-5-1.
   (a) Write the expression for the said laser material for $x = 0.37$.
   (b) Estimate the bandgap energy in electronvolts.
   (c) Calculate the emitting wavelength in micrometers.
   (d) Identify the type of light.

8-6   Compound Lasers

8-6-1. The Nd^{+3}:YAG (trivalent neodymium: yttrium-aluminum-garnet) is a very useful laser in optical communications. It emits laser light at 1.064 μm.
   (a) Estimate its bandgap energy in electronvolts.
   (b) Identify the type of light.

8-8   Laser Modulators

8-8-1. When a voltage is applied to a Pockels-cell laser modulator, the light beam will be rotated by certain degrees, depending on the magnitude of the voltage and the nature of the dielectric crystal as shown in Eq. (8-8-1). Derive the equation.

8-8-2. The output light intensity from a Pockels-cell laser modulator is expressed by Eq. (8-8-3). Derive the equation.

8-8-3. The changing phase angle for a traveling-wave electrooptic modulator is shown in Eq. (8-8-14). Verify the equation.

8-8-4. The crystal KD*P is often used for the Pockels-cell laser modulator and it has a refractive index n of 1.52. Its electrooptic constant r_{63} is 26.4×10^{-12} m/V. The half-wave voltage is 3.72 kV.
(a) Determine the wavelength and the frequency of light.
(b) Identify the type of light.
(c) Find the applied voltage for a phase change of 90°.

8-8-5. The crystal KDP in a Pockels-cell laser modulator has a refractive index n of 1.51 and an electrooptic constant r_{63} of 10.60×10^{-12} m/V.
(a) Compute the half-wave voltage at a wavelength of 0.5461 μm.
(b) Identify the type of light.
(c) Calculate the output light intensity in terms of the input light intensity for $V = V_{\lambda/2}$.

8-8-6. A Kerr cell uses the liquid nitrobenzene with a Kerr constant K of 2.44×10^{-12} m/V². Its dimensions are $L = 10$ cm and $d = 2$ cm. Calculate the applied voltage for a phase change of 90°.

8-8-7. A magnetooptic laser modulator uses a flint glass as the refractor. The flint glass has a verdet constant V of 3.17×10^4 minutes of arc per meter per tesla. The magnetic flux density is 0.2 tesla and the cell length L is 10 cm. Determine the light rotation at a wavelength of 5893 Å. (*Note*: 1 tesla = 1 weber per square meter = 10^4 gausses.)

8-8-8. A maximum retardation requires a maximum power as shown in Eq. (8-8-3c). Start from Eqs. (3-2-3), (8-8-3a), and (8-8-3b) and derive Eq. (8-8-3c).

8-8-9. A KD*P Pockels cell has the following parameters:

Relative dielectric constant	$\epsilon_r = 50$
Electrooptic constant	$r_{63} = 26.4 \times 10^{-12}$ m/V
Wavelength of light source	$\lambda_0 = 0.54$ μm
Crystal cross section	$A = 2 \times 10^{-4}$ m²
Crystal length	$L = 50$ mm
Crystal refractive index	$n_0 = 1.52$
Maximum retardation angle	$\theta_m = 20°$
Finite modulation bandwidth	$\Delta f = 0.2$ GHz

Calculate the required maximum power in watts.

Chapter 9

Infrared Devices and Systems

9-0 INTRODUCTION

Infrared (IR) radiation is an electromagnetic radiation generated by the vibration and rotation of atoms and molecules within any material at temperatures above absolute zero—that is, 0°K or −273°C. In recent years there has been an increasing emphasis on the research, design, development, and deployment of various infrared devices and systems for military applications at night or during the day when vision is diminished by fog, haze, smoke, or dust. Infrared systems are defined as those that sense the passive infrared radiation emitted by some target or source and process it to the point that a visual image of that target or source is formed.

The history of infrared discovery is an interesting one. In 1800 Sir William Herschel discovered infrared radiation while working for the British Royal Navy [1, 2, 3] but at that time he did not use the term infrared. Herschel referred to the new portion of the radiation by such names as invisible rays, radiant heat, dark heat, and the rays that occasion heat. Sir Herschel found that the heating effect increased as he moved the thermometer toward the red from the blue end of the spectrum. In 1829 Nobili made the first thermocouple. The thermocouple was an improved thermometer based on the thermoelectric effect discovered by Seebeck in 1821 [4]. In 1833 Melloni invented a thermopile, which consisted of a number of thermocouples connected in series. The thermopile was more sensitive than the thermocouple and it could detect radiant heat from a person at a distance of 10 m [5]. In 1901

Langley and Abbot developed an improved bolometer that could detect radiant heat from a cow at a distance of 400 m [6]. During World War I an infrared search system that could detect aircraft at a distance of 1.6 km and people at a distance of 300 m was used. In 1917 Case constructed the first photoconductive sensor by using thallous sulfide [7]. Many sensitive infrared detectors, such as photon detectors and image converters, were developed during World War II. The sniperscope, which consisted of an image converter and an illuminator mounted on a carbine, allowed a soldier to fire accurately at night at targets as far away as 60 m. In the late 1950s the Sidewinder and Falcon heat-seeking infrared-guided missiles were developed. Subsequently infrared devices and systems were installed in the Walleye, Redeye, and Chaparral missiles and A-6E aircraft. Furthermore, infrared techniques could be used in the altitude stabilization of space vehicles, the measurement of planetary temperatures, earth mapping, and the early detection of cancer [8]. The pioneer and fundamental work on infrared thermal imaging systems resulted from the efforts of many dedicated scientists and engineers, such as Hudson [4, 9], Jones [10], and Johnson [11].

9-1 INFRARED SPECTRUM

All bodies radiate energy throughout the infrared spectrum when their temperatures are above absolute zero. If the radiating source is hot enough (above 1000°K), some of the emitted energy may be visible to the human eye in the 0.4- to 0.7-μm range. Energy emitted at wavelengths between 0.75 and 1000 μm is defined as *infrared radiation*. It should be noted that wavelengths between 100 and 1000 μm are ultramicrowaves or millimeter waves. Figure 9-1-1 shows the infrared spectrum [4].

The infrared spectrum can be further subdivided into four divisions as shown in Table 9-1-1. The first three divisions include spectral intervals in which the earth's atmosphere is relatively transparent, the so-called atmospheric windows. It is these windows that will be used by any infrared sensor that must look through the earth's atmosphere. The extreme infrared, which is often called the *ultramicrowave*, is generally used only for laboratory applications where the instrument can be evacuated because the atmosphere is essentially opaque [9]. Typical targets of interest have peak emittances at a wavelength of about 10 μm; a good atmospheric window exists between 8 and 14 μm; and scattering is much lower at a wavelength of 10 μm. It is obvious, then, why the 8- to 12-μm band has been consistently chosen for thermal imaging at ranges longer than 900 m. Shorter wavelengths of 3 to 5 μm, for instance, can readily be applied to thermal imaging at ranges shorter than 900 m. Figure 9-1-2 show the transmittance of the atmosphere for a 1.8-km horizontal path at sea level in the infrared range.

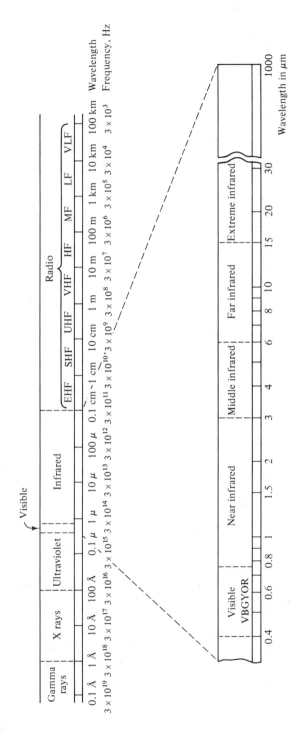

Figure 9-1-1 Infrared spectrum (After R. D. Hudson, Jr. [4]; reprinted by permission of John Wiley & Sons, Inc.)

TABLE 9-1-1   INFRARED SPECTRUM

Division	Wavelength (μm)	Frequency (Hz)
Near infrared (NIR)	0.7–3	4.29×10^{14}–1×10^{14}
Middle infrared (MIR)	3–6	1×10^{14}–5×10^{13}
Far infrared (FIR)	6–15	5×10^{13}–2×10^{12}
Extreme infrared (XIR)	15–1000	2×10^{12}–3×10^{11}

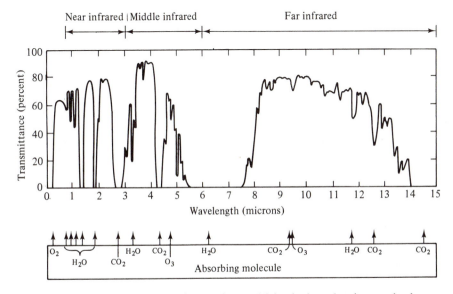

Figure 9-1-2   Transmittance of atmosphere at 1.8 km horizontal path at sea level (Adapted from H. A. Gebbie et al. [12], from R. D. Hudson, Jr. [4]; reprinted by permission of John Wiley & Sons, Inc.)

9-2 INFRARED RADIATION

The fundamental law of physics states that all bodies at temperatures above absolute zero in Kelvin—that is, $T > 0°$K or $-273°$C—emit radiation. The amount of the infrared energy emitted depends on the absolute temperature, nature of the body, and wavelength of the radiation. The emitting body may be a blackbody, which absorbs all incident radiation, or a graybody, which absorbs a portion of the incident radiation.

9-2-1   Spectral Radiant Emittance

The spectral radiant emittance of a blackbody is given by Planck's law as

$$W(\lambda) = \frac{C_1}{\lambda^5}\left[\exp\left(\frac{C_2}{\lambda T}\right) - 1\right]^{-1} = \frac{2\pi hc^2}{\lambda^5}\left[\exp\left(\frac{hc}{\lambda kT}\right) - 1\right]^{-1} \qquad \text{watts/cm}^2/\mu\text{m}$$

$$(9\text{-}2\text{-}1)$$

where $C_1 = 2\pi hc^2 = 3.7415 \times 10^4$ watts-$(\mu\text{m})^4/\text{cm}^2$
        $h = 6.6256 \times 10^{-34}$ watts-sec^2 is Planck's constant
        $c = 3 \times 10^{10}$ cm/sec is the velocity of light in vacuum
        $C_2 = ch/k = 1.4388 \times 10^4$ μm-°K
        $k = 1.38 \times 10^{-23}$ watt-sec/°K is Boltzmann's constant
        λ = the wavelength in micrometers
        T = the absolute temperature in degrees Kelvin

9-2-2   Radiant Emittance

The total radiant emittance into a hemisphere from a blackbody at a given absolute temperature can be obtained by integrating Eq. (9-2-1) over wavelength limits extending from zero to infinity. That is,

$$W = \int_0^\infty W(\lambda)\,d\lambda = \frac{2\pi^5 k^4}{15c^2 h^3}T^4 = \sigma T^4 \qquad \text{watts/cm}^2 \qquad (9\text{-}2\text{-}2)$$

where $\sigma = 5.6697 \times 10^{-12}$ watts/cm^2/°K^4. Equation (9-2-2) is well known as the Stefan–Boltzmann or Boltzmann law and it indicates the radiant flux in watts per unit area of a source.

9-2-3   Irradiance

When the radiant flux from a source arrives at a detector, the irradiance H incident on the surface of a detector is less than the radiant flux at the source due to the atmospheric absorption and molecular scattering in the transmission path. The irradiance H is expressed in watts per square centimeter and it can be measured at the detector site.

9-2-4   Maximum Spectral Radiant Emittance

Differentiating Eq. (9-2-1) with respect to wavelength λ and equating the derivative equal to zero result in Wien's displacement law. That is,

$$\lambda_m T = a \qquad \mu\text{m-°K} \qquad (9\text{-}2\text{-}3)$$

where $a = 2898$ μm-°K
        λ_m = the wavelength in micrometers for maximum spectral radiant emittance
        T = the absolute temperature in degrees Kelvin

The maximum wavelength for peak radiation from a 300°K source, for example, can be determined from Wien's displacement law as about 10 μm.

Example 9-2-1   Computations of Infrared Radiations

The jet engine of an aircraft emits infrared radiation at a wavelength of 10 μm and at a temperature of 200°C.

(a) Calculate the spectral radiant emittance of the jet engine in watts per square centimeter per micrometer and watts per square meter per meter.

(b) Compute the radiant emittance in watts per square centimeter.

(c) Find the temperature in degrees Kelvin for maximum radiant emittance if the wavelength is fixed at 10 μm and determine the maximum radiant emittance in watts per square centimeter.

(d) Determine the wavelength for maximum spectral radiant emittance if the temperature is fixed at 200°C and find the maximum spectral radiant emittance in watts per square centimeter per micrometer.

Solution.   (a) Given: $\lambda = 10$ μm and $T = 273 + 200 = 473$°K. Using Eq. (9-2-1), we obtain

$$W(\lambda) = \frac{3.7415 \times 10^4}{10^5} \left[\exp\left(\frac{1.4388 \times 10^4}{10 \times 473} \right) - 1 \right]^{-1}$$

$$= 18.70 \times 10^{-3} \text{ watt/cm}^2/\mu m$$

$$W(\lambda) = \frac{2\pi h c^2}{\lambda^5} \left[\exp\left(\frac{ch}{\lambda kT} \right) - 1 \right]^{-1}$$

$$= \frac{2 \times 3.1416 \times 6.625 \times 10^{-34} \times (3 \times 10^8)^2}{(10 \times 10^{-6})^5}$$

$$\times \left[\exp\left(\frac{3 \times 10^8 \times 6.625 \times 10^{-34}}{10 \times 10^{-6} \times 1.38 \times 10^{-23} \times 473} \right) - 1 \right]^{-1}$$

$$= 18.72 \times 10^7 \text{ watts/m}^2/m$$

(b) The radiant emittance is

$$W = \sigma T^4 = 5.6697 \times 10^{-12} \times (473)^4 = 0.2838 \text{ watt/cm}^2$$

(c) Using Wien's displacement law, we have

$$\lambda T_m = a = 2898$$

$$T_m = \frac{2898}{10} = 289.8°K$$

$$W = \sigma T^4 = 5.6697 \times 10^{-12} \times (289.8)^4 = 0.04 \text{ watt/cm}^2$$

(d) From Eq. (9-2-3), we have

$$\lambda_m = \frac{a}{T} = \frac{2898}{473} = 6.13 \ \mu m$$

$$W(\lambda) = \frac{3.7415 \times 10^4}{(6.13)^5} \left[\exp\left(\frac{1.4388 \times 10^4}{6.13 \times 473} \right) - 1 \right]^{-1}$$

$$= 0.030 \text{ watt/cm}^2/\mu m$$

9-2-5 Radiant Intensity and Radiance

The radiant intensity J and radiance N of a blackbody in a hemisphere are expressed as

$$J = \frac{WA}{\pi} = NA \qquad \text{watts/sr} \tag{9-2-4}$$

and

$$N = \frac{W}{\pi} = \frac{J}{A} \qquad \text{watts/cm}^2\text{/sr} \tag{9-2-5}$$

where A is the area of a radiating surface in square centimeters.

It should be noted that a hemisphere has a solid angle of 2π. The unit of a solid angle is expressed in steradians, which is often written sr. Here, however, it is assumed that the detector is a Lambertian receiver that has an effective integrated solid angle of 1π for a hemisphere. If the source is an isotropic radiator, the radiation will emit uniformly throughout the entire solid angle of 4π. If the source is small compared with the field of view of the infrared system—that is, a point source—the irradiance will vary with the distance but not the angle about the radiator. If the source is large compared with the system field of view—namely, an extended source—the irradiance will be constant. This situation can be explained by Lambert's cosine laws, which state that the radiant intensity in any direction propagating from any point of a surface is a function of the cosine of the angle θ between the said direction and the normal line to the surface at that point. In other words, the maximum radiation is in the direction normal to the surface and zero radiation is in the tangential direction. This is why a scanning detector always receives the same amount of radiation regardless of the change of angle θ between the detector's line of sight and the normal line to the radiating surface. As the radiating area viewed by the detector is increased, the angle θ is also increased and the value of $\cos\theta$ is decreased. Consequently, the total irradiance is constant.

9-2-6 Emissivity

The emissivity of a thermal radiator is a measure of its radiation efficiency. It is defined as

$$\text{Emissivity} = \frac{\text{total radiant emittance of a graybody}}{\text{total radiant emittance of a blackbody at the same temperature}}$$

$$\epsilon = \frac{W'}{W} = \frac{\int_0^\infty \epsilon(\lambda)W(\lambda)\,d\lambda}{\int_0^\infty W(\lambda)\,d\lambda} = \frac{1}{\sigma T^4}\int_0^\infty \epsilon(\lambda)W(\lambda)\,d\lambda \tag{9-2-6}$$

The emissivity ϵ is therefore an indication of the graybody of the thermal radiator. In other words, the lower the emissivity the grayer is the radiator and the higher the emissivity the blacker is the body. The blackbody has an emissivity of unity. Table 9-2-1 lists the emissivity of commonly used materials in total normal radiation.

TABLE 9-2-1 EMISSIVITY OF COMMONLY USED MATERIALS

Material	Temperature (°C)	Emissivity ϵ
Metals and their Oxides		
Aluminum:		
polished sheet	100	0.05
sheet as received	100	0.09
anodized sheet, chromic acid process	100	0.55
vacuum deposited	20	0.04
Brass:		
highly polished	100	0.03
rubbed with 80-grit emery	20	0.20
oxidized	100	0.61
Copper:		
polished	100	0.05
heavily oxidized	20	0.78
Gold: highly polished	100	0.02
Iron:		
cast, polished	40	0.21
cast, oxidized	100	0.64
sheet, heavily rusted	20	0.69
Magnesium: polished	20	0.07
Nickel:		
electroplated, polished	20	0.05
electroplated, no polish	20	0.11
oxidized	200	0.37
Silver: polished	100	0.03
Stainless steel:		
type 18-8, buffed	20	0.16
type 18-8, oxidized at 800°C	60	0.85
Steel:		
polished	100	0.07
oxidized	200	0.79
Tin: commercial tin-plated sheet iron	100	0.07
Other Materials		
Brick: red common	20	0.93
Carbon:		
candle soot	20	0.95
graphite, filed surface	20	0.98
Concrete	20	0.92
Glass: polished plate	20	0.94
Lacquer:		
white	100	0.92
matte black	100	0.97
Oil, lubricating (thin film on nickel base):		
nickel base alone	20	0.05
film thickness of 0.001, 0.002, 0.005 in.	20	0.27, 0.46, 0.72
thick coating	20	0.82
Paint, oil: average of 16 colors	100	0.94
Paper: white bond	20	0.93
Plaster: rough coat	20	0.91
Sand	20	0.90
Skin, human	32	0.98
Soil:		
dry	20	0.92
saturated with water	20	0.95
Water:		
distilled	20	0.96
ice, smooth	-10	0.96
frost crystals	-10	0.98
snow	-10	0.85
Wood: planed oak	20	0.90

(After R. D. Hudson, Jr. [4]; reprinted by permission of John Wiley & Sons, Inc.)

9-2-7 Kirchhoff's Law

Kirchhoff [13] discovered that at a given temperature the ratio of radiant emittance to absorptance is a constance for all materials and that it is equal to the radiant emittance of a blackbody at that temperature [4]. That is,

$$\frac{W'}{\alpha} = W \tag{9-2-7}$$

where α is the absorptance of a graybody. It is evident from Eqs. (9-2-6) and (9-2-7) that the emissivity of any graybody at a given temperature is numerically equal to its absorptance at that temperature. That is,

$$\epsilon = \alpha \tag{9-2-8}$$

Table 9-2-2 shows the values of the absorptance α, the emissivity ϵ, and the ratio α/ϵ for several materials.

TABLE 9-2-2 ABSORPTANCE α AND EMISSIVITY ϵ OF MATERIALS

Material	α	ϵ	$\dfrac{\alpha}{\epsilon}$
Aluminum:			
polished and degreased	0.387	0.027	14.35
foil, dull side, crinkled and smoothed	0.223	0.030	7.43
foil, shiny side	0.192	0.036	5.33
sandblasted	0.42	0.21	2.00
oxide, flame sprayed, 0.001 in. thick	0.422	0.765	0.55
anodized	0.15	0.77	0.19
Fiberglass:	0.85	0.75	1.13
Gold: plated on stainless steel and polished	0.301	0.028	10.77
Magnesium: polished	0.30	0.07	4.30
Paints:			
Aquadag, 4 coats on copper	0.782	0.490	1.60
aluminum	0.54	0.45	1.20
Microbond, 4 coats on magnesium	0.936	0.844	1.11
TiO_2, gray	0.87	0.87	1.00
TiO_2, white	0.19	0.94	0.20
Rokide A	0.15	0.77	0.20
Stainless steel: type 18-8, sandblasted	0.78	0.44	1.77

(After R. D. Hudson, Jr. [4]; reprinted by permission of John Wiley & Sons, Inc.)

9-3 INFRARED RADIATION SOURCES

A *blackbody* is defined as any object that completely absorbs all incident radiation. Conversely, the radiation emitted by a blackbody at a given temperature is maximum. Therefore a blackbody is an ideal radiator and absorber of radiation for all wavelengths at all temperatures and its emissivity is equal to unity. Because the blackbody is a theoretical thermal radiator, it is commonly used as a standard source to calibrate all other infrared devices. Objects with emissivity less than unity are called *graybodies*; the great majority of radiating objects belong to this group.

In general, infrared radiation sources can be classified into two groups: natural and artificial (Table 9-3-1). In addition to the radiation from a target or object, a certain amount of background radiation will be present. The blackgroup radiation appears in the infrared detection system as unwanted noise and it must be filtered out for proper detection.

TABLE 9-3-1   INFRARED RADIATION SOURCES

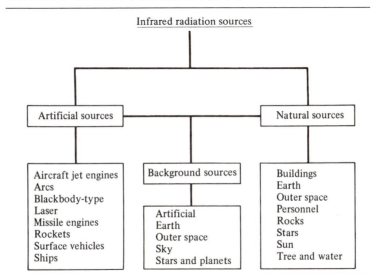

9-3-1   Artificial Sources

This type of infrared radiation source includes the controlled sources, such as blackbody-type sources, and the active sources, such as aircraft and rocket engines.

Most blackbody-type sources used as standard sources for the calibration of infrared devices are of the cavity type with an opening of 1.27 cm or less and they operate in the temperature range of 400 to 1300°K. The spectral radiant emittances of a blackbody as calculated from Planck's law in Eq. (9-2-1) in terms of temperature and wavelength are shown in Fig. 9-3-1 [4]. The dashed curve in Fig. 9-3-1 indicates the locus of these maxima as computed from Wien's displacement law in Eq. (9-2-3).

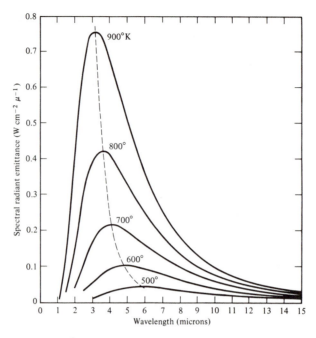

Figure 9-3-1 Spectral radiant emittance of a blackbody (From R. D. Hudson, Jr. [4]; reprinted by permission of John Wiley & Sons, Inc.)

Example 9-3-1 Spectral Radiant Emittance of a Blackbody

A blackbody emits radiation at 900°K.

(a) Determine the wavelength in micrometers for a maximum spectral radiant emittance.

(b) Find the peak spectral radiation in watts per square centimeter per micrometer.

Solution. (a) Using Eq. (9-2-3), we see that the wavelength for peak radiation is

$$\lambda_m = \frac{a}{T} = \frac{2898}{900} = 3.22 \ \mu m$$

(b) Then from Eq. (9-2-1) the peak spectral radiation is

$$W(\lambda) = \frac{3.7415 \times 10^4}{(3.22)^5} \left[\exp\left(\frac{1.4388 \times 10^4}{3.22 \times 900} \right) - 1 \right]^{-1}$$

$$= 0.76 \ watt/cm^2/\mu m$$

Aircraft jet engines, rockets, and missiles are powerful active sources of infrared radiation. The prime sources of infrared radiation on aircraft and missiles are the hot metal of a jet tail pipe or an engine exhaust manifold and the jet plume. Figure 9-3-2 shows the exhaust temperature contours of the turbojet engine JT4A being used on the Boeing 707 [4]. At supersonic speeds the skin of an aircraft or missile becomes an infrared radiation source because of aerodynamic heating. Figure 9-3-3 shows the equilibrium surface temperature caused by aerodynamic

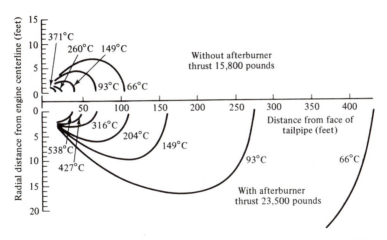

Figure 9-3-2 Exhaust temperature contours of a Boeing 707 (After R. D. Hudson, Jr. [4]; reprinted by permission of John Wiley & Sons, Inc.)

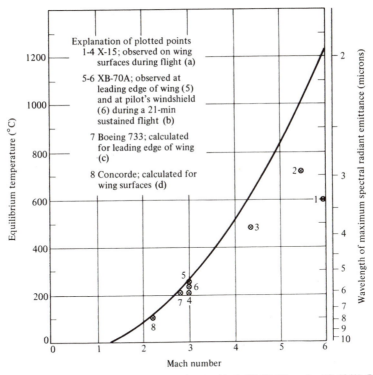

(a) Various articles on the X–15 program, *Aviation Week*: **75**, 52 (November 20, 1961); **75**, 60 (November 27, 1961); **77**, 35 (August 13, 1962); **78**, 38 (June 10, 1963).

(b) C.M. Plattner, XB–70A flight research – part 2. *Aviation Week* **84**, 60 (June 13, 1966).

(c) C.M. Plattner, Variable-sweep wing keynotes Boeing 733 SST proposal. *Aviation Week*, **80**, 36 (May 4, 1964).

(d) Size, Speed, Safety of SST are debated. *Aviation Week*, **80**, 29 (June 1, 1964).

Figure 9-3-3 Surface temperature of SST and Concorde (From R. D. Hudson, Jr. [4]; reprinted by permission of John Wiley & Sons, Inc.)

heating in altitudes above 11.28 km and laminar flow for the Boeing 733 [that is, Supersonic Transport (SST)] and the British-French Concorde [4]. A large aircraft may emit several kilowatts of infrared energy, but the human body emits only about 2 watts.

The laser beam provides coherent sources of extremely high radiance in the portion of the spectrum extending from the ultraviolet to microwaves. The first application of the laser in the infrared portion of the spectrum was for military operations and communication systems.

Surface vehicles may radiate enough infrared energy to be considered targets. The paint used on such vehicles usually has an emissivity of 0.85 or greater. The exhaust pipes and mufflers may radiate several times as much energy as the rest of the vehicle because of their high temperature.

Example 9-3-1A   Radiant Emittance of a Jet Engine

A certain jet engine radiates infrared energy at a temperature of 100°C at a distance of 100 m. Determine the radiant emittance of the engine at that location.

Solution.   From Eq. (9-2-2) the radiant emittance is

$$W = \sigma T^4 = 5.67 \times 10^{-12} \times (373)^4$$

$$= 109.75 \text{ mW/cm}^2$$

$$= 1.098 \text{ kW/m}^2$$

9-3-2 Natural Sources

This type of infrared radiation source includes the terrestial sources, such as rocks, trees, earth, and water, the celestial sources, such as the sun, sky, stars, and planets, and the buildings on the ground. The sun radiates a total radiant emittance of 26.9 W/m^2 normal to the earth's surface. In other words, the sun radiates as a 5900°K blackbody. Of this solar radiation, approximately 596 mW/m^2 is absorbed by the earth's surface from the sun at wavelengths longer than 3 μm.

During the daytime the infrared radiation from the surface of the earth is a combination of reflected and scattered solar energy and thermal emission from the earth itself. At night, when the sun has gone, the spectral distribution becomes that of a graybody at the ambient temperature of the earth. Figure 9-3-4 illustrates the spectral radiance of typical terrain objects as observed during the daytime [4].

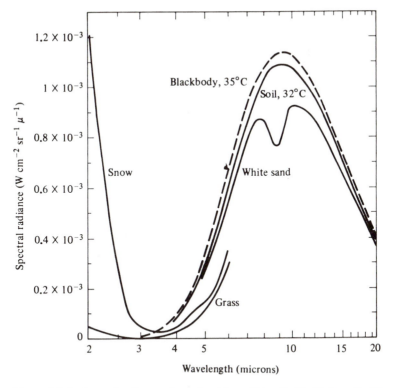

Figure 9-3-4 Spectral radiance of terrain objects (After R. D. Hudson, Jr. [4]; reprinted by permission of John Wiley & Sons, Inc.)

Example 9-3-2 Spectral Radiance of Soil

A certain soil has a temperature of 32°C.

(a) Calculate its emitting wavelength for maximum spectral radiant emittance.

(b) Compute its peak spectral radiance in watts per square centimeter per micrometer.

Solution. (a) From Eq. (9-2-3) the wavelength for peak radiance is

$$\lambda_m = \frac{2898}{305} = 9.50 \ \mu\text{m}$$

(b) From Eq. (9-2-1) the peak spectral radiance is

$$W(\lambda) = \frac{3.7415 \times 10^4}{(9.50)^5} \left[\exp\left(\frac{1.4388 \times 10^4}{9.5 \times 305} \right) - 1 \right]^{-1}$$

$$= 0.484 \times 7.02 \times 10^{-3}$$

$$= 3.40 \times 10^{-3} \ \text{watt/cm}^2/\mu\text{m}$$

9-4 INFRARED OPTICAL COMPONENTS

In a radar system the antenna acts as a transmitter and receiver for the electromagnetic energy. But in an infrared system the optics collect and transmit the infrared radiant flux. Therefore the optical components in an infrared system are simply analogous to the antennas in a radar system. The commonly used lenses and mirrors in an infrared system are shown in Fig. 9-4-1.

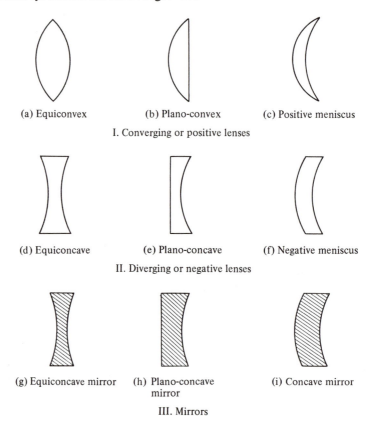

(a) Equiconvex   (b) Plano-convex   (c) Positive meniscus

I. Converging or positive lenses

(d) Equiconcave   (e) Plano-concave   (f) Negative meniscus

II. Diverging or negative lenses

(g) Equiconcave mirror   (h) Plano-concave   (i) Concave mirror
                              mirror

III. Mirrors

Figure 9-4-1 Lenses and mirrors

9-4-1 Focal Points and Focal Lengths

The axis in a lens is a straight line through the geometrical center of the lens and normal to the two faces at the points of intersection (see Fig. 9-4-2). The primary focal point F lies on the axis and is defined for a positive lens as the point from which diverging rays are refracted by the lens into a parallel beam. The secondary focal point F' is defined by applying the principle of reversibility to the same lens. The distance between the center of a lens and either of its focal points is called its *focal length*.

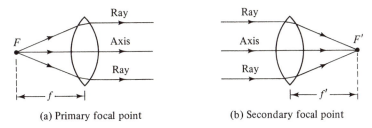

(a) Primary focal point   (b) Secondary focal point

Figure 9-4-2   Focal point and focal length

9-4-2 Magnification and Conjugate

In any optical device the ratio between the transverse dimension of the final image and the corresponding dimension of the original object is defined as the *magnification* of the lens. If an object is placed at the position previously occupied by its image, it will be imaged at the position previously occupied by the object. Here the object and image are said to be *conjugate*. The conjugate plane for a collimator is at an infinite distance.

9-4-3 Telescope

In an infrared system the infrared radiation from an object scene is collected by a telescope and brought out in a collimated beam with a small pupil at the scan mirror. A terrestrial-type telescope with a two-lens erecting system is used in the infrared system.

9-4-4 Field of View and Stops

The field of view (FOV) in geometrical optics determines how much of the surface of a broad object can be seen through an optical system. It is often subdivided into narrow field of view (NFOV), mediate field of view (MFOV), and wide field of view (WFOV), depending on the size of the aperture. In order to control the brightness of images, an aperture stop is placed between the lens and the focal plane to limit the incident bundles of rays; the field stop in front of the focal plane determines the extent of the object, or the field, that will be produced in the image.

9-5 INFRARED DETECTORS

The purpose of an infrared detector in the infrared imaging system is to collect irradiance energy from a target and convert it into some other measurable form, such as an electrical current or photographic image. At wavelengths of 8 to 12 μm and at video frequencies of 0.01 to 5 MHz, there are only a few detectors with sufficient sensitivity for high-performance infrared imaging systems. These detectors are mercury-doped germanium (Ge : Hg), mercury-cadmium telluride (Hg : Cd, Te), and lead-tin telluride (Pb : Sn, Te). The quality and characteristics of a detector are usually described by three parameters:

1. Responsivity (R)
2. Specific detectivity ($D*$)
3. Time constant (τ)

9-5-1 Responsivity (R)

The *responsivity* (R) is a measure of merit for detectors and it is defined as the ratio of the detector output over input. That is,

$$R(f) = \frac{V_s}{HA_d} \qquad \text{volts/watt} \tag{9-5-1}$$

where V_s = signal voltage in volts (root-mean-square, rms) is the fundamental component of the signal

H = incident radiant flux in watts per square centimeter is the rms value of the fundamental component of the irradiance on the detector

A_d = sensitive area of the detector in square centimeters

9-5-2 Specific Detectivity (D*)

The responsivity, as shown in Eq. (9-5-1), indicates only the behavior of the detector signal and does not give any information about the amount of noise in the output of the detector that will ultimately obscure the signal. In order to know the signal-to-noise ratio at the output of the detector, it is necessary to define another parameter, detectivity (D). Before doing so, it is desirable to define the noise equivalent power (NEP) at the detector output as

$$\text{NEP} = \frac{HA_d}{V_s/V_n} = \frac{HA_dV_n}{V_s} = \frac{V_n}{R(f)} \qquad \text{watts} \tag{9-5-2}$$

where V_n is the noise voltage in rms volts at the detector output.

The *specific detectivity* ($D*$) is defined as directly proportional to the square root of the product of the detector area and the noise bandwidth and inversely proportional to the noise equivalent power [10]. That is,

$$D*(f) = \frac{(A_d \times \Delta f)^{1/2}}{\text{NEP}} \qquad \text{cm-Hz}^{1/2}\text{/watt} \tag{9-5-3}$$

where Δf is the noise bandwidth in hertz.

The *detectivity* (D) of a detector is related to the specific detectivity ($D*$) by the following equation [10]:

$$D = D*(A_d \times \Delta f)^{-1/2} = \frac{1}{\text{NEP}} \qquad \text{watt}^{-1} \tag{9-5-4}$$

This relationship was verified by Jones [10] because $DA_d^{1/2} = $ constant is well known through extensive theoretical and experimental studies. It should be noted that the quantity of the specific detectivity (D^*) refers to an electrical noise bandwidth of 1 Hz and a detector area of 1 cm². It is customary to indicate D^* by two numbers in parentheses. The first number shows the temperature of the blackbody and the second indicates the spatial frequency. For instance, D^* (400°K, 800) means a value of D^* measured with a 400°K blackbody at a spatial frequency of 800 Hz.

9-5-3 Time Constant (τ)

The *time constant* (τ) of a detector is defined as the time required for the detector output to reach 63% of its final value after a sudden change in the irradiance. The responsive time constant (τ_r) of a detector [10] is defined as

$$\tau_r = \frac{\frac{1}{4}R_m^2}{\int_0^\infty [R(f)]^2 \, df} \qquad \text{seconds} \qquad (9\text{-}5\text{-}5)$$

where R_m is the maximum value of $R(f)$ with respect to frequency.

Similarly, the detection time constant (τ_d) of a detector is expressed [10] as

$$\tau_d = \frac{\frac{1}{4}D_m^{*2}}{\int_0^\infty [D^*(f)]^2 \, df} \qquad \text{seconds} \qquad (9\text{-}5\text{-}6)$$

where D_m^* is the maximum value of $D^*(f)$ with respect to frequency.

9-5-4 Mercury-Doped Germanium (Ge:Hg)

As noted, all infrared thermal imaging systems are designed to operate at wavelengths of 8 to 12 μm. The response of the detector is then limited to that wavelength band for high performance. Various detectors were constructed with germanium (Ge) as a host lattice. Such impurities as gold (Au) yielded a response to about 10 μm, copper (Cu) to 30 μm, and mercury (Hg) to 14 μm. For the band range of 8 to 14 μm, Ge:Hg was found most effective [14] because it had a detectivity of 4×10^{10} cm-Hz$^{1/2}$/W and a detection time constant of 1 ns. The disadvantage of the Ge:Hg detector is that it requires cooling to 25 to 30°K. The cooling requirement and the cost have prevented Ge:Hg from being continuously used in the infrared imaging system.

9-5-5 Mercury-Cadmium Telluride (Hg:Cd,Te)

Mercury-cadmium telluride (Hg:Cd,Te) is an alloy consisting of a mixture of the compounds HgTe and CdTe. Its spectral response varies from 9.5 to 12 μm and its impedance ranges from 20 to 110 Ω. The detectivity is 1×10^{10} cm-Hz$^{1/2}$/W and the detection time constant is 0.05 ns [14].

9-5-6 Lead-Tin Telluride (Pb:Sn,Te)

Lead-tin telluride detectors are available only in the photovoltaic mode. They have a detectivity of 1×10^{10} cm-Hz$^{1/2}$/W and a detection time constant of 0.1 ns. The operating wavelengths vary from 8 to 12 μm. Figure 9-5-1 shows the characteristic properties of the foregoing mentioned three detectors [14].

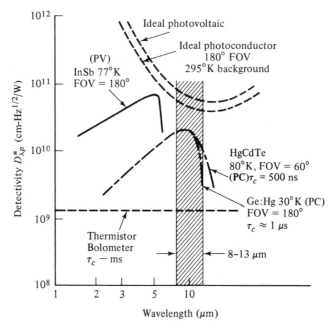

Figure 9-5-1 Specific detectivity (D^*) of leading detectors (From J. J. Richter [14]; reprinted by permission of the IEEE, Inc.)

In addition, many other infrared detectors are available, such as

1. Thermal detectors—thermocouple, thermopile, and bolometer
2. Photon or quantum detectors—photoelectric detector, photoconductive (PC) detector, and photovoltaic (PV) or *p-n* junction detector
3. Imaging detectors—infrared film, Vidicon, and photothermionic image converter (Thermicon)

The Vidicon is a small television-type camera tube in which an electron beam scans a photoconductive target. The Thermicon is based on the thermal variation of photoemission and produces the scene image on its retina. Table 9-5-1 lists the properties of several infrared photoconductive detectors [8]. Figure 9-5-2 illustrates the infrared detectors versus the infrared spectrum [9].

TABLE 9-5-1 INFRARED PHOTOCONDUCTIVE DETECTORS

Material	Maximum temperature for background limited operation	Long wavelength cutoff (50%) (μm)	Peak wavelength (μm)	Absorption coefficient (cm^{-1})	Quantum efficiency	Resistance (Ω)	D^* Peak ($cm\text{-}Hz^{1/2}/W$)	Approximate response time (seconds)
InAs		3.6	3.3	$\sim 3 \times 10^3$			3×10^{11}	5×10^{-7}
InSb	110	5.6	5.3	$\sim 3 \times 10^3$	0.5–0.8	10^3–10^4	6×10^{10} -1×10^{11}	5×10^{-6}
Ge : Au	60	9	6	~ 2	0.2–0.3	4×10^5	$3 \times 10^9 - 10^{10}$	3×10^{-8}
Ge : Au(Sb)	60	9	6			10^4	6×10^9 7×10^9 -4×10^{10}	1.6×10^{-9}
Ge : Hg	35	14	11	~ 3	0.2–0.6	1.4×10^4	4×10^{10}	$\begin{cases} 3 \times 10^{-8} \\ -10^{-9} \end{cases}$
		14	10.5	~ 4	0.62	1.2×10^5		
Ge : Hg(Sb)	35	14	11			5×10^9	1.8×10^{10}	$3 \times 10^{-10} - 2 \times 10^{-9}$ $3 \times 10^{-10} - 3 \times 10^{-9}$
Ge : Cu	17	27	23	~ 4	0.2–0.6	2×10^4	2.4×10^{10}	$3 \times 10^{-4} - 10^{-8}$
Ge : Cu(Sb)	17	27	23	$\sim 10^3$		2×10^5	2×10^{10}	$4 \times 10^{-9} - 1.3 \times 10^{-7}$ $< 2.2 \times 10^{-9}$
Hg : Cd, Te x = 0.2		14	12		0.05–0.3	60–400	10^{10}	$< 10^{-8}$
Pb : Sn, Te x = 0.17–0.2		11	10	$\sim 10^4$		20–200	6×10^{10}	$< 4 \times 10^{-4}$
						42	3×10^8	1.5×10^{-8}
		15	14			52	1.7×10^{10}	1.2×10^{-4}

(After Levinstein and Mudar [8]; reprinted by permission of the IEEE, Inc.)

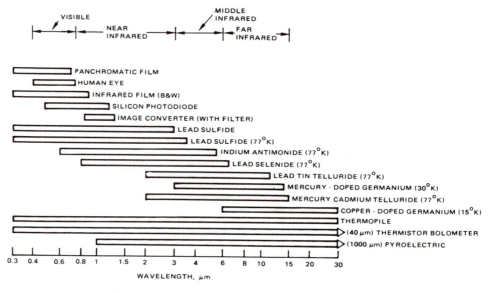

Figure 9-5-2 Infrared detectors versus spectral ranges. Operating temperature is 300°K if not indicated (From R. D. Jr. and J. W. Hudson [9]; reprinted by permission of the IEEE, Inc.)

Example 9-5-5 Merit Figures of Infrared Detectors

A certain infrared detector has the following parameters:

Infrared wavelength	$\lambda_0 = 8\ \mu m$
Detecting area	$A_d = 100 \times 10^{-4}\ cm^2$
Bandwidth	$\Delta f = 1\ Hz$
Incident radiant flux	$H = 1\ W/cm^2$
Signal voltage	$V_s = 1\ mV$
Noise voltage	$V_n = 0.25\ pV$

(a) Determine the responsivity $R(f)$.
(b) Find the noise equivalent power (NEP).
(c) Calculate the detectivity D.
(d) Compute the specific detectivity D^*.

Solution. (a) Using Eq. (9-5-1), we find the responsivity as

$$R(f) = \frac{V_s}{HA_d} = \frac{10^{-3}}{1 \times 10^{-2}} = 0.1\ V/W$$

(b) Then from Eq. (9-5-2) the noise equivalent power is

$$NEP = \frac{V_n}{R(f)} = \frac{0.25 \times 10^{-12}}{0.1} = 2.5\ pW$$

(c) According to Eq. (9-5-4), the detectivity is

$$D = \frac{1}{NEP} = \frac{1}{2.5 \times 10^{-12}} = 4 \times 10^{11}\ W^{-1}$$

(d) And from Eq. (9-5-3) the specific detectivity is

$$D^* = D\left(A_d \times \Delta f\right)^{1/2} = 4 \times 10^{11} \times \left(100 \times 10^{-4} \times 1\right)^{1/2}$$
$$= 4 \times 10^{10} \text{ cm-Hz}^{1/2}/\text{W}$$

9-6 INFRARED SYSTEMS

An *infrared system* consists of the optical unit, the detector array, the signal processors, and the display indicators (see Fig. 9-6-1).

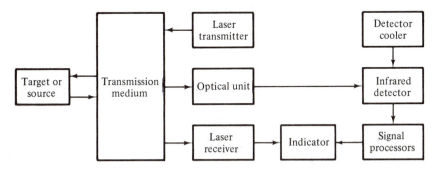

Figure 9-6-1   Block diagram of an infrared system

Because a simple infrared system is passive, it can be used for the detection, recognition, and identification of a target by sensing the radiation emitted by the target but it provides no information on the distance to the target. If an infrared system has some illuminating devices like the laser transmitter and receiver built in, the entire infrared system may become active and have the capability of determining the range to the target. The target is the object of interest for which the infrared system is designed and built. The radiated energy varies with the temperature of the target, its emissivity, and the viewing angle. The atmosphere is not a very favorable transmission medium. If the operating wavelength is chosen in the range of the atmospheric window as shown in Fig. 9-1-2, however, the atmospheric absorptance may be reduced to a minimum. The optical unit collects the radiation emitted by the target and delivers it to the detector just like an antenna in the radar system. Before reaching the detector, the information radiation from the target may be recovered by an optical demodulator from the unwanted emission in the background. The detector then converts the information radiation into an electrical signal. Because a wide field of view (WFOV) and high resolution require a large number of detector elements, the forward-looking infrared (FLIR) unit uses the newest detection techniques, based on discrete detector arrays to scan the object space and generate a video signal. Finally, the signal processor amplifies the electrical signal and sends the coded target information to the indicator for display. In addition, the infrared detector must be cooled down by a cryogenic cooler to a specified operating level of temperature—30°K for the Ge : Hg detector—for high performance. This step is necessary because the internal radiation emitted by the detector itself must be reduced to a minimum. Otherwise it is difficult, if not impossible, to separate the two radiations with equal magnitude from the target and the detector itself.

The performance of an infrared imaging system is described by using the following three parameters:

1. Noise equivalent differential temperature (NEΔT)
2. Minimum resolvable temperature (MRT)
3. Modulation transfer function (MTF)

9-6-1 Noise Equivalent Differential Temperature (NEΔT)

The signals from the detectors in an infrared imaging system respond to the variation in the irradiance at the entrance pupil of the optical system as the detectors are scanned across the target scene. Small differences in irradiance or temperature can be detected as an ac video signal while the high ambient irradiance or temperature is presented as a dc level or white noise. Then the noise signal is subtracted by a single RC lowpass filter. The *Noise Equivalent Differential Temperature* (NEΔT) is defined as the temperature difference required at the input of the detector to produce a peak signal-to-noise (rms) ratio of unity at the detector preamplifier output [10]. That is,

$$\text{NE}\Delta T = \frac{V_n}{V_s}\Delta T \qquad \text{degrees in C} \qquad (9\text{-}6\text{-}1)$$

where V_s = ac signal voltage in volts (peak to peak)
V_n = noise voltage in volts (rms), and
ΔT = target temperature difference in degrees centigrade

In Eq. (9-6-1) it is assumed that the reflectance of a collimator is 100%. The peak signal voltage from a target can be measured by an oscilloscope. The noise voltage can be recorded by turning off the scanner, covering the optical aperture with a flat black cover or an opaque, and reading the noise on an rms voltmeter.

9-6-2 Minimum Resolvable Temperature (MRT)

The *Minimum Resolvable Temperature* (MRT) is a measurement that is sensitive to the thermal imaging system–human observer combination. Its output is a function that describes the minimum temperature difference to resolve the various spatial-frequency patterns projected into the input port of the infrared system under test. A square 4-bar pattern, 7-to-1 aspect ratio, is currently the standard pattern used for such measurements.

Johnson's imaging model. One of the earliest attempts to relate the threshold resolution with the visual discrimination of images of real scenes theoretically is attributed to Johnson [11]. The basic experimental scheme was to move a real scene object, such as a car, out in range until it could only barely be discerned on a detector's display at a given discrimination level, such as detection, recognition, or identification. Then the real scene object was replaced by a bar pattern of contrast similar to that of the scene object. The number of bars per minimum object dimension in the pattern was then increased until the bars could just barely be individually resolved [15]. Figure 9-6-2 shows Johnson's imaging model.

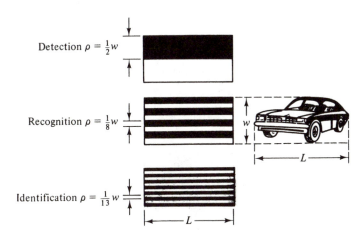

Figure 9-6-2 Johnson's imaging model

In Fig. 9-6-2 the car is replaced by a bar pattern. According to Johnson's hypothesis, if the car is to be barely detected, the bar pattern width ρ should be equal to one-half the car's minimum dimension. For simple recognition, the bar width should be one-eighth whereas for identification the bar width should be one-thirteenth the car's minimum dimension. The length of the bars was un-specified, but it was assumed that the length is to be equal to the car's longest dimension.

In the laboratory measurement the bars are normally oriented vertically, but 45° and horizontal orientations are sometimes also used. The observer is allowed to adjust the display Brightness and Contrast controls for the most resolvable pictures at some minimum temperature ΔT. This process is repeated for other spatial frequencies. It is important that the ambient temperature during the measurement not be changed by more than $\pm 0.20°C$ from its initial values. The smaller the resolvable temperature, the better is the infrared system.

The characteristic spatial frequency f of a 4-bar pattern, 7-to-1 aspect ratio, is defined as

$$f = \frac{1}{2\theta} = \frac{L}{2\rho} \qquad \text{cycles/mrd} \qquad (9\text{-}6\text{-}2)$$

where L = focal length of a collimator in inches

ρ = bar width in mils is the smallest resolution dimension of the target that is to be viewed

$\theta = \arctan(\rho/L)$ is the angle subtended by the target (or bar) width in milliradians

Figure 9-6-3 shows schematically the relationship between the bar width and the focal length of a collimator. In practice, the minimum resolvable temperature is usually plotted against some normalized spatial frequency f_0 instead of the characteristic spatial frequency f.

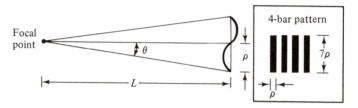

Figure 9-6-3   Relationship between bar width and focal length

Example 9-6-2   Spatial Angle of a 4-bar Pattern or a Target

A 4-bar pattern has a bar width of 10 mils and the focal length of the collimator is 100 in.

(a) Determine the angle subtended by the bar width in milliradians (mrd).

(b) Find the characteristic spatial frequency in cycles per milliradian (cycles/mrd).

Solution.   (a) Using Fig. 9-6-3, we obtain the angle subtended by the 4-bar pattern as

$$\theta = \arctan\left(\frac{\rho}{L}\right) \approx \frac{\rho}{L} \qquad \text{for } \rho \ll L$$

$$= \frac{10 \times 10^{-3}}{100} = 0.1 \text{ mrd}$$

(b) Then from Eq. (9-6-2) the characteristic spatial frequency is

$$f = \frac{100}{2 \times 10 \times 10^{-3}} = 5 \text{ cycles/mrd}$$

9-6-3   Modulation Transfer Function (MTF)

According to electrical system theory, the total system transfer function is the product of the individual system's element transfer functions. The purpose of an infrared thermal imaging system is to collect the thermal energy emitted by a target and convert it to an electrical current or voltage for display. Because an infrared imaging system is composed of optical, physical, chemical, mechanical, electrical, and electronic elements, the system is more complicated than any ordinary electrical or electronic system. Therefore it is extremely difficult, if not impossible, to calculate analytically the infrared system's transfer function.

In Fourier theory any analytical function can be transferred to a Fourier series. The *Modulation Transfer Function* (MTF) of an infrared system may be defined as the modulus of the Fourier transform of the one-dimensional spatial impulse response of the system. For a 4-bar pattern, 7-to-1 aspect ratio, the object consists of alternate light and dark bands that vary sinusoidally. The distribution of brightness, which is a spatial function, can be resolved into a Fourier transform as

$$B(x) = B_0 + B_1 \cos 2\pi f x \tag{9-6-3}$$

where f = the spatial frequency of the brightness variation in cycles per milliradian
B_0 = the dc or average level of brightness
B_1 = the magnitude of the variable brightness
x = the spatial coordinate in milliradians

Figure 9-6-4 shows the energy response of a 4-bar pattern. For simplicity, the third and higher harmonic terms are neglected. The modulation of the object pattern at the input port of the system is given by Smith [16] as

$$M_0 = \frac{(B_0 + B_1) - (B_0 - B_1)}{(B_0 + B_1) + (B_0 - B_1)} = \frac{B_1}{B_0} \tag{9-6-4}$$

where $(B_0 + B_1)$ is the maximum brightness and $(B_0 - B_1)$ is the minimum brightness. When the object brightness pattern passes through the infrared system, the image pattern will be affected by the system transfer function as shown in Fig. 9-6-5.

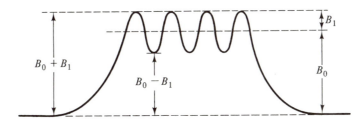

Figure 9-6-4   Object energy response of a 4-bar pattern

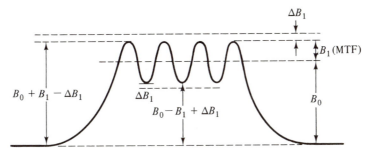

Figure 9-6-5   Image energy response of a 4-bar pattern

The modulation of the image pattern at the output port of the system is given by Smith [16] as

$$M_i = \frac{[B_0 + B_1(\text{MTF})] - [B_0 - B_1(\text{MTF})]}{[B_0 + B_1(\text{MTF})] + [B_0 - B_1(\text{MTF})]} = \frac{B_1}{B_0}(\text{MTF}) = M_0(\text{MTF})$$

(9-6-5)

Therefore the modulation transfer function (MTF) of an infrared thermal imaging system is expressed as

$$\text{MTF} = \frac{M_i}{M_0}$$

(9-6-6)

In practice, the quantities of the image modulation M_i and the object modulation M_0 are unknown. The maximum and minimum image patterns can be measured in the infrared system laboratory from an oscilloscope as shown in Fig. 9-6-5. Then Eq. (9-6-5) can be expressed as

$$\text{MTF} = \frac{M_i}{M_0} = \frac{B_0}{B_i} M_i = \frac{B_0}{B_i} \cdot \frac{(B_0 + B_1 - \Delta B_1) - (B_0 - B_1 + \Delta B_1)}{(B_0 + B_1 - \Delta B_1) + (B_0 - B_1 + \Delta B_1)} = \frac{B_1 - \Delta B_1}{B_1}$$

(9-6-7)

where ΔB_1 = the differential portion decreased by the effect of the system transfer function

B_1 = the magnitude of the variable object brightness

$(B_0 + B_1 - \Delta B_1)$ = the maximum image pattern

$(B_0 - B_1 + \Delta B_1)$ = the minimum image pattern

In conclusion, the higher the modulation transfer function, the better is the infrared imaging system.

9-7   INFRARED SYSTEM MEASUREMENT AND APPLICATIONS

The infrared thermal imaging system is currently based on discrete detectors and optical scanning mechanisms for image display. The basic principles of operation and measurement are shown in Fig. 9-7-1.

System Operation

The infrared thermal energy emitted by a target is collected by the telescope of the infrared system and brought out in a collimated beam with a small pupil at the scan mirror. The scan mirror provides the scan motion to generate one angular dimension of the sensor field of view. The scanned energy at the smaller aperture is collected by the detector lens and focused on the infrared detectors. The signals from the detectors and preamplifier are processed through the data processors for image display.

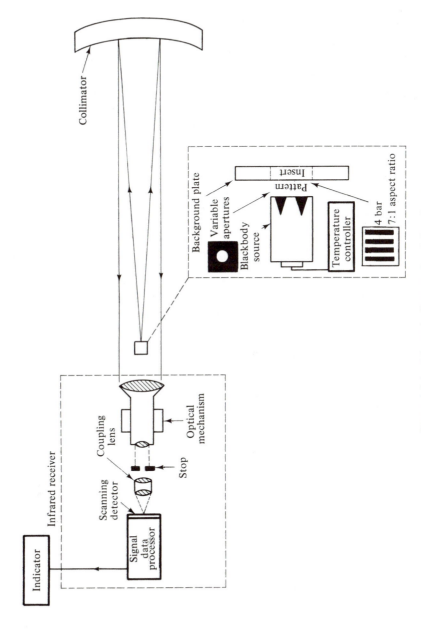

Figure 9-7-1 Schematic diagram of an infrared system

Collimator

Background plate

Variable apertures

Insert

Pattern

Blackbody source

Temperature controller

4 bar

7:1 aspect ratio

Coupling lens

Optical mechanism

Stop

Scanning detector

Infrared receiver

Signal data processor

Indicator

System Measurement

A square 4-bar pattern, 7-to-1 aspect ratio, is inserted into the plate slot. The temperature of the source is reduced until a match is obtained between the background and target. When a differential temperature ΔT is increased, the 4-bar image is barely distinguished. The differential temperature ΔT value is the target temperature difference for the tested pattern. The same procedure is repeated for various spatial frequencies. When the best image of each spatial frequency appears on the oscilloscope, the maximum, minimum, and average levels of brightness are recorded. The noise voltage or power can be measured by covering the aperture of the infrared receiver with an opaque [17–19].

Infrared Applications

The infrared thermal imaging system has unlimited application potential in many areas. Some current applications are listed here:

Military Applications.   These include military fire control at night or during the day when the vision is diminished due to fog, smoke, or haze; detection and tracking of ships, aircraft, missiles, surface vehicles, and personnel; submarine detection and range finding. It has been demonstrated that a missile can be fired accurately at a target in night by using an infrared device as far away as 2 km.

Medical Applications.   These include early detection and identification of cancer, obstacle detection for the blind, location of blockage in a vein, and early diagnosis of incipient stroke.

Scientific Applications.   These include satellite and space communications, environmental survey and control, detection of life and vegetation on other planets, and measurement of lunar and planetary temperature.

Industrial Applications.   These include aircraft landing aid, traffic counting, forest fire detection, and natural source detection.

REFERENCES

[1] Herschel, W. Investigation of the powers of the prismatic colours to heat and illuminate objects: With remarks that prove the different refrangibility of radiant heat. *Phil. Trans. Roy. Soc. London*, pt. II, **90** (1800), 255.

[2] Herschel, W. Experiments on the refrangibility of the invisible rays of sun. *Phil. Trans. Royal Soc. London*, pt. II, **90** (1800), 284.

[3] Herschel, W. Experiments on the solar, and on the terrestrial rays that occasion heat. *Phil. Trans. Royal Soc. London*, pt. II, **90** (1800), 293, 437.

[4] Hudson, R. D., Jr. *Infrared System Engineering*. New York: John Wiley & Sons, 1968, pp. 6, 21, 36, 40, 43, 45, 92, 102, 107, 115.

[5] Barr, E. S. The infrared pioneers. II. Macedonia Melloni. *Infrared Phys.* (1962), p. 67.

[6] Barr, E. S. The infrared pioneers. III. Pierpont Langley. *Infrared Phys.*, **3** (1963), 195.

[7] Case, T. W. Notes on the change of resistance of certain substances in light. *Phys. Rev.*, **9** (1917), 305–310.

[8] Levinstein, H., and J. Mudar. Infrared detectors in remote sensing. *Proc. IEEE.*, **63**, no. 1 (January 1975), 6–14.

[9] Hudson, R. D., Jr., and J. W. Hudson. The military applications of remote sensing by infrared. *Proc. IEEE.*, **63**, no. 1 (January 1975), 104–128.

[10] Jones, R. C. Phenomenological description of the response and detecting ability of radiation detectors. *Proc. IRE.*, vol. 47, no. 9 (September 1959).

[11] Johnson, J. Analysis of image forming systems. *Proc. of Image Intensifier Symposium*. Ft. Belvoir, Va., AD220160, October 1958.

[12] Gebbie, H. A., et al. Atmospheric transmission in the 1- to 14-μm region. *Proc. Roy. Soc.*, **A206** (1951), 87.

[13] Planck, M. *Theory of Heat Radiation*. New York: Dover, 1960. (A reprint of the 1910 edition)

[14] Richter, J. J. Infrared thermal imaging. *Proc. IEEE. Southeast Region 3 Conference*. Orlando, Fla., April 29, 1974.

[15] Rosell, F. A. Levels of visual discrimination for real scene objects vs. bar pattern resolution for aperture and noise limited imagery. *Proc. National Aerospace and Electronics Conference*. pp. 327–334. Dayton, Ohio, January 10–12, 1975.

[16] Smith, F. D. Optical image evaluation and the transfer function. *Applied Optics*, **2**, no. 4 (April 1963), 335–350.

[17] Liao, Samuel Y. System performance of the detecting and ranging set (DRS) for the Navy A-6E TRAM aircraft. Report for Hughes Aircraft Company, El Segundo, Calif., August 1978.

[18] Wood, J. T. Test and evaluation of thermal imaging systems. Northeast Electronics Research and Engineering Meeting. Record part 3, November 1973.

[19] Liao, Samuel Y. Capability and reliability of forward-air-controller (FAC) receiver subsystem for the Navy A-6E TRAM aircraft. Hughes Aircraft Company, El Segundo, Calif., August 1981.

SUGGESTED·READINGS

1. Hudson, Richard D., Jr. *Infrared System Engineering*. New York: John Wiley & Sons, 1969.

2. *IEEE Proceedings*, vol. 63, no. 1 (January 1975). Special issue on infrared technology for remote sensing.

3. Liao, Samuel Y. *Microwave Devices and Circuits*, Chapter 6. Englewood Cliffs, N.J.: Prentice-Hall, Inc., 1980.

PROBLEMS

9-2   Infrared Radiation

9-2-1. A military aircraft emits infrared radiation in a wavelength of 4 μm at a temperature of 1100°K.

  (a) Calculate the spectral radiant emittance of the aircraft in watts per square centimeter per micrometer.

  (b) Find the radiant emittance in watts per square centimeter.

  (c) Estimate the temperature in degrees Kelvin for the maximum spectral radiant emittance if the wavelength is fixed at 4 μm and determine the maximum spectral radiant emittance in watts per square centimeter per micrometer.

  (d) Determine the wavelength for maximum spectral radiant emittance if the temperature is fixed at 1100°K and compute the maximum radiation in watts per square centimeter per micrometer.

9-2-2. The irradiance H is the radiation incident on the surface of a detector and it is usually less than the radiant flux at the source due to the atmospheric absorption and the molecular scattering in the transmission path. If the loss is considered 30% and the radiant flux from the source is 20 W/cm^2, determine the irradiance H at a detector.

9-3   Infrared Radiation Sources

9-3-1. The blackbody source is usually used as the standard one for the calibration of infrared devices. It emits radiation into a hemisphere at 600°C with an emitting area of 150 cm^2.

  (a) Compute the radiant emittance of the blackbody.

  (b) Calculate the irradiance of the blackbody.

  (c) Find the radiant intensity.

  (d) Determine the radiance.

  (e) Determine the wavelength for maximum spectral radiant emittance.

9-5   Infrared Detectors

9-5-1. The quality and characteristics of an infrared detector are usually described by three parameters: responsivity (R), specific detectivity (D^*), and the time constant (τ). Describe the quality and characteristics of the three commonly used infrared detectors:

  (a) Mercury-doped germanium, Ge : Hg (photoconductive detector)

  (b) Mercury-cadmium telluride, Hg : Cd, Te (photoconductive detector)

  (c) Lead-tin telluride, Pb : Sn, Te (photovoltaic detector)

9-5-2. A certain infrared detector has the following parameters:

Detecting area	$A_d = 200 \times 10^{-4}$ cm^2
Signal voltage	$V_s = 2$ mV
Noise voltage	$V_n = 0.5$ pV
Incident radiant flux	$H = 1.5$ W/cm^2
Bandwidth	$\Delta f = 1$ Hz
Infrared wavelength	$\lambda_0 = 10$ μm

  (a) Determine the responsivity $R(f)$.

  (b) Find the noise equivalent power (NEP).

  (c) Calculate the detectivity D.

  (d) Compute the specific detectivity D^*.

9-6   Infrared System

9-6-1. An infrared imaging system is usually described by three parameters:
  (a) The modulation transfer function (MTF)
  (b) The minimum resolvable temperature (MRT)
  (c) The noise equivalent differential temperature (NEΔT)
  Describe the three parameters in detail.

9-7   Infrared System Measurements

9-7-1. One of the earliest experimental attempts to relate the threshold resolution to the visual discrimination of images of real scenes is attributed to Johnson. Explain Johnson's imaging model for thermal imaging measurements.

9-7-2. A square 4-bar pattern, 7-to-1 aspect ratio, is currently used as the standard pattern for measurement of the minimum resolvable temperature (MRT). The bar width of the 4-bar pattern is 5 mils and the focal length of the collimator is 50 in.
  (a) Find the angle subtended by the bar width in milliradians.
  (b) Determine the characteristic spatial frequency f_0 in cycles per milliradians.

9-7-3. The modulation transfer function (MTF), minimum resolvable temperature (MRT), and noise equivalent differential temperature (NEΔT) are the three most important parameters for measuring an infrared imaging system. Explain how to measure them in detail.

9-7-4. The modulation transfer function (MTF) of an infrared thermal imaging system is the product of the individual system's element transfer functions. In an infrared imaging system laboratory the differential brightness ΔB_1 decreased by the effect of the system transfer function, the magnitude of the variable object brightness B_1, and the dc or average level of the brightness B_0 responding to the characteristic spatial frequency f_0 are 0.1, 0.4, and 1, respectively. (Refer to Figs. 9-6-4 and 9-6-5.)
  (a) Determine the modulation of the object pattern at the input port of the system.
  (b) Find the modulation transfer function of the system at the characteristic spatial frequency f_0.

9-7-5. In an infrared imaging system laboratory the temperature of the source is reduced to such a level that a match is obtained between the background and target. The differential temperature ΔT, the noise voltage, and the ac signal voltage were measured to be 2°C, 2 V (rms), and 10 V (peak to peak), respectively.
  (a) Determine the noise-to-signal ratio for the system.
  (b) Calculate the noise equivalent differential temperature (NEΔT).

Chapter 10

Light-Emitting Diodes and Liquid Crystal Displays

10-0 INTRODUCTION

A *quantum wave* is defined as an electromagnetic radiation that is in the wavelength range of infrared, visible (optical), ultraviolet, and x rays and that travels in a vacuum with a speed of 2.998×10^8 m/sec. In general, solid-state quantum wave devices can be divided into two groups.

Group I. Light-Emitting Diodes and Liquid Crystal Displays. This group includes two types: active devices and passive devices. Active devices are those radiant devices that convert electric energy into a light wave; so they are referred to as *light-emitting diodes* (LEDs), such as infrared LEDs and visible LEDs. Passive devices are *liquid crystal displays* (LCDs) whose operations depend on ambient light for displays; so they are referred to as light-converting devices, such as digital display and infrared display.

Group II. Radiation Detectors. This group consists of *radiation detectors* that absorb the quantum radiation and convert it to electric energy. They are subdivided into photon detectors, thermal detectors, and photothermic detectors.

Light-emitting diodes and liquid crystal displays are often used commercially in digital watches and calculator displays and militarily in microwave electronic circuits. LCDs, however, are more effective and less expensive than LEDs. Radiation detectors are widely used both in military weapons and in commercial detec-

346

tions. In this chapter and Chapter 11 we investigate light-emitting diodes, liquid crystal displays, and the radiation-detecting devices, respectively. Table 10-0-1 lists the commonly used quantum wave devices.

TABLE 10-0-1   COMMONLY USED QUANTUM WAVE DEVICES

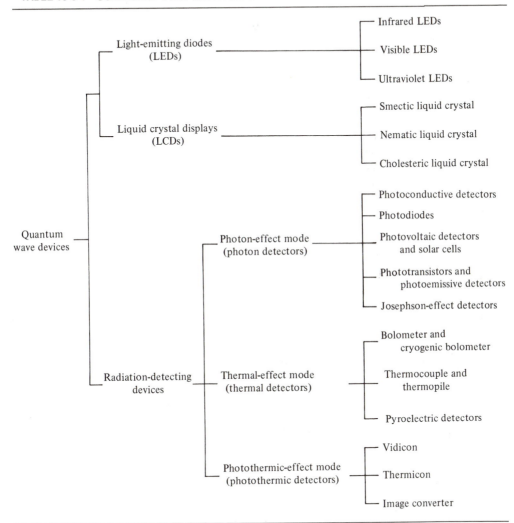

10-1   *LIGHT-EMITTING DIODES (LEDs)*

Light-emitting diodes (LEDs) are those compound semiconductor *p-n* junction diodes that emit spontaneous radiation under forward bias. The radiation energy determines the radiation frequency. According to solid-state theory, there are three transition processes for interaction between a photon and an electron in a semiconductor: absorption, spontaneous emission, and stimulated emission. When an excited electron in the conduction band returns to the valence band due to instability, the electron will give off a photon of energy in $h\nu$ by spontaneous emission. If an electron in the conduction band is stimulated by a photon of energy, the excited electron will return to the ground state and will emit two photons of energy by the stimulated-emission process. The relationship between the photon and the bandgap energy is given by

$$E_g = E_c - E_v = h\nu \qquad \text{eV} \qquad (10\text{-}1\text{-}1)$$

or

$$\lambda_0 = \frac{1.242}{E_g} \qquad \mu m$$

where E_g = bandgap energy in electronvolts

$\quad h = 6.625 \times 10^{-34}$ J-sec is Planck's constant

$\quad \nu$ = frequency in hertz

$\quad E_c$ = conduction band minimum (excited state)

$\quad E_v$ = valence band maximum (ground state)

$\quad \lambda_0$ = free space wavelength in micrometers

Equation (10-1-1) shows that the bandgap energy of a compound semiconductor determines the radiation frequency.

Example 10-1-0   Radiation of a GaAs LED

The binary compound gallium-arsenide diode (GaAs) has a bandgap energy of 1.43 eV.
(a) Determine its emitting wavelength.
(b) Identify its emitted light.

Solution.   (a) From Eq. (10-1-1) the emitted wavelength is

$$\lambda_0 = \frac{hc}{E_g} = \frac{6.625 \times 10^{-34} \times 3 \times 10^8}{1.43 \times 1.6 \times 10^{-19}} = 0.87 \, \mu m$$

(b) The emitted light is an infrared at 0.87 μm.

If the photon energy is less than the bandgap energy, no radiation occurs. In other words, the radiation is cut off when the wavelength in free space is greater than a certain value, as defined by

$$\lambda_{0c} > \frac{1.242}{E_g} \qquad \mu m \qquad \text{for cutoff} \qquad (10\text{-}1\text{-}2)$$

The radiation light of an LED is emitted randomly from the junction in all directions; so it is incoherent in phase and time. The LED light source is very useful in optical-fiber communications, however. This aspect was described earlier in Section 7-5.

10-1-1 LED Materials

Various compound semiconductor *p-n* junction diodes emit light radiation at optical frequencies in the infrared, visible light, or ultraviolet region of the electromagnetic spectrum. The wavelength of the light radiation is determined by the photon energy of the bandgap. Table 10-1-1 lists the commonly used materials of LEDs. It can be seen that LED materials include binary, ternary, and quaternary semiconductor compounds.

TABLE 10-1-1   MATERIALS OF LIGHT-EMITTING DIODES

Material		Bandgap E_g (eV)	Wavelength λ_0 (μm)	Light wave		
Scientific name	Symbol			Infrared	Visible	Ultra-violet
Indium antimonide	InSb	0.18	6.90	√		
Germanium	Ge	0.80	1.55	√		
Gallium-indium arsenide-phosphide	$Ga_xIn_{1-x}As_yP_{1-y}$ ($x = 0.28$, $y = 0.60$)	1.00	1.24	√		
Silicon	Si	1.12	1.11	√		
Gallium-indium arsenide-phosphide	$Ga_xIn_{1-x}As_yP_{1-y}$ ($x = 0.17$, $y = 0.34$)	1.14	1.10	√		
Gallium arsenide	GaAs	1.43	0.87	√		
Cadmium telluride	CdTe	1.50	0.83	√		
Cadmium selenide	CdSe	1.70	0.73	√		
Gallium arsenide-phosphide	$GaAs_{1-x}P_x$ ($x = 0.4$)	1.80	0.69		Red	
Gallium-aluminum arsenide	$Ga_xAl_{1-x}As$ ($x = 0.6$)	1.80	0.69		Red	
Gallium phosphide-zinc oxide	GaP:ZnO	1.91	0.65		Red	
Gallium arsenide-phosphide	$GaAs_{1-x}P_x$ ($x = 0.65$)	2.07	0.60		Orange	
Gallium arsenide-phosphide	$GaAs_{1-x}P_x$ ($x = 1.0$)	2.14	0.58		Yellow	
Gallium phosphide-nitride	GaP:N	2.22	0.56		Green	
Gallium phosphide	GaP	2.26	0.55		Green	
Zinc telluride	ZnTe	2.30	0.54		Green	
Cadmium sulphide	CdS	2.43	0.51		Green	
Zinc selenide	ZnSe	2.67	0.47		Blue	
Silicon carbide	SiC	2.90	0.43		Violet	
Gallium nitride	GaN	3.11	0.40			√
Zinc sulphide	ZnS	3.60	0.35			√

Generally the notation used is $AB_{1-x}C_x$ (or $D_xE_{1-x}F$) for ternary compounds, where A (or D and E) is the element of group III, B and C (or F) the elements of group V, and *x* the percentage of element atoms concerned. The symbol

$A_x B_{1-x} C_y D_{1-y}$ is used for quaternary compounds; here A and B refer to the elements of group III, C and D the elements of group V, and x and y the fractions of the element atoms, respectively.

Consider a ternary compound, such as gallium arsenide-phosphide, $GaAs_{1-x}P_x$. If $x = 0.4$, then $GaAs_{0.6}P_{0.4}$ will emit a photon energy of 1.80 eV at a red light of 0.69 μm. Several other ternary semiconductor compounds that emit light radiation are GaInP, GaSbP, and InAlAs.

10-1-2 Principles of Operation

Four processes can excite photon light from semiconductor junction diodes:

Intrinsic Process.   If the powder of a semiconductor, such as ZnS, is embedded in a dielectric like plastic or glass and subjected to an alternating electric field, light is emitted by the impact ionization of accelerated electrons from the trapping center.

Avalanche Process.   When a *p-n* junction is reverse biased into avalanche breakdown, light is generated by the impact ionization, as described earlier in Section 6-3-2.

Tunneling Process.   When a sufficiently large reverse bias is applied to a metal–semiconductor barrier, light is injected by the radiative recombination of holes with electrons.

Injection Process.   When a forward bias is applied to a *p-n* junction, the injection of minority carriers across the junction can radiate light. This process is an important one and is described later.

(1) Injection process.   Luminescence is the general property of light emission and it can be divided into four types, depending on the input energy and excitation methods.

1. Photoluminescence. The carriers are excited by photon absorption and the light radiation is emitted by the recombination of the excited carriers.
2. Cathodoluminescence. The carriers are excited by a high-energy electron beam or cathode ray.
3. Radioluminescence. The carriers are excited by other fast particles or high-energy radiation.
4. Electroluminescence. The carriers are excited by a forward-biased voltage or current.

As for electron behavior inside a semiconductor, electroluminescence is generated by different processes, such as intrinsic, avalanche, tunneling, and injection processes. When a forward bias is applied to the junction of an LED, minority carriers are injected into the region of the semiconductor. As a result, an optical radiation is emitted. In this section we are particularly concerned with the injection electroluminescence of LEDs.

A necessary condition for radiative recombination in most types of LEDs is the injection of minority carriers into the region of the semiconductor in which recombination may occur. The Fermi level E_F in a semiconductor is an important parameter and it is defined as the energy level at which the electron occupation probability is 50%. For a highly doped n-type semiconductor, the Fermi level E_F may lie above the conduction band minimum E_c. Similarly, for a p-type semiconductor, E_F is below the valence band maximum E_v. When two adjacent regions of a semiconductor are doped p and n types, electrons initially flow from the n region to the p region and holes flow from the p region to the n region until a sufficient electrostatic potential is established across the interfacial region to prevent the further net flow of charges. Under equilibrium conditions (zero bias) Fermi levels in both regions are equal. As a result, a depletion region W is created as shown in Fig. 10-1-1.

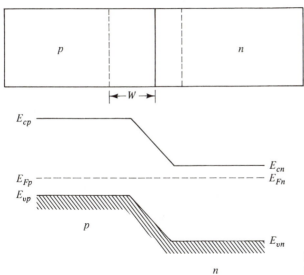

Figure 10-1-1 Injection of minority carriers

(2) Voltage-current equation.   The magnitude of current flowing through a semiconductor p-n junction diode determines the light intensity of an LED. In general, there are three currents in a p-n junction diode: tunnel current, space-charge recombination current, and diffusion current. In most practical LEDs the tunnel current is negligible, the space-charge recombination current is small when the bias voltage is high enough, but the diffusion current will always dominate at a sufficiently high bias.

When a forward-biased voltage V is applied across a p-n junction, the potential energy of an electron in the n-type region is increased and the barrier height $(V_0 - V)$ is decreased. As soon as an overlap of electrons and holes occurs, their recombination takes place via the injection process. As a result, electrons from the n-type side of the material will diffuse across the junction to the p side and recombine with the majority carriers (holes). On the average, the minority electrons will have a lifetime τ_e and will travel a diffusion length L_e before recombining. The diffusion length is given by [1]

$$L_e = \sqrt{D_e \tau_e} \quad \text{in centimeters} \tag{10-1-3}$$

where $D_e = \mu_e \dfrac{kT}{e}$ is the electron diffusion coefficient or Einstein's relationship in centimeters squared per second

$\mu_e =$ electron mobility in centimeters squared per volt-second

Similarly, holes will diffuse across the junction into the n-type side and both diffusion currents will contribute to the total junction current.

The voltage-current equation is then given by

$$I = I_0 \left[\exp\left(\frac{eV}{kT} \right) - 1 \right] \tag{10-1-4}$$

where $I_0 = eA \left(p_n \dfrac{D_p}{L_p} + n_p \dfrac{D_n}{L_n} \right)$ is the reverse saturation current

$e =$ 1.6×10^{-19} coulomb is the magnitude of electron charge

$V =$ applied voltage in volts

$\dfrac{kT}{e} =$ 26 mV at room temperature is the volt equivalent of temperature

$T =$ absolute temperature in degrees Kelvin

$k =$ 1.38×10^{-23} watt-second per degrees Kelvin is Boltzmann's constant

$L = \sqrt{D\tau}$ is the diffusion length in centimeters

$D = \mu \dfrac{kT}{e}$ is the diffusion coefficient or Einstein's relationship in centimeters squared per second

$\tau =$ lifetime of electron or hole in seconds

$\mu =$ electron or hole mobility in centimeters squared per volt-second

$A =$ cross-sectional area of the junction in centimeters squared

$p_n =$ equilibrium concentration of minority holes in the n-type material

$n_p =$ equilibrium concentration of minority electrons in the p-type material

GaAs (gallium arsenide) is a direct-bandgap material, but Si (silicon) and Ge (germanium) are indirect-bandgap semiconductors. Typically bulk GaAs has a background impurity level of 10^{16} cm^{-3}. This condition is due to contamination by silicon and gives rise to n-type material with a resistivity of 1 Ω-cm. At room temperature the intrinsic carrier density n_i of GaAs is low, 10^6 cm^{-3} compared to the device's high doping levels. However, n_i increases rapidly with temperature, doubling for every 10°C. Figure 10-1-2 shows the intrinsic carrier densities of GaAs, Si, and Ge as a function of reciprocal absolute temperature T [7].

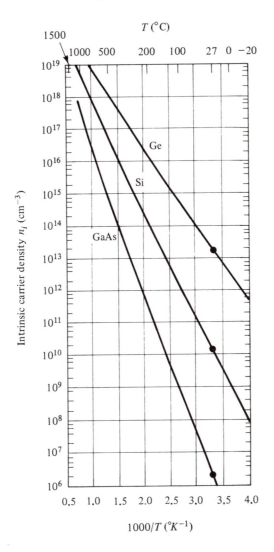

Figure 10-1-2 Intrinsic carrier densities n_i of GaAs, Si, and Ge as a function of reciprocal absolute temperature T (From S. M. Sze [7]; reprinted by permission of John Wiley & Sons, Ltd.)

At high temperatures, thermal generation is the dominant factor of carrier generation. This thermal effect becomes equal to the background concentration at the intrinsic temperature T_i. Figure 10-1-3 shows the curves of intrinsic temperature T_i for GaAs, Si, and Ge as a function of background concentration [7]. It has been shown experimentally that the carrier concentration is relatively temperature independent below T_i, and that above T_i it rises exponentially with temperature.

The mobilities of carriers in semiconductors decrease with temperature, the effective mass, and the impurity concentration as shown in Appendix 7.

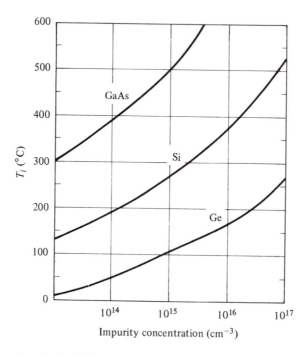

Figure 10-1-3 Intrinsic temperature T_i for GaAs, Si, and Ge as a function of background concentration (From S. M. Sze [7]; reprinted by permission of John Wiley & Sons, Ltd.)

Example 10-1-2   Currents of a LED

The light intensity of a LED depends on the magnitude of the current generated by the incident photons. An *n*-type GaAs has an intrinsic concentration n_i of 10^{10} cm^{-3} at a temperature of 416°K. The electron carrier concentration N_d is 10^{18} cm^{-3}, the cross section A of the absorbing area is 1 cm^2, and the carrier lifetime τ_p is equal to τ_n in 1 nsec.

(a) Determine the diffusion coefficients of the carriers.
(b) Find the diffusion lengths of the carriers.
(c) Calculate the reverse saturation current.
(d) Compute the current flow in the GaAs LED.

Solution.   (a) From Eq. (10-1-4) the carrier diffusion coefficients are

$$D_n = \mu_n \frac{kT}{e} = 2600 \times \frac{1.38 \times 10^{-23} \times 416}{1.6 \times 10^{-19}}$$

$$= 93.29 \text{ cm}^2/\text{sec}$$

$$D_p = \mu_p \frac{kT}{e} = 180 \times \frac{1.38 \times 10^{-23} \times 416}{1.6 \times 10^{-19}}$$

$$= 6.46 \text{ cm}^2/\text{sec}$$

(b) From Eq. (10-1-4) the carrier diffusion lengths can be computed as

$$L_n = \sqrt{D_n \tau_n} = \sqrt{93.3 \times 10^{-9}} = 3.05 \times 10^{-4} \text{ cm} = 3.05 \; \mu\text{m}$$

$$L_p = \sqrt{D_p \tau_p} = \sqrt{6.46 \times 10^{-9}} = 0.80 \times 10^{-4} \text{ cm} = 0.80 \; \mu\text{m}$$

(c) Also from Eq. (10-1-4) the equilibrium concentrations are

$$p_n = n_p = \frac{n_i^2}{N_d} = \frac{(10^{10})^2}{10^{18}} = 100 \text{ cm}^{-3}$$

Then the reverse saturation current is

$$I_0 = 1.6 \times 10^{-19} \times 1 \times \left(10^2 \times \frac{6.46}{0.8 \times 10^{-4}} + 10^2 \times \frac{93.29}{3.05 \times 10^{-4}}\right)$$

$$= 1.6 \times 10^{-19} \times 10^6 (8.08 + 30.59)$$

$$= 6.19 \ \mu\mu\text{A}$$

(d) Finally the current flowing in the GaAs LED is

$$I = I_0 \left[\exp\left(\frac{eV}{2kT}\right) - 1\right]$$

$$= 6.19 \times 10^{-12} \times \left[\exp\left(\frac{1.43}{2 \times 36 \times 10^{-3}}\right) - 1\right]$$

$$= 6.19 \times 10^{-12} \times [\exp(19.93) - 1]$$

$$= 6.19 \times 10^{-12} \times [4.52 \times 10^8 - 1]$$

$$= 2.798 \text{ mA}$$

(3) Recombination rate.   Under thermal equilibrium conditions the rate R_{vc} for electrons being thermally excited from the valence band to the conduction band is equal to the rate R_{cv} for electrons returning from the conduction band to the valence band. Then they are related by

$$R_{vc} = R_{cv} = Bn_0 p_0 \qquad (10\text{-}1\text{-}5)$$

where $B = 7.21 \times 10^{-10}$ cm^3/sec is the recombination constant for GaAs
n_0 = electron concentration in equilibrium
p_0 = hole concentration in equilibrium

If electrons are injected into the p-type region at a rate R_n, the steady-state excess electrons on the p side are

$$\Delta n_p = \frac{R_n}{B(n+p)} = R_n \tau_e = n_p \left[\exp\left(\frac{eV}{\eta kT}\right) - 1\right] \qquad (10\text{-}1\text{-}6)$$

where $\tau_e = 10^{-9}$ sec for GaAs with hole concentration of 10^{18} cm^{-3} is the lifetime of the excess electron.
Similarly, for excess holes on the n side, it is

$$\Delta p_n = \frac{R_p}{B(n+p)} = R_p \tau_p = p_n \left[\exp\left(\frac{eV}{\eta kT}\right) - 1\right] \qquad (10\text{-}1\text{-}7)$$

(4) Quantum efficiency.   The quantum efficiency of a LED usually refers to the internal quantum efficiency of the device, for the overall efficiency (often referred to as the external efficiency) is much lower than the internal efficiency due to various losses from the inside to the outside of the device.

The probability for the radiative recombination rate R_r is a reciprocal of the radiative lifetime τ_r. Similarly, the probability of the nonradiative recombination rate R_{nr} is the reciprocal of the nonradiative lifetime τ_{nr}. For a given input excitation energy, the quantum efficiency η_q is defined as the ratio of the radiative recombination rate R_r to the total recombination rate R, where R is the sum of R_r and R_{nr}. Thus the quantum efficiency is expressed in terms of their lifetimes as

$$\eta_q = \frac{R_r}{R} = \frac{\tau_{nr}}{\tau_{nr} + \tau_r} = \frac{\tau}{\tau_r} = 1 - \frac{\tau}{\tau_{nr}} \tag{10-1-8}$$

where $R_r =$ radiative recombination rate
$R_{nr} =$ nonradiative recombination rate
$R = R_r + R_{nr}$ is the total recombination rate
$\tau_r =$ radiative lifetime
$\tau_{nr} =$ nonradiative lifetime
$\tau = \dfrac{\tau_r \tau_{nr}}{\tau_r + \tau_{nr}}$ is the minority-carrier lifetime

The total recombination rate and lifetime are related to the n-type and p-type semiconductors, respectively, by

$$R = \frac{n - n_0}{\tau} \tag{10-1-9}$$

and

$$R = \frac{p - p_0}{\tau} \tag{10-1-10}$$

where n_0 and p_0 are the electron and hole concentrations under the thermal equilibrium condition and n and p are the electron and hole concentrations in the optical excitation condition.

In order to achieve a high quantum efficiency, the nonradiative recombination rate must be made as low as possible so as to increase τ_{nr}. In other words, the radiative lifetime must be as small as possible. This effect can be achieved by using a perfectly clean fabrication process and very low imperfections in the crystal.

Example 10-1-2A   Recombination Rate in GaAs

The intrinsic carrier density of a GaAs material is 10^6 cm^{-3} and the hole concentration of the p-type GaAs is 10^{18} cm^{-3} at room temperature.
(a) Determine the recombination rate under a thermal equilibrium condition.
(b) Estimate the lifetime of the excessive electrons in the p side.

Solution.   (a) From Eq. (10-1-5) the recombination rate under an equilibrium condition is

$$R_{vc} = R_{cv} = Bn_0 p_0 = 7.21 \times 10^{-10} \times 10^6 \times 10^6$$

$$= 721 \text{ sec}^{-1} \text{ cm}^{-3}$$

(b) The lifetime of the excessive electrons is

$$\tau_e = \frac{1}{B(n+p)} \simeq \frac{1}{Bp} = \frac{1}{7.21 \times 10^{-10} \times 10^{18}}$$

$$= 1.39 \text{ nsec} \qquad \text{for } p \gg n$$

10-1-3 Infrared LEDs

The infrared wavelength ranges from 0.70 to 1000 μm as shown in Section 9-1. Some commonly used infrared LEDs are listed in Table 10-1-1. At present, the binary compound gallium arsenide (GaAs) is one of the most important LEDs with the highest quantum efficiency. This LED has a direct bandgap energy of 1.43 eV and it emits an infrared at a wavelength of 0.87 μm. This infrared source is very useful in optoisolators and communications. Figure 10-1-4 shows the bandgap energy and electron mobility of the GaAs LED as a function of temperature [2].

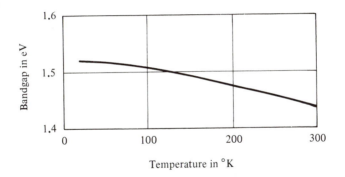

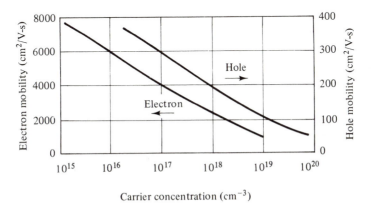

Figure 10-1-4 Bandgap energy and electron mobility of GaAs LED (From C. H. Gooch [2]; reprinted by permission of John Wiley & Sons, Ltd.)

Another very useful infrared source is the quaternary compound gallium-indium arsenide-phosphide, $Ga_xIn_{1-x}As_yP_{1-y}$. If $x = 0.28$ and $y = 0.6$, the LED $Ga_{0.28}In_{0.72}As_{0.6}P_{0.4}$ has an indirect bandgap energy of 1.00 eV and it emits an infrared at 1.24 μm. If $x = 0.17$ and $y = 0.34$, the infrared radiation is in the wavelength of 1.10 μm. These infrared sources are suitable for use in optical-fiber communications because the fiber losses are low at a wavelength of 1.3 μm.

10-1-4 Visible LEDs

Visible light wavelengths range between 0.43 and 0.70 μm as shown in Section 9.1. Some widely used visible LEDs are tabulated in Table 10-1-1.

The binary compound gallium phosphide, GaP, has an indirect bandgap energy of 2.26 eV and it emits a green light at a wavelength of 0.55 μm. Its efficiency is about 0.1%. Figure 10-1-5 shows the bandgap energy and electron mobility of the GaP LED as a function of temperature [2].

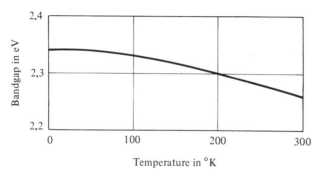

(a) Bandgap energy of GaP LED

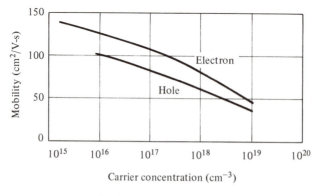

(b) Electron and hole mobility in
      GaP at 300°K

Figure 10-1-5 Bandgap energy and mobility of GaP LED (After C. H. Gooch [2]; reprinted by permission of John Wiley & Sons, Ltd.)

The ternary compound gallium arsenide-phosphide, $GaAs_{1-x}P_x$, emits a red, orange, or a green light, depending on the percentage of x for the element concerned (see Table 10-1-1). The bandgap energy and efficiency of $GaAs_{1-x}P_x$ LEDs are shown in Fig. 10-1-6 [2]. If $x = 0.40$, the $GaAs_{0.6}P_{0.4}$ is a red LED. If $x = 1.0$, it is a yellow LED. Its quantum efficiency is about 0.1%. These red LEDs are commonly used as displays in calculators and digital watches.

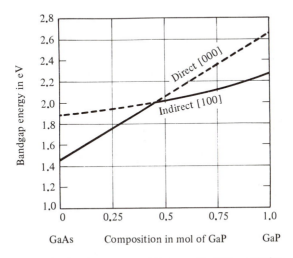

(a) Bandgap energy of $GaAs_{1-x}P_x$ LEDs at 300°K

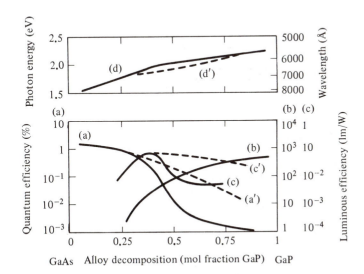

(a) Quantum efficiency
(b) Luminous efficiency of radiation
(c) Luminous efficiency of diode
(d) Photon energy curves (a'), (c') and (d') represent results obtained with nitrogen doping

(b) Quantum efficiency and photon energy of $GaAs_{1-x}P_x$

Figure 10-1-6 Properties of $GaAs_{1-x}P_x$ LEDs (After C. H. Gooch [2]; reprinted by permission of John Wiley & Sons, Ltd.)

The red LED, such as $GaAs_{1-x}P_x$, which has a direct bandgap, is fabricated on GaAs substrates and the other colored LEDs (orange, yellow, and green) are grown on GaP substrates. Their structures are shown in Fig. 10-1-7 [3]. It can be seen that radiation photons are emitted in all directions and only a small fraction of these photons can reach the eyes of an observer.

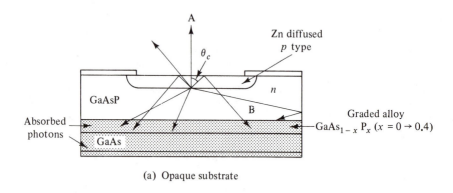

(a) Opaque substrate

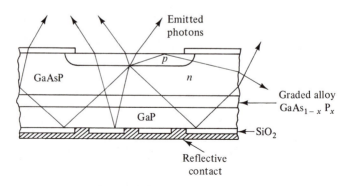

(b) Transparent substrate

Figure 10-1-7 Structures of $GaAs_{1-x}P_x$ LEDs (From G. Gage et al. [3]; reproduced by permission of McGraw Hill Book Company.)

10-1-5 Optical Applications

Light-emitting diodes possess a number of special features, such as small size, high reliability, and high luminescence. Consequently, they are used in modulators, displays, lamps, communications, high-speed logic circuits, and optical radar systems.

Visible LEDs are commonly used as displays in calculators and as optoisolators. Figure 10-1-8 shows a basic format for LED displays [1]. The 14 segments and the 5×7 matrix array are used to display alphameric formats from A to Z and 0 to 9. The 7 segments and the 3×5 array are used in numeric displays.

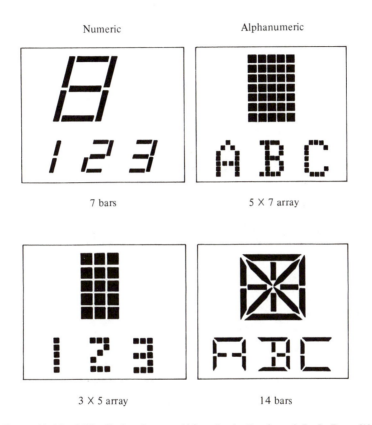

Figure 10-1-8 LED display formats (After A. A. Bergh and P. J. Dean [1]; reprinted by permission of Oxford University Press.)

Visible LED can also be used in optoisolators, where the input signal is isolated from the output signal, because there is no feedback from the output to the input signal. Figure 10-1-9 shows a diagram for an optoisolator [7]. When an input electrical signal is applied to the LED, the photon light emitted from the LED is detected by the photodiode and the light is then converted back to an electrical signal flowing through the load resistor.

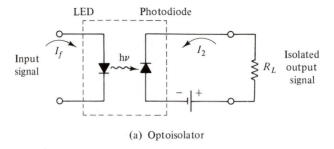

(a) Optoisolator

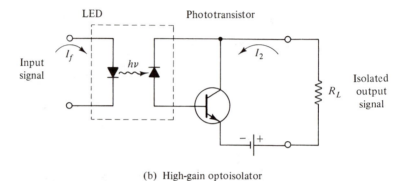

(b) High-gain optoisolator

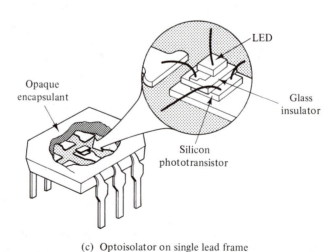

(c) Optoisolator on single lead frame

Figure 10-1-9 Circuit diagrams for optoisolators (After S. M. Sze [7]; reprinted by permission of John Wiley & Sons, Ltd.)

Radar and communications systems are often operated at optical frequencies. An optical radar system consists of a transmitter and a receiver (see Fig. 10-1-10) [2]. The modulator of the transmitter drives the LED and an optical system is used to collimate the radiation light for propagation. In the receiver unit an optical detector converts the light radiation to the electrical signal for electronic processes and displays.

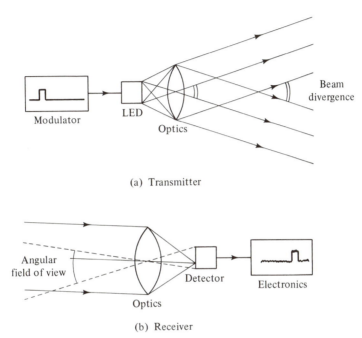

(a) Transmitter

(b) Receiver

Figure 10-1-10   Block diagram of an optical radar system (After C. H. Gooch [2]; reprinted by permission of John Wiley & Sons, Ltd.)

10-2 *LIQUID CRYSTAL DISPLAYS (LCDs)*

All light-emitting diodes (LEDs) are active devices, but the liquid crystal displays (LCDs) (or light-converting devices) are passive devices, for their operation depends on the presence of ambient light at a level adequate for human vision.

Although liquid crystals have been in existence for a century, it was only in the 1970s that their unique optical, electrooptic, and thermal properties were exploited for use in digital (or alphameric) displays, such as in digital watches, analog displays, image converters, and matrix-type picture screens.

The power consumption of a LCD depends on the surface area of its active part and it is extremely low. The liquid crystal usually has a very high resistivity, in the range of 10^8 to 10^{10} Ω-cm. For an electric field-effect display, a typical power dissipation is 10 μW/cm². For a 4-digit 7-segment display, the power consumption is 15 mW for LCDs compared to 550 mW for LEDs.

The LCDs are particularly useful in applications where high switching speeds and compatibility with integrated circuits are minor considerations whereas low power consumption and high light ambients are major ones.

10-2-1 Types of Liquid Crystals

Liquid crystals consist of liquid organic compounds with sugar-shaped polar molecules and spontaneous anisotropy. There are three basic types of liquid crystals: smectic, nematic, and cholesteric structures (see Fig. 10-2-1).

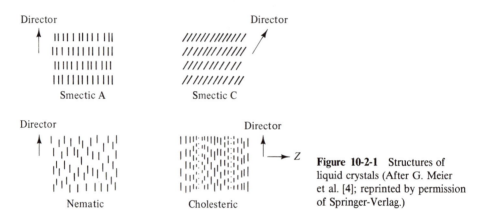

Figure 10-2-1 Structures of liquid crystals (After G. Meier et al. [4]; reprinted by permission of Springer-Verlag.)

(1) Smectic Liquid Crystals.   All smectic liquid crystals have a layer structure. The centers of gravity of the elongated molecules are aligned in the equidistance plane. The location of the centers of gravity within the plane may be random or regular. The long axes of the molecules are parallel to a preferred direction that may be normal to the planes (smectic A) or tilted by an angle (smectic C). Up to now smectic liquid crystals are little used.

(2) Nematic Liquid Crystals.   These liquid crystals have long-range orientational order and their molecular axes align parallel to a preferred direction. The centers of gravity are distributed at random and the molecules rotate freely about their long axes. Due to their physical structures, nematic liquid crystals are uniaxial with respect to all physical properties. The axis of symmetry is identical with the preferred axis of the structure. Nematic liquid crystals are very useful for electro-optical applications.

(3) Cholesteric Liquid Crystals.   These liquid crystals lack long-range translational order. On a local scale, cholesteric and nematic orderings are similar. On a larger scale, however, the cholesteric director **n** follows a helical path as shown in Fig. 10-2-2(a).

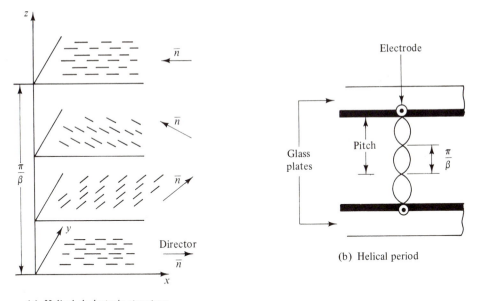

(a) Helical cholesteric structure

(b) Helical period

Figure 10-2-2   Helical rotation of cholesteric order

The cholesteric director **n** can be expressed in three directions as

$$n_x = \cos(\beta z + \phi) \tag{10-2-1}$$

$$n_y = \sin(\beta z + \phi) \tag{10-2-2}$$

$$n_z = 0 \tag{10-2-3}$$

where z = helical axis
    β = phase angle
    ϕ = arbitrary phase angle

Thus the structure of a cholesteric liquid crystal is periodic with a spatial period L given by

$$L = \frac{\pi}{|\beta|} \tag{10-2-4}$$

The nominal pitch P of a helical structure is equal to 2π. Therefore the spatial period L equals one-half of a pitch as shown in Fig. 10-2-2(b). The sign of the phase constant β may be positive or negative, corresponding to a right or left helix. When the phase constant β is zero, the period L is infinite. In this case, the cholesteric liquid crystal is operating in nematic structure or mode. When the period L is zero (or $\beta = \infty$), the cholesteric liquid crystal is in the typical cholesteric phase.

When the period L is approaching optical wavelengths, the periodicity results in strong scattering light. If the wavelength of the scattered light is in the visible light range, the cholesteric liquid crystal will exhibit a bright-colored light.

10-2-2 Operational Principles of Liquid Crystals

When a liquid crystal is inserted in an electric field, it experiences a combination of elastic, dielectric, and electrical forces. The elastic force produces splay, twist, and bend effects in the liquid crystal. The dielectric and electrical forces introduce electrooptic deformations of the liquid crystal. These effects play an important role in liquid crystal applications.

(1) Elastic effect. Liquid crystals exhibit elastic properties in an electric field. Any distortion of the undisturbed state requires a certain amount of energy because elastic torques attempt to maintain the original configuration. Three elastic deformations, such as splay, twist, and bend, of a nematic liquid crystal are shown in Fig. 10-2-3. The letters k_{11}, k_{22}, and k_{33} in the figure represent the elastic constants for splay, twist, and bend deformations, respectively. The values of the elastic constants are 10^{-10} to 10^{-12} N (newton) (or 10^{-5} to 10^{-7} dyne). (*Note:* 1 newton = 10^5 dynes.)

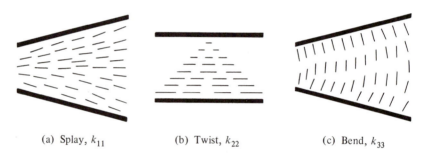

(a) Splay, k_{11} (b) Twist, k_{22} (c) Bend, k_{33}

Figure 10-2-3 Elastic deformations of a nematic liquid crystal (After G. Meier et al. [4]; reprinted by permission of Springer-Verlag.)

(2) Dielectric effect. In general, most crystals are dielectric in nature and do not carry electric current. The liquid crystals do carry current, however, when a certain level of electric voltage is applied to them; so they have a sizable electrooptic effect. The liquid crystals are uniaxial dielectric types. An uniaxial dielectric crystal is defined as $\epsilon_{r11} = \epsilon_{r22} \neq \epsilon_{r33}$ in Section 2-2-1. If the nematic axis of an uniaxial nematic liquid crystal is chosen in the z direction, the relative dielectric tensor of the nematic liquid crystal can be expressed as

$$\hat{\epsilon} = \begin{bmatrix} \epsilon_{\perp} & 0 & 0 \\ 0 & \epsilon_{\perp} & 0 \\ 0 & 0 & \epsilon_{\parallel} \end{bmatrix} \tag{10-2-5}$$

Then the relative dielectric anisotropy is given by

$$\Delta\epsilon = \epsilon_{\parallel} - \epsilon_{\perp} \tag{10-2-6}$$

where $\epsilon_\perp$ = relative dielectric constant perpendicular to the nematic axis (or in the x
       and y directions)
   $\epsilon_\|$ = relative dielectric constant parallel to the nematic axis (or in the z
       direction)

The value of $\Delta\epsilon$ is the differential relative dielectric constant between the parallel
and perpendicular directions and it varies from negative 10 to positive 10.

There are three configurations for dielectric-effect deformations in a nematic
liquid crystal: parallel orientation, perpendicular orientation, and twist orientation
as shown in Fig. 10-2-4 [4]. The optic axis of a liquid crystal can be uniformly
aligned by a special treatment of the electrode surfaces.

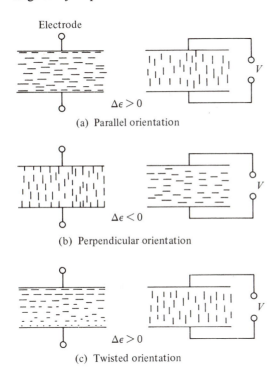

(a) Parallel orientation

(b) Perpendicular orientation

(c) Twisted orientation

Figure 10-2-4 Dielectric deformations in a
nematic liquid crystal (After G. Meier et al.
[4]; reprinted by permission of
Springer-Verlag.)

Parallel Orientation.   The undisturbed optic axis of the liquid crystal is
aligned parallel to the surface of the electrodes as shown in Fig. 10-2-4(a). The
relative dielectric anisotropy $\Delta\epsilon$ is assumed to be positive. When a voltage is applied
across the crystal, the electric field force exerts a torque on each molecule so that the
molecule tends to align its optic axis along the field and perpendicular to the
electrodes. Below a critical threshold voltage V_{th}, the restoring elastic torque is large
enough to prevent deformation. The threshold voltage V_{th} in cgs units is given by

$$V_{\text{th}} = \pi\sqrt{\frac{k_{11}}{\epsilon_0 \,\Delta\epsilon}} \quad \text{stat-volts (cgs)} \tag{10-2-7}$$

where $k_{11} =$ splay elastic constant in dynes
$\quad \epsilon_0 = $ 1 esu (electrostatic unit) is the free-space permittivity
$\quad \Delta\epsilon = \epsilon_{\parallel} - \epsilon_{\perp}$ is the relative dielectric anisotropy

The threshold voltage V_{th} in mks units can be written

$$V_{th} = \sqrt{\frac{\pi k_{11}}{4\epsilon_0 \Delta\epsilon}} \quad \text{volts (mks)} \tag{10-2-8}$$

where $k_{11} = $ splay elastic constant in newtons
$\quad \epsilon_0 = 8.854 \times 10^{-12}$ F/m is the free-space permittivity

Several conversion factors are available in changing from one unit system to another. For instance, 1 F/m $= 4\pi \times 10^{-7} c^2$ esu, where $c = 3 \times 10^8$ m/sec, 1 V $= 10^6/c$ stat-V, 1 stat-V $= 3 \times 10^2$ V, 1 newton $= 10^5$ dynes, and a factor of 4π comes from changing cgs units to mks units via Coulomb's law.

Example 10-2-2   Computation of Threshold Voltage for LCD

A nematic liquid crystal has a splay elastic constant k_{11} of 8×10^{-11} newton (N) and its relative dielectric anisotropy $\Delta\epsilon$ is 5. Compute the threshold voltages from Eqs. (10-2-7) and (10-2-8) and compare the results.

Solution.   1. From Eq. (10-2-7) we have

$$V_{th} = 3.1416 \times \left(\frac{8 \times 10^{-11} \times 10^5}{1 \times 5} \right)^{1/2}$$

$$= 3.97 \times 10^{-3} \text{ stat-volt}$$

2. From Eq. (10-2-8) we obtain

$$V_{th} = \left[\frac{3.1416}{4} \times 8 \times \frac{10^{-11}}{8.854 \times 10^{-12} \times 5} \right]^{1/2}$$

$$= 1.19 \text{ volts}$$

3. 3.97×10^{-3} stat-volt is equivalent to 1.19 volts. So the results are the same.

Typical values for k_{11} are from 10^{-10} to 5×10^{-12} N. Threshold voltages for the parallel orientation usually vary from 3 to 6 V.

Perpendicular Orientation.   The optic axis of the nematic liquid crystal is aligned perpendicular to the electrodes when the relative dielectric anisotropy is negative as shown in Fig. 10-2-4(b). The threshold voltage in cgs units is expressed as

$$V_{th} = \pi \sqrt{\frac{k_{33}}{\epsilon_0 |\Delta\epsilon|}} \quad \text{stat-volts (cgs)} \tag{10-2-9}$$

where k_{33} is the bend elastic constant in dynes. The threshold voltage in mks units can be given by

$$V_{th} = \sqrt{\frac{\pi k_{33}}{4\epsilon_0 |\Delta\epsilon|}} \qquad \text{volts (mks)} \qquad (10\text{-}2\text{-}10)$$

where k_{33} is the bend elastic constant in newtons. Typical values for k_{33} are from 10^{-10} to 3×10^{-12} N. Threshold voltages for the perpendicular orientation usually vary from 3 to 6 V.

Twist Cell. The nematic liquid crystal is continuously twisted 90° by the specially treated surface force of the electrodes so that the optic axis of the liquid crystal is aligned parallel to the electrodes as shown in Fig. 10-2-4(c). As a result, the preferred directions on the two electrodes are at right angles to each other. The relative dielectric anisotropy $\Delta\epsilon$ is assumed to be positive. The threshold voltage for deformation in cgs units becomes

$$V_{th} = \frac{\pi}{2} \left(\frac{4k_{11} + k_{33} - 2k_{22}}{\epsilon_0 \Delta\epsilon} \right)^{1/2} \qquad \text{stat-volts (cgs)} \qquad (10\text{-}2\text{-}11)$$

where k_{22} is the twist elastic constant in dynes. The threshold voltage in mks units can be expressed as

$$V_{th} = \frac{\pi}{4} \left(\frac{4k_{11} + k_{33} - 2k_{22}}{\pi\epsilon_0 \Delta\epsilon} \right)^{1/2} \qquad \text{volts (mks)} \qquad (10\text{-}2\text{-}12)$$

where k_{22} is the twist elastic constant in newtons. Typical values for k_{22} are from 10^{-10} to 7×10^{-12} N. Threshold voltages for the twisted cell usually vary from 1 to 5 V.

Example 10-2-2A   Threshold Voltage of Twisted Liquid Crystals

A twist cell has the following parameters:

Relative dielectric anisotropy	$\Delta\epsilon = +4.5$
Splay elastic constant	$k_{11} = 5 \times 10^{-11}$ newton (N)
Twist elastic constant	$k_{22} = 7 \times 10^{-11}$ N
Bend elastic constant	$k_{33} = 8 \times 10^{-10}$ N

Determine the threshold voltage of the twist cell.

Solution. From Eq. (10-2-12) the threshold voltage is

$$V_{th} = \frac{3.1416}{4} \left(\frac{4 \times 5 \times 10^{-11} + 8 \times 10^{-10} - 2 \times 7 \times 10^{-11}}{3.1416 \times 8.854 \times 10^{-12} \times 4} \right)^{1/2}$$

$$= 2.18 \text{ volts}$$

(3) Electrooptic effect. A new electrooptic property of the liquid crystal has been discovered in a twisted nematic liquid crystal cell. A homogeneously aligned film of a nematic liquid crystal with a positive relative dielectric anisotropy is sandwiched between two transparent conducting electrodes as shown in Fig. 10-2-5(a). The thickness of the nematic liquid crystal film is in the range of 10 to 20 μm. Part (b) of Fig. 10-2-5 illustrates its equivalent circuit.

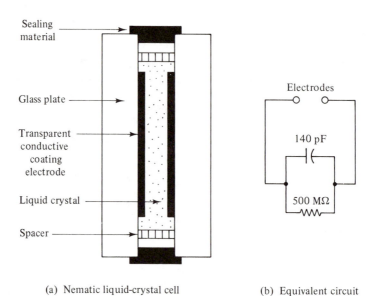

(a) Nematic liquid-crystal cell (b) Equivalent circuit

Figure 10-2-5 Schematic diagram of a nematic liquid-crystal cell with its equivalent circuit

There are two types of twisted nematic liquid crystal displays, depending on the orientations of the two glass plates or polarizers: two crossed polarizers and two parallel polarizers.

Two Crossed Polarizers. A 90° twist is imposed on the liquid crystal by turning one of the two glass plates in its own plane about an axis perpendicular to the nematic crystal film as shown in Fig. 10-2-6(a). The polarization vector of light transmitted through this cell follows the twisted structure. When an electric voltage is applied, a deformation occurs above a threshold voltage given by

$$V_{th} = \left[\frac{\pi^2 k_{11} + \phi^2 (k_{33} - 2k_{22})}{4\pi\epsilon_0 \Delta\epsilon} \right]^{1/2} \text{ volts (mks)} \qquad (10\text{-}2\text{-}13)$$

where ϕ is the twisted angle. When the twisted angle ϕ is 90°, Eq. (10-2-13) is identical to Eq. (10-2-12).

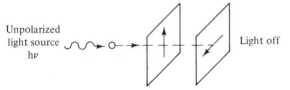

(a) Two crossed polarizers

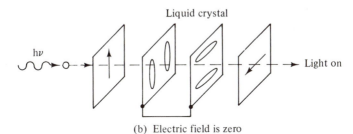

(b) Electric field is zero

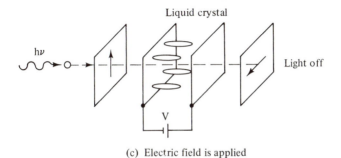

(c) Electric field is applied

Figure 10-2-6   Schematic diagrams of a nematic liquid-crystal cell with two crossed polarizers

Example 10-2-2B   Threshold Voltage of Crossed Polarizers

A twisted nematic liquid crystal cell between two crossed polarizers has the following parameters:

Relative dielectric anisotropy	$\Delta\epsilon = +4$
Splay elastic constant	$k_{11} = 5 \times 10^{-10}$ N
Twist elastic constant	$k_{22} = 5 \times 10^{-11}$ N
Bend elastic constant	$k_{33} = 5 \times 10^{-10}$ N
Twisted angle	$\phi = 45°$

Determine the threshold voltage of the twisted cell.

Solution.   Using Eq. (10-2-13), we see that the threshold voltage is

$$V_{th} = \left[\frac{(3.14)^2 \times 5 \times 10^{-10} + (3.14/4)^2 \times (5 \times 10^{-10} - 2 \times 5 \times 10^{-11})}{4 \times 3.1416 \times 8.854 \times 10^{-12} \times 4} \right]^{1/2}$$

$$= 3.41 \text{ volts}$$

Applying an electric field to a liquid crystal cell causes the liquid crystal to deform. If the relative dielectric anisotropy is positive ($\Delta\epsilon > 0$), the director tends to align in the direction of the electric field. If the relative dielectric anisotropy is negative ($\Delta\epsilon < 0$), the director aligns perpendicular to the applied electric field. The relative dielectric anisotropy plays an important role in electrooptic applications of the liquid crystal.

Linearly polarized light that is propagating perpendicular to the twisted nematic cell is rotated by 90° as it passes through the liquid crystal when there is no applied voltage as shown in Fig. 10-2-6(b). When the applied voltage is larger than the threshold voltage, the electric field causes the nematic director to become untwisted and to align parallel to the applied field. Consequently, light is extinguished as shown in Fig. 10-2-6(c).

Two Parallel Polarizers.   Light between two parallel polarizers is extinguished when the electric field is off, but when the electric field is switched on, the molecules in the liquid crystal will orient themselves perpendicular to the glass plates and light will be transmitted. This case is illustrated in Fig. 10-2-7.

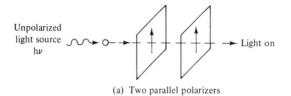

(a) Two parallel polarizers

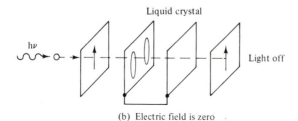

(b) Electric field is zero

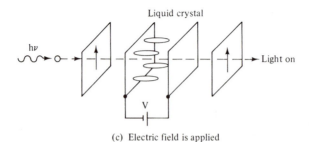

(c) Electric field is applied

Figure 10-2-7   Schematic diagrams of a nematic liquid-crystal cell with two parallel polarizers

Light transmission or extinction depends on the orientation of the two polarizers; so either a black-on-white or a white-on-black display can be achieved.

In effect, these two switching modes are complementary to each other. Figure 10-2-8 shows the light transmission for a twisted nematic liquid crystal with either two crossed or two parallel polarizers [5]. Curve *A* indicates the light transmission for two crossed polarizers whereas curve *B* is for two parallel polarizers.

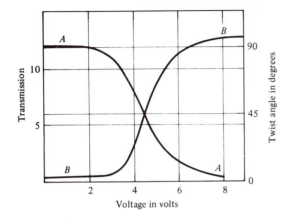

Figure 10-2-8 Light transmission for a twisted nematic liquid crystal (After E. B. Priestley [5]; used by permission of author and Plenum Press.)

10-2-3 *Modes of Operation*

There are three basic modes of operation for liquid crystal displays: transmissive mode, reflective mode, and storage mode.

(1) Transmissive mode.   The transmissive mode of a liquid crystal display (LCD) is an operation in which the reviewed light is transmitted through the liquid crystal from an incident light source. Figure 10-2-9 shows the optimum design of a scattered-light transmissive-mode display with an added display film [4]. The louver

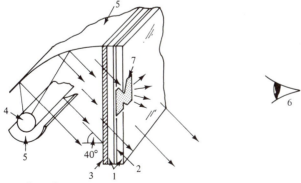

Legend:

1 Glass plate with electrodes   5 Reflector
2 Liquid crystal   6 Observer
3 Light control film   7 Activated liquid crystal
4 Fluorescent lamp

Figure 10-2-9 Transmissive mode of LCD (From G. Meier et al. [4]; used by permission of the author and Springer-Verlag.)

angle is 40°. The light from a fluorescent lamp is collimated by two reflectors and passes through the liquid crystal for observation. The light unaffected by the liquid crystal display is invisible to the observer.

(2) Reflective mode. The reflective mode of a liquid crystal display is an operation in which the light is reflected by a reflector at the liquid crystal back through the device into the viewing field as shown in Fig. 10-2-10 [4]. The reflector is a retroreflective foil about 40 μm in diameter with part of its surface mirrored coated. In the upper part of the cell the liquid crystal is in the reflective mode. A light ray r_1 is scattered at the liquid crystal in the forward direction, reflected at the retroreflector, and again scattered when passing through the cell a second time. The ray r_2 incident on a nonscattering point of the liquid crystal layer is not reflected; in this case, the incident and reflected rays are coincident.

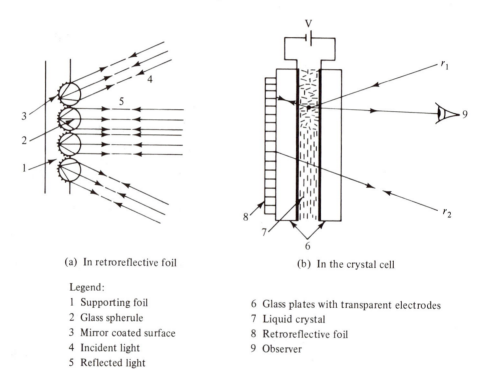

(a) In retroreflective foil   (b) In the crystal cell

Legend:
1 Supporting foil   6 Glass plates with transparent electrodes
2 Glass spherule   7 Liquid crystal
3 Mirror coated surface   8 Retroreflective foil
4 Incident light   9 Observer
5 Reflected light

Figure 10-2-10 Reflective mode of LCD (From G. Meier et al. [4]; used by permission of author and Springer-Verlag.)

(3) Storage mode. The storage mode of a LCD is an operation in which a quasi-permanent pattern remains when the applied electric field is off. This optical storage mode was reported in mixtures of nematic and cholesteric materials with negative relative dielectric anisotropy. With no applied voltage, the sample was in a relatively clear state. When an electric field was applied to a certain level, a dynamic scattering was induced. When the voltage was off, a quasi-permanent state remained [5].

10-2-4 Applications of Liquid Crystal Displays

Liquid crystal displays have many uses, such as digital displays in digital watches and infrared displays.

(1) Digital or alphanumeric displays.   Liquid crystal displays are effective and inexpensive devices. For example, a 7- or 8×5-segment device is used to display numbers or letters as shown in Fig. 10-2-11 [4]. The type of transmissive or reflective mode depends on the particular back electrode. Digital liquid crystal displays are being used in digital watches, calculators, electronic clocks, and data output equipment.

(a) 7-segment display   (b) 14-segment display   (c) 8×5-segment display

Figure 10-2-11   Schemes of digital liquid-crystal displays (From G. Meier et al. [4]; used by permission of the author and Springer-Verlag.)

(2) Infrared display.   When the wavelength of the transmitted light is in the infrared range, the cholesteric liquid crystal will exhibit infrared light. An infrared display unit using a cholesteric crystal is shown in Fig. 10-2-12 [4]. Window 2, which consists of sodium chloride, is transparent to infrared. Window 3, which consists of plastic glass, is transparent to visible light. An image converter is located between the two windows. A polyester foil (about 2 μm) on the right surface of the image converter is designed to absorb the infrared radiation and distribute the infrared heat to the gold or nickel layer. A cholesteric layer (about 10 μm) on the left side of the image converter is used to display the infrared. If the infrared radiation from a target is focused by an optic system onto the image converter, the target can be displayed with illumination by a light source. This type of liquid crystal display may have potential applications in the military field.

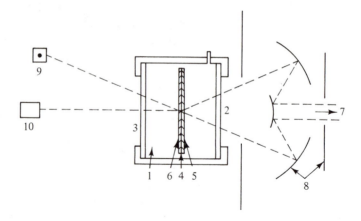

Legend:

1 Chamber	6 Cholesteric layer
2 NaCl window (IR translucent)	7 Object
3 Glass window (translucent for visible light)	8 IR optic (mirrors and diaphragm)
4 Supporting	9 Light source
5 Heat-absorbing layer (Au, Ni)	10 Observer

Figure 10-2-12 Infrared display with cholesteric layer (From G. Meier et al. [4]; used by permission of the author and Springer-Verlag.)

(3) Other applications. LCDs can be used in traffic and industrial control systems, in advertising displays, in medical diagnoses, and in scientific areas.

REFERENCES

[1] Bergh, A. A., and P. J. Dean. *Light-Emitting Diodes.* Oxford: Clarendon Press, 1976, pp. 39, 548.

[2] Gooch, C. H. *Injection Electroluminescent Devices.* New York: John Wiley & Sons, 1973, pp. 79, 80, 89, 91, 95, 96.

[3] Gage, G., et al. *Optoelectronics Applications Manual.* New York: McGraw-Hill Book Company, 1977.

[4] Meier, G., et al. *Applications of Liquid Crystals.* New York: Springer-Verlag, 1975, pp. 12, 86, 104, 106, 108.

[5] Priestley, E. B., ed. *Introduction to Liquid Crystals.* New York: Plenum Press, 1975, pp. 247, 255.

[6] Thurmond, C. D. The standard thermodynamic function of the formation of electrons and holes in Ge, Si, GaAs, and GaP. *J. Electrochem. Soc.*, **122** (1975), 1133.

[7] Sze, S. M. *Physics of Semiconductor Devices.* 2nd ed. New York: John Wiley & Sons, 1981, pp. 19, 20, 698.

SUGGESTED READINGS

1. Bergh, A. A., and P. J. Dean. Light-emitting diodes. *IEEE Proc.* **60**, no. 2 (February 1972), 156–223.

2. Chandrasekhar, S. *Liquid Crystals*. New York: Academic Press, 1978.

3. Gennes, P. G. *Physics of Liquid Crystals*. Oxford: Clarendon Press, 1974.

4. *IEEE Proceedings*. vol. 60, no. 2 (February 1972). Special issue on light-emitting diodes.

5. *IEEE Transactions on Electron Devices*. Special issues on LEDs and LCDs.

   vol. ED-26, no. 8, August 1979.

   vol. ED-28, no. 4, April 1981.

   vol. ED-28, no. 6, June 1981.

6. Kallard, Thomas, ed. *Liquid Crystal Devices*. New York: Optosonic Press, 1973.

7. Liebert, L., ed. *Liquid Crystals*. New York: Academic Press, 1978.

8. Milnes, A. G. *Semiconductor Devices and Integrated Electronics*, Chapter 14. New York: Van Nostrand Reinhold Company, 1980.

9. Pankove, J. I., ed. *Electroluminescence*. Berlin: Springer-Verlag, 1977.

10. Sze, S. M. *Physics of Semiconductor Devices*, Chapter 12. 2nd ed. New York: John Wiley & Sons, 1981.

PROBLEMS

10-1 LEDs

10-1-1. The photon energy $h\nu$ must be equal to or greater than the bandgap energy E_g in order for the stimulated-emission process to take place. Verify that $\lambda = 1.242/E_g$.

10-1-2. A few LEDs are listed as follows: GaAs, GaP, $GaAs_{0.6}P_{0.4}$, and ZnSe.
(a) Calculate the emitting wavelength at room temperature for each LED.
(b) Identify the emitted light for each LED.

10-1-3. An n-type GaAs LED operates at 300°K and has a cross section A of 2×10^{-3} cm^2.
(a) Calculate the voltage equivalent of temperature in millivolts.
(b) Compute the diffusion coefficient.
(c) Determine the diffusion length for an electron lifetime of 2 nsec.

10-1-4. A $GaAs_{0.5}P_{0.5}$ LED is given.
(a) Compute the emitting wavelength in micrometers.
(b) What light does the LED emit?

10-1-5. Compare the characteristics and applications of LEDs with those of semiconductor junction lasers.

10-1-6. A certain infrared LED emits infrared at a wavelength of 0.87 μm.
(a) Estimate the bandgap energy in electronvolts for the LED.
(b) Identify the type of LED.

10-1-7. Why are infrared LEDs so useful for optical-fiber communications?

10-1-8. A certain visible LED emits green light at a wavelength of 0.55 μm.
     (a) Estimate the bandgap energy in electronvolts of the LED.
     (b) Identify the type of LED.

10-1-9. A popular visible LED GaAsP emits red light at a wavelength of 0.69 μm.
     (a) Estimate the bandgap energy in electronvolts for the LED.
     (b) Determine the fractions of As and P.
     (c) Find the quantum efficiency and luminous efficiency.

10-1-10. The electron concentration of an n-type GaAs is 10^{19} cm^{-3} and the intrinsic carrier density of a GaAs is 10^6 cm^{-3}.
     (a) Determine the recombination rate under the equilibrium condition.
     (b) Estimate the lifetime of the excessive holes in the n side.

10-2 LCDs

10-2-1. All equations for LCDs in the literature are expressed in esu (electrostatic unit) for the cgs system. In order to change the equations from the cgs to the mks system, it is necessary to use certain conversion factors.
     (a) Write Coulomb's law in esu (cgs) and compute it for $R = 1$ cm.
     (b) Write Coulomb's law in the mks system and calculate it for $R = 1$ m.
     (c) Compare the results.

10-2-2. The threshold voltage in stat-volts (cgs units) for the splay elastic effect of a nematic liquid crystal is given by Eq. (10-2-7). Verify Eq. (10-2-8) in volts for mks units.

10-2-3. The threshold voltage in stat-volts (cgs units) for the bend elastic effect of a nematic liquid crystal is expressed by Eq. (10-2-9). Verify Eq. (10-2-10) in volts for the mks system.

10-2-4. The threshold voltage in stat-volts (cgs system) for the twist elastic effect of a nematic liquid crystal is shown in Eq. (10-2-11). Verify Eq. (10-2-12) in volts for the mks system.

10-2-5. A nematic liquid crystal display has the following parameters:

$$k_{11} = 6 \times 10^{-12} \text{ N} \qquad k_{33} = 7.5 \times 10^{-12} \text{ N}$$
$$k_{22} = 4 \times 10^{-12} \text{ N} \qquad \Delta\epsilon = 0.5$$

     (a) Calculate the threshold voltage in stat-volts for a 90°-twisted nematic liquid crystal cell.
     (b) Compute the threshold voltage in volts for the same liquid crystal cell.
     (c) Compare the results.

10-2-6. A certain liquid crystal display has the following parameters:

$$k_{11} = 5.85 \times 10^{-12} \text{ N} \qquad \epsilon_{\parallel} = 11.2$$
$$k_{22} = 2.40 \times 10^{-12} \text{ N} \qquad \epsilon_{\perp} = 10.0$$
$$k_{33} = 7.70 \times 10^{-12} \text{ N} \qquad \phi = 45°$$

     (a) Compute the threshold voltage in stat-volts.
     (b) Calculate the threshold voltage in volts.
     (c) Compare the results.

Chapter 11

Radiation-Detecting Devices

11-0 INTRODUCTION

Radiation-detecting devices are those which detect quantized radiation and convert it into electrical energy. The two basic types of radiation-detecting devices are photon detectors and thermal detectors. Incident radiation changes the electrical properties in each detector. Both photon and thermal detectors are quantum detectors, since radiation is quantized. Still another type of radiation detector is a combination of the first two and it is called a photothermic detector or imaging detector. All three radiation detectors are listed in Table 10-0-1 of Chapter 10.

11-1 OPERATIONAL MECHANISMS

According to modern solid-state theory, all solids are divided into two thermodynamic mechanisms: electronic and lattice. The electronic mechanism is characterized by three energy bands: the valence band, the forbidden band, and the conduction band, as described previously. The radiation is absorbed directly by the electronic system of the solid and produces changes in the electrical properties. The unit of radiation for the electronic mechanism in a solid is a photon. The lattice of a solid is composed of atoms or molecules and the lattice mechanism is characterized by lattice vibrations. The radiation is absorbed by the lattice, thereby producing heat changes in the lattice. The change in temperature of the lattice causes a change in the electronic system. The unit of radiation for the lattice mechanism is a phonon.

From quantum theory the unit of a photon is $h\nu$ and the unit of a phonon is kT. Because both units are quantized, the quantum condition is determined by

$$h\nu = kT \qquad \text{eV} \qquad (11\text{-}1\text{-}1)$$

where $h = 6.6256 \times 10^{-34}$ W-sec^2 (or J-sec) is Planck's constant
         ν = frequency in hertz
         $k = 1.38 \times 10^{-23}$ W-sec/°K (or J/°K) is Boltzmann's constant
         T = absolute temperature in degrees Kelvin

The quantum-condition temperature is given by

$$T_{\mathrm{D}} = \frac{h\nu}{k} \simeq \frac{\nu}{20} \qquad °\text{K} \qquad (11\text{-}1\text{-}2)$$

where ν = frequency in gigahertz
         T_{D} = Debye temperature in degrees Kelvin

If $\nu = 1000$ GHz, for example, the Debye temperature T_{D} is 50°K.

At very low temperatures kT energy is so small compared to the $h\nu$ of most radiation energies that there are few phonons. At somewhat higher temperatures there are many low-frequency phonons but few high-frequency ones. At high temperatures kT becomes greater than $h\nu$ for even the highest-frequency radiations. Therefore the quantum region is at

$$T \leq \frac{h\nu}{k} \qquad °\text{K} \qquad (11\text{-}1\text{-}3)$$

and the classic region is at

$$T > \frac{h\nu}{k} \qquad °\text{K} \qquad (11\text{-}1\text{-}4)$$

11-2 FIGURES OF MERIT

Six figures of merit are used to describe the quality and performance of radiation detectors:

1. responsivity R
2. time constant τ
3. noise equivalent power (NEP)
4. detectivity D, specific detectivity D^*, and background-limited detectivity D^{**}
5. quantum efficiency η_d, and
6. photocurrent gain G

1 Responsivity R

The *responsivity* is a measure of the dependence of the signal output of a detector on the input radiant power and it is commonly expressed as voltage responsivity in volts per watt or current responsivity in amperes per watt. The voltage responsivity is defined as

$$R(T,f) = \frac{V_s}{P} = \frac{V_s}{HA_d} \qquad \text{volts/watt} \qquad (11\text{-}2\text{-}1)$$

where V_s = rms signal voltage at the output of a detector of area A_d measured at frequency f in response to incident radiation power P (rms) at temperature T

P = incident radiation rms power

H = irradiance in watts per centimeter squared

A_d = sensitive area of the detector in centimeters squared

f = modulation frequency in hertz

T = absolute temperature of the blackbody in degrees Kelvin

Sometimes the responsivity can be expressed as $R(\lambda, f)$, where the output signal is measured at a frequency f in response to the monochromatic radiation of a wavelength λ modulated at frequency f.

The current responsivity is defined as

$$R(T,f) = \frac{\eta_d q}{h\nu} = \frac{\eta_d \lambda_0}{1.242} \qquad \text{amperes/watt} \qquad (11\text{-}2\text{-}2)$$

where η_d = quantum efficiency of the detector

q = charge of the carrier in coulombs

$h\nu$ = photon energy in electronvolts

λ_0 = free-space wavelength in micrometers

The voltage or current sensitivity of a microwave solid-state detector is identical to the voltage or current responsivity defined in Eqs. (11-2-1) and (11-2-2).

Alternatively, the responsivity of a radiation detector can be expressed in terms of modulation frequency f and time constant τ. That is,

$$R(f) = \frac{R(0)}{\left[1 + (\omega\tau)^2\right]^{1/2}} \qquad (11\text{-}2\text{-}3)$$

where $R(0)$ = responsivity at zero frequency

τ = time constant of the detector in seconds

$\omega = 2\pi f$

f = modulation frequency in hertz

The cutoff frequency f_c is defined as the frequency at which the responsivity falls to a value $R(0)/\sqrt{2}$ and is thus given by $f_c = 1/(2\pi\tau)$.

2 Time Constant τ

The time constant τ, also known as the response time, is a measure of the radiation detector's speed of response and it is defined as

$$\tau = \frac{1}{2\pi f_{3\,\text{dB}}} \qquad \text{seconds} \qquad (11\text{-}2\text{-}4)$$

where f_{3dB} is the frequency at which the signal power is 3 dB below the value at zero frequency—that is, the voltage is 0.707 that of the final value or the power is 0.5 that of the final power. In practice, the time constant τ of a detector is the time required for the detector output to reach to 63% of its final value or drop to 37% from its peak value.

3 Noise Equivalent Power (NEP)

The *noise equivalent power* (NEP) is defined as the rms incident radiant power falling on the detector that is required to produce an rms signal voltage or current equal to the rms noise voltage or current at the detector output. It is expressed as

$$\text{NEP} = \frac{HA_d}{V_s/V_n} = \frac{HA_dV_n}{V_s} = \frac{V_n}{R(T, f)} \qquad \text{watts} \qquad (11\text{-}2\text{-}5)$$

where V_n is the rms noise voltage at the detector output.

The post-detector electrical bandwidth, also called the noise bandwidth, should be specified. The NEP varies in proportion to the square root of the electrical bandwidth. The smaller the bandwidth, the lower is the NEP and the better is the detector sensitivity.

4 Detectivity D, Specific Detectivity D*, and Background-Limited Detectivity D**

Detectivity D. The sensitivity of a detector is described by its detectivity. The *detectivity D* is defined as the signal-to-noise ratio per unit incident radiation power and it is, then, the reciprocal of the noise equivalent power (NEP). That is,

$$D = \frac{V_s/V_n}{HA_d} = \frac{1}{\text{NEP}} \qquad \text{watt}^{-1} \qquad (11\text{-}2\text{-}6)$$

Specific detectivity D*. The *specific detectivity D**, also known as normalized detectivity, is a reciprocal of the NEP, normalized to a detector area of 1 cm² and an electrical bandwidth of 1 Hz. Then assuming that the detector noise varies with $A^{1/2}$ and $B^{1/2}$, the specific detectivity is expressed as

$$D^* = \frac{V_s/V_nB^{1/2}}{HA_d^{1/2}} = \frac{(A_dB)^{1/2}}{\text{NEP}} = D(A_dB)^{1/2} \qquad \text{cm-(Hz)}^{1/2}/\text{watt}$$

$$(11\text{-}2\text{-}7)$$

where A_d = detector area in centimeters squared (note that for a square detector, $A^{1/2}$ is its linear dimension)

B = post-detection electrical bandwidth in hertz

Specific detectivity D^* is usually specified by two parameters, such as the source temperature and the modulation frequency. For instance, $D^*(400°\text{K}, 900)$

means a value of D^* measured with a 400°K blackbody at a spatial frequency of 900 Hz.

The significant feature of D^* is that the value of D^* is independent of the size of the detector and the NEP for a truly background-limited detector. Therefore the detectivity of a detector can be determined if the background irradiance and the detector area are specified.

The value of D^* for a specific detector will depend on the wavelength of the signal radiation and the frequency at which it is modulated. Figure 11-2-1 shows the

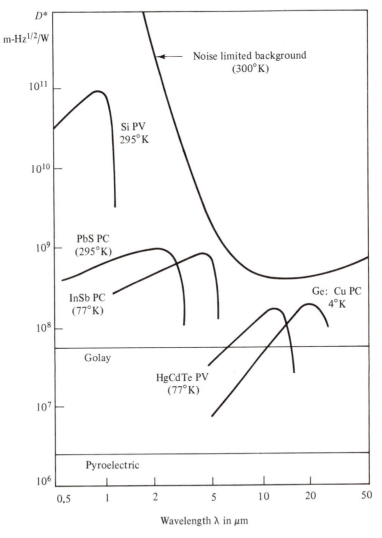

Figure 11-2-1 Specific detectivity D^* as a function of wavelength (After J. Wilson and J. F. B. Hawkes [16]; reprinted by permission of Prentice-Hall, Inc.)

curves of D^* as a function of wavelength for several photodetectors, where PC stands for photoconductive and PV for photovoltaic.

Background-limited detectivity D^{**}.   The detectivity D^{**} (dee-double-star) is a figure of merit that includes the reference conditions of D^* and a reference to a field of view having a solid angle of π steradians (which is equal to a hemisphere). Thus D^{**} is defined as

$$D^{**} = \left(\frac{\Omega}{\pi}\right)^{1/2} D^* = D^* \sin\theta \qquad \text{cm-(Hz-ster)}^{1/2}/\text{watt} \qquad (11\text{-}2\text{-}8)$$

where $\Omega = \pi \sin^2\theta$ is the effective solid angle
        θ = a half angle subtended by the detector aperture

For a detector without shield, $\theta = 90°$ and $\Omega = \pi$; then $D^{**} = D^*$. Figure 11-2-2 shows a limited background for a detector.

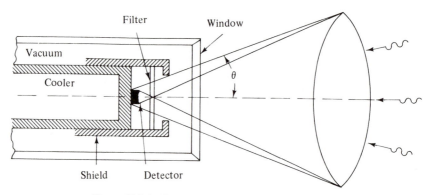

Figure 11-2-2   Detector with limited-background view

5 Quantum Efficiency

When a photon is absorbed by a detector, its quantum energy is transferred to a single electron within the surface. Therefore the quantum efficiency of a radiation detector is defined as

$$\eta_d = \frac{\text{number of electrons collected}}{\text{number of incident electrons}} \qquad (11\text{-}2\text{-}9)$$

The quantum efficiency for an ideal detector is unity. In other words, an ideal detector absorbs the incident radiation and converts it into a voltage or current that is proportional to the total radiation power incident on its surface.

6 Photocurrent Gain G

The photocurrent gain G of a detector is defined as the number of charge carriers flowing between the two contact electrodes of a detector per second for each photon absorbed per second. That is,

$$G = \frac{I_p}{I_{ph}} = \frac{\tau}{\tau_t} \qquad (11\text{-}2\text{-}10)$$

where I_p = photocurrent flowing between the electrodes
    I_{ph} = photocurrent
      τ = carrier lifetime or recombination time
      τ_t = carrier transit time

Table 11-2-1 lists some typical values of figures of merit for several commonly used radiation detectors.

TABLE 11-2-1   FIGURES OF MERIT OF COMMONLY USED RADIATION DETECTORS

Detectors	Time constant τ (seconds)	D^{**} [cm-(Hz-ster)$^{1/2}$/W]	Gain G	Operating temperature (°K)
Avalanche photodiodes	10^{-10}		10^2–10^4	300
pin photodiodes	10^{-8}–10^{-12}		10^0	300
Photovoltaic detectors	10^{-6}–10^{-7}	10^8–10^{11}		12–100
Photoconductive detectors	10^{-3}–10^{-8}	10^9–10^{10}	10^6	4.2–300
Bolometers	10^{-3}	10^8–10^{10}		300
Thermocouples	10^{-3}–10^{-5}	10^9		300
Pyroelectric detectors	10^{-1}	10^9		300

Example 11-2-1   Merit Figures of a Photon Detector

A certain photon detector has the following parameters:

Detecting area $A_d = 200 \times 10^{-6}$ m^2
Wavelength $\lambda_0 = 1 \ \mu$m
Quantum efficiency $\eta_d = 20\%$
Bandwidth $B = 1$ Hz
Noise current $I_n = 10$ pA

(a) Determine the responsivity.
(b) Find the noise equivalent power.
(c) Calculate the detectivity.
(d) Compute the specific detectivity.

Solution.   (a) Using Eq. (11-2-2), we obtain the responsivity as

$$R(f) = \frac{\eta_d \lambda_0}{1.242} = \frac{0.2 \times 1}{1.242} = 161.00 \text{ mA/W}$$

(b) Then from Eq. (11-2-5) the noise equivalent power is

$$\text{NEP} = \frac{I_n}{R(f)} = \frac{10 \times 10^{-12}}{161.0 \times 10^{-3}} = 62 \text{ pW}$$

(c) The detectivity is found by using Eq. (11-2-6).

$$D = \frac{1}{\text{NEP}} = \frac{1}{62 \times 10^{-12}} = 1.61 \times 10^{10} \text{ W}^{-1}$$

(d) Finally, from Eq. (11-2-7) the specific detectivity is

$$D^* = D(A_d B)^{1/2} = 1.61 \times 10^{10}(200 \times 10^{-2} \times 1)^{1/2}$$

$$= 2.28 \times 10^{10} \qquad \text{cm-Hz}^{1/2}/\text{W}$$

11-3   MODES OF OPERATION

All radiation detectors can be classified into three modes, depending on their radiation effects. (See Table 10-0-1 in Chapter 10.)

1 Photon-Effect Mode

The photon effects are characterized by changes in certain properties of a material due to the interactions of the incident photon with the electrons. The photon effects can be subdivided into two types: internal and external.

Internal photon effects.   The internal photon effects are those in which the photoexcited carriers (electrons or holes) remain within the sample. They include photodiodes, photovoltaic detectors, photoconductive detectors, and others.

External photon effects.   The external photon effect, also known as the photoemission effect, is one in which an incident photon causes the emission of an electron from the surface of the absorbing material. The photoemissive detector is an example.

2 Thermal-Effect Mode

The thermal effects are characterized by changes in certain properties of a material resulting from temperature changes caused by the heating effect of the incident radiation. Bolometers and thermocouples are included in this group. Generally they are wavelength independent.

3 Photothermic-Effect Mode

The photothermic effect is characterized by a combination of photon and thermal effects. The Vidicon, for instance, uses photoconductivity as the detection phenomenon, but its spatial variation is read out by an electron beam.

11-4 *PHOTON DETECTORS*

Photon detectors are those solid-state devices that operate under the influence of photon effects. Based on their physical structures, they can be divided into junction type and bulk type. The biased photodiodes, such as the *pin* photodiodes and avalanche photodiodes (APD), and the unbiased photovoltaic detectors are the junction type; the photoconductive detectors are the bulk type.

Photon detectors have a small size, minimum noise, low biasing voltage, high sensitivity, high reliability, and fast response time. Therefore they are very useful in optical-fiber communications systems.

Basically, if a photon of sufficient energy excites an electron from a nonconducting state into a conducting state, the photoexcited electron will generate current or voltage in the detector. According to Eq. (10-1-1), the electronic excitation requires that the incident photon energy must be equal to or greater than the electronic excitation energy. In other words, the excitation condition is

$$E_{exc} \leq h\nu \tag{11-4-1}$$

or

$$E_{exc} \leq \frac{1.242}{\lambda_0} \tag{11-4-2}$$

where E_{exc} = electronic excitation energy in electronvolts
λ_0 = free space wavelength in micrometers

11-4-1 *Photoconductive Detectors*

The photoconductive detectors are microwave devices that are constructed by a single crystal of semiconductor material; its two ends are fixed with ohmic contacts as shown in Fig. 11-4-1.

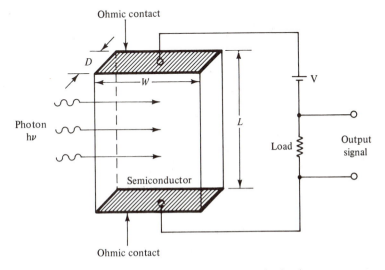

Figure 11-4-1 Schematic diagram of photoconductive detector

The incident photon energy creates free carriers in the crystal and changes the conductivity of the material:

$$\sigma = q(n\mu_n + p\mu_p) \qquad (11\text{-}4\text{-}3)$$

where q = carrier charge
n = electron concentration
μ_n = electron mobility
p = hole concentration
μ_p = hole mobility

This type of detector can be used for automatic light control in homes and office buildings to turn light on at dawn and off at dark. Also, they are useful in optical signaling systems.

There are two types of excitation in a semiconductor (see Fig. 11-4-2). The intrinsic excitation can occur only from the valence band to the conduction band to create an electron-hole pair whereas the extrinsic excitation can take place either from a discrete crystal-defect (dopant) energy level in the forbidden band to conduction band or from the valence band to the forbidden band to generate a conduction electron or hole.

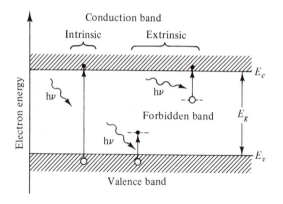

Figure 11-4-2 Processes of photoexcitations (After S. M. Sze [3]; reprinted by permission of John Wiley & Sons, Ltd.)

Intrinsic photoconductive detectors.   In the intrinsic mode photoconduction is produced by absorption of a photon across the energy bandgap that creates a free electron and hole simultaneously; the cutoff wavelength is determined by Eq. (10-1-2). The absorption coefficient α is very large due to the large number of available electrons in the valence and conduction bands. Here α is 10^4 cm^{-1} for photoexcitation near the bandgap energy.

The total current of electrons and holes in an intrinsic photoconductive detector is

$$I = \sigma EWD = qWD(n\mu_n + p\mu_p)E \qquad (11\text{-}4\text{-}4)$$

where W = width of the crystal
D = thickness of the crystal
E = applied electric field across the crystal when the conductivity is measured

The photocurrent is given by

$$I_{\text{ph}} = qWD\left(\Delta n \mu_n + \Delta p \mu_p\right)E \tag{11-4-5}$$

where $\Delta n = n - n_0$ is the photoexcited electron concentration
    $n_0 =$ electron concentration in equilibrium
    $\Delta p = p - p_0$ is the photoexcited hole concentration
    $p_0 =$ hole concentration in equilibrium

If there is no trapping, the photoexcited electron and hole concentrations may be expressed as

$$\Delta n = \Delta p = g\tau \tag{11-4-6}$$

where $g = \dfrac{\eta P_{\text{in}}}{h\nu}$ is the average generation rate of the optical flux in photons per
    second   (11-4-7)
    $\eta =$ quantum efficiency
    $P_{\text{in}} =$ incident radiation signal power
    $\tau =$ carrier lifetime

Extrinsic photoconductive detectors.   It is often desirable to reduce the excitation energy level so as to improve the photoconductor performance and minimize the internal noise figure. As a result, the extrinsic photoconductor is used. In this mode a photon is absorbed at the impurity dopant ionization energy level and then a free electron is created in an n-type photoconductor or a free hole in a p-type photoconductor. If the incident photon energy is insufficient to induce the intrinsic photoconduction, it still can create an extrinsic photoconduction. Table 11-4-1 lists ionization energies E_i of selected impurities in Ge and Si that correspond

TABLE 11-4-1   IONIZATION ENERGY AND WAVELENGTH OF IMPURITY IN GERMANIUM AND SILICON

Semiconductor	Impurity	Ionization energy in eV	Wavelength λ_0 in μm
Ge	Au	0.15	8.3
Ge	Hg	0.09	14
Ge	Cd	0.06	21
Ge	Cu	0.041	30
Ge	Zn	0.033	38
Ge	B	0.0104	120
Si	In	0.155	8
Si	Ga	0.0723	17
Si	Bi	0.0706	18
Si	Al	0.0685	18
Si	As	0.0537	23
Si	P	0.045	28
Si	B	0.0439	28
Si	Sb	0.043	29

(After R. J. Keyes [1]; reprinted by permission of Springer-Verlag.)

to their cutoff wavelengths [1]. The cutoff wavelength is determined by the approximate ionization energy for the impurity. The absorption coefficient α is very small, about 1 to 10 per centimeter.

The total currents of the n-type and p-type modes are given, respectively, by

$$I_n = qWDn\mu_n E = qWDnv_{dn} \tag{11-4-8}$$

and

$$I_p = qWDp\mu_p E = qWDpv_{dp} \tag{11-4-9}$$

where $v_d = \mu E$ is the drift velocity

The photocurrents are

$$I_{nph} = qWD\Delta n\mu_n E \tag{11-4-10}$$

and

$$I_{pph} = qWD\Delta p\mu_p E \tag{11-4-11}$$

Example 11-4-1   Photocurrent of a Photoconductive Detector

A typical GaAs photoconductive detector has the following parameters:

Incident radiation power	$P_{in} = 20$ mW
Efficiency	$\eta = 90\%$
Carrier lifetime	$\tau = 1$ nsec
Infrared signal frequency	$\nu = 3.45 \times 10^{14}$ Hz
Applied electric field	$E = 400$ V/cm
Crystal width	$W = 100$ cm
Crystal thickness	$D = 10$ cm

(a) Compute the generation rate.
(b) Calculate the photoexcited carriers.
(c) Determine the photocurrent.

Solution.   (a) From Eq. (11-4-6) the generation rate is

$$g = \frac{\eta P_{in}}{h\nu} = \frac{0.90 \times 20 \times 10^{-3}}{6.625 \times 10^{-34} \times 3.45 \times 10^{14}}$$

$$= 7.88 \times 10^{16} \text{ sec}^{-1}$$

(b) The photoexcited carriers are

$$\Delta n = \Delta p = g\tau = 7.88 \times 10^{16} \times 10^{-9}$$

$$= 7.88 \times 10^7 \text{ carriers/cm}^3$$

(c) Then from Eq. (11-4-5) the photocurrent is

$$I_{ph} = qWD(\Delta n\mu_n + \Delta p\mu_p)E$$

$$= 1.6 \times 10^{-19} \times 10^2 \times 10 \times 7.88 \times 10^7 \times (8500 + 400) \times 400$$

$$= 44.88 \text{ mA}$$

11-4-2 Photodiodes

Photodiodes are usually reverse biased with relatively large biasing voltages in order to reduce the transit time and the diode capacitance for high-speed performance. The photoinduced carriers in the vicinity of the junction modify the current-voltage characteristic so that the radiation level can be measured. Figure 11-4-3 shows a schematic diagram for a *p-n* junction photodiode with its voltage-current characteristic curves.

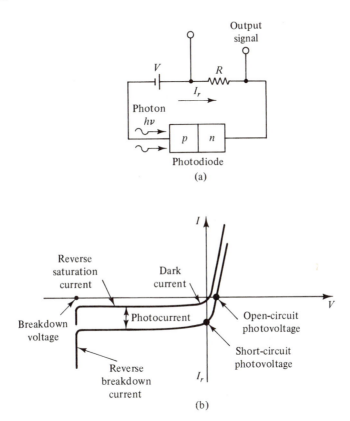

Figure 11-4-3   Schematic diagram of a photodiode

When the junction is reverse biased, the observed photosignal is a photocurrent rather than a photovoltage. The dark current due to the thermal generation of electron-hole pairs in the depletion layer is given by

$$I_d = I_{rs}\left[\exp\left(\frac{qV}{nkT}\right) - 1\right] \tag{11-4-12}$$

where I_{rs} = reverse-saturation current
       n = constant of the order of unity
       q = charge
       V = applied voltage
       k = Boltzmann's constant
       T = absolute temperature

The average photocurrent due to the optical signal is

$$I_{ph} = \frac{q\eta P_{in}}{h\nu} \qquad (11\text{-}4\text{-}13)$$

The response time of the detector in optical detection applications is very critical. When a photodiode detector is exposed to a light signal of incident pulses, the photogenerated minority carriers must diffuse to the junction and be swept across the depletion layer to the other side in a time much shorter than the pulse width. Therefore it is often desirable to increase the width W of the depletion layer so that most of the photons are absorbed within W rather than in the neutral p and n regions. In addition, a wide W results in a small junction capacitance and eventually reduces the RC time constant of the detector circuit. The *pin* photodiode and the avalanche photodiode are two of the most common photodetectors with a large wide depletion layer.

(1) ***pin* photodiodes.**   The *pin* photodiode has an intrinsic *i*-region sandwiched between a *p*- and an *n*-region as shown in Fig. 11-4-4.

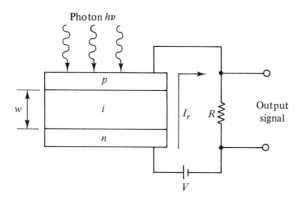

Figure 11-4-4 *pin* photodiode circuit

When the diode is reverse biased, the depletion regions occur at the junctions of both the *p*- and *n*-regions and the effective depletion layer width is increased by the effect of the *i*-region. Thus the depletion layer capacitance is much less than that of an ordinary *p-n* diode and the response time to the modulated light signal is much shorter. This type of photodiode is very useful in optical-fiber communications and star-tracking systems.

Currents.   When the *pin* photodiode is reverse biased, the applied voltage appears almost entirely across the *i*-region. If the carrier lifetime within the *i*-region is long enough compared to the drift time, most of the photogenerated carriers will be collected by the *p*- and *n*-regions. When the incident photons are absorbed, the electron-hole pairs are released within the *i*-region. While the electrons drift to the *n* region and holes to the *p* region under reverse bias, both are added to the reverse current. Under steady-state conditions the total current density through the reverse-biased depletion layer is given by [2]

$$J_{\text{tot}} = J_{\text{dr}} + J_{\text{diff}} \qquad (11\text{-}4\text{-}14)$$

where J_{dr} = drift current density per unit area
J_{diff} = diffused current density per unit area

The drift current density is due to carriers generated inside the depletion region and it is expressed as

$$J_{\text{dr}} = q\Phi_0(1 - e^{-\alpha w}) \qquad (11\text{-}4\text{-}15)$$

where $\Phi_0 = \dfrac{P_{\text{in}}(1 - R)}{Ah\nu}$ is the incident photon flux per unit area per second
P_{in} =   incident radiation signal power
R =   reflection coefficient
A =   detecting area
W =   width of *i*-region
α =   absorption of ionization coefficient

The diffusion current density is due to the carriers generated outside the depletion layer and diffusing into the reverse-biased junction and it is given by

$$J_{\text{diff}} = q\Phi_0 \frac{\alpha L_p}{1 + \alpha L_p} e^{-\alpha w} + q p_0 \frac{D_p}{L_p} \qquad (11\text{-}4\text{-}16)$$

where L_p = hole diffusion length
D_p = hole diffusion coefficient in area per second
p_0 = hole concentration in equilibrium

Then the total current density is

$$J_{\text{tot}} = q\Phi_0 \left(1 - \frac{e^{-\alpha w}}{1 + \alpha L_p}\right) + q p_0 \frac{D_p}{L_p} \qquad (11\text{-}4\text{-}17)$$

Under normal operating conditions the last term is very small and it may be neglected.

(2) Avalanche photodiodes (APD).

If the amplitude of the reverse bias across a photodiode is increased to the breakdown-voltage level, avalanche multiplication occurs, thus resulting in a much larger current gain than in the other photodiodes. The avalanche photodiodes have a low noise figure over a wide bandwidth and their typical gain-bandwidth product is 100 GHz. They are widely used with lasers in optical-fiber communications systems.

   Multiplication and Current.   For equal absorption coefficients ($\alpha_n = \alpha_p$), the multiplication factor is expressed by [3]

$$M = \frac{1}{1 - \alpha_n W} \qquad (11\text{-}4\text{-}18)$$

If both leakage and photocurrents are subjected to avalanche multiplication, the combined multiplication factor is given by

$$M = \frac{I}{I_u} = \frac{I_{ph} - I_d}{I_{uph} - I_{ud}} = \frac{1}{1 - (V_r/V_b)^n} \qquad (11\text{-}4\text{-}19)$$

where I = total multiplied current
I_u = total unmultiplied current
I_{ph} = multiplied photocurrent
I_d = multiplied dark current
I_{uph} = unmultiplied photocurrent
I_{ud} = unmultiplied dark current
V_r = reverse bias voltage
V_b = breakdown voltage
n = a constant

The multiplication factor M can have a value between 1 and 100, depending on V_r. Figure 11-4-5 shows an equivalent circuit for an avalanche photodiode.

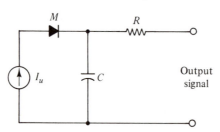

Figure 11-4-5   Equivalent circuit for an avalanche photodiode

Example 11-4-2   Photocurrent of an Avalanche Photodiode
   A typical avalanche photodiode has the following parameters:

Incident radiation power	$P_{in} = 50$ mW
Efficiency	$\eta = 95\%$
Red-light frequency	$\nu = 4.5 \times 10^{14}$ Hz
Breakdown voltage	$V_b = 35$ V
Reverse bias voltage	$V_r = 34$ V
Dark current	$I_d = 10$ nA
Constant	$n = 2$

(a) Compute the multiplication factor M.
(b) Calculate the generation rate g.
(c) Determine the average photocurrent I_{ph}.

Solution.   (a) From Eq. (11-4-19) the multiplication factor is

$$M = \left[1 - \left(\frac{V_r}{V_b}\right)^2\right]^{-1} = \left[1 - \left(\frac{34}{35}\right)^2\right]^{-1} = 16.67$$

(b) The generation rate is

$$g = \frac{\eta P_{\text{in}}}{h\nu} = \frac{0.95 \times 50 \times 10^{-3}}{6.625 \times 10^{-34} \times 4.5 \times 10^{14}}$$

$$= 1.59 \times 10^{17} \text{ sec}^{-1}$$

(c) From Eq. (11-4-13) the average photocurrent is

$$I_{\text{ph}} = qg = 1.6 \times 10^{-19} \times 1.59 \times 10^{17}$$

$$= 25.44 \text{ mA}$$

11-4-3 Photovoltaic Detectors and Solar Cells

(1) Photovoltaic detectors.   A *photovoltaic detector* is a microwave solid-state device that is operated under the intrinsic photoconductive effect without bias and it requires an internal potential barrier with a built-in electric field to separate the photoexcited electron-hole pairs [4]. The physical structure of a photovoltaic detector consists of a *p-n* junction formed in an intrinsic semiconductor as shown in Fig. 11-4-6.

When the incident photons impinge on the detector, the induced electron-hole pairs are separated by the junction electric field and a photovoltage is generated as shown at point *A* in Fig. 11-4-7. Figure 11-4-7 shows several different modes of operation for photodiodes. Curve 1 indicates the *V-I* curve for a photodetector without incident radiation and curve 2 is for detection with incident radiation. The photovoltaic mode operates at point *A* for open-circuit voltage in a high impedance without bias. The photoconductive mode operates at point *C* and the avalanche photodiode mode is at point *D*.

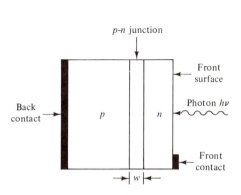

Figure 11-4-6 Diagram of a *p-n* junction photovoltaic detector

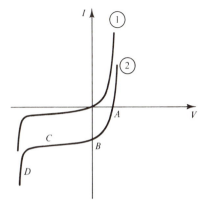

Figure 11-4-7 *V-I* characteristic curves for photodiodes

The photovoltaic detector has no bias supply and its circuit is much simpler than that of the other detectors. Its detectivity is usually 40% greater than in a photoconductive mode. Because the photovoltaic detector operates only in intrinsic photoexcitation, its bandgap energy must be equal to or less than the incident photon energy and its cutoff wavelength is determined by Eq. (10-1-2).

(2) Solar cells. Solar energy is a huge energy reserve for human beings. Every second the sun emits electromagnetic radiation of 4×10^{20} Joules with a net mass less of 4×10^3 kg in the ultraviolet-to-middle infrared regions (0.2 to 3 μm). At sea level, with the sun at the zenith, the solar power density is about 1 kW/m². The total mass of the sun is now approximately 2×10^{30} kg and its mass for energy conversion is about 5×10^{20} kg. The total life of the sun is approximately 10 billion years and its active life has about 3 billion years to go. In the future solar radiation will be the major energy source to be explored and used.

Solar cells are semiconductor junction devices that collect solar energy from the sun. When the incident solar radiation produces electron-hole pairs in the device, a voltage is generated by the separation of the barrier electric field. A current output may be taken from a resistive load. The energy-wavelength equation as shown in Eq. (10-1-2) is applied and it is

$$\lambda_0 = \frac{1.242}{h\nu} \qquad \mu m \qquad (11\text{-}4\text{-}20)$$

Semiconductor junction devices are the most common solar energy collectors, but the photon of energy that is less than the bandgap energy will not be absorbed. Bandgap energies for selected semiconductors were tabulated in Table 10-1-1. A number of *p-n* junctions made of Ge, Si, GaAs, and GaP are commonly used as solar cells. Figure 11-4-8 shows a schematic structure of silicon *p-n* junction solar cell with its equivalent diagram [3].

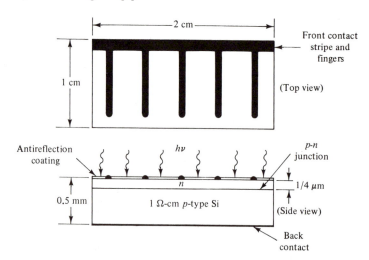

Figure 11-4-8 Schematic structure of a Si *p-n* junction solar cell with equivalent diagram (After S. M. Sze [3]; reprinted by permission of John Wiley & Sons, Ltd.)

The electric field of the junction-depletion region causes the induced electrons in the p-region to move to the n-region and the induced holes move to the opposite direction as shown in Fig. 11-4-9(a).

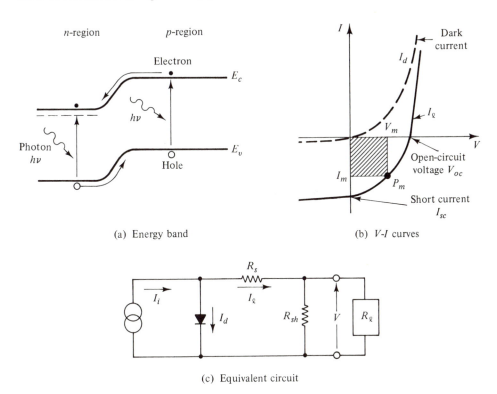

(a) Energy band

(b) V-I curves

(c) Equivalent circuit

Figure 11-4-9 Solar cell   (a) energy band   (b) V-I curves, and   (c) equivalent circuit

In an ideal case, the current flowing through the load is given by

$$I_\ell = I_i - I_s\left[\exp\left(\frac{qV}{kT}\right) - 1\right] \tag{11-4-21}$$

where $I_s =$ diode saturation current

$I_d = I_s\left[\exp\left(\frac{qV}{kT}\right) - 1\right]$ is the dark current due to the thermal effect

$I_i =$ illuminated current by solar radiation

The open-circuit voltage is expressed as

$$V_{oc} = \frac{kT}{q}\ell\mathrm{n}\left(\frac{I_i}{I_s} + 1\right) \approx \frac{kT}{q}\ell\mathrm{n}\left(\frac{I_i}{I_s}\right) \qquad \text{for } \frac{I_i}{I_s} \gg 1 \tag{11-4-22}$$

The output power of the solar cell is

$$P = VI_\ell = V\left\{I_i - I_s\left[\exp\left(\frac{qV}{kT}\right) - 1\right]\right\} \tag{11-4-23}$$

As shown in Fig. 11-4-9(b), the curve of the load current I_ℓ passes through the fourth quadrant; thus the power can be extracted from the device. By choosing a load resistor properly, about 80% of the power of $V_{oc}I_{sc}$ can be collected.

Solar cells are commonly used for satellites in space and they will be used to generate electrical power for industrial and commercial purposes in the future.

Example 11-4-3   Output Power of Silicon Solar Cells

A single silicon solar cell with an area of 4 cm^2 has an illuminated current I_i of 100 mA and a diode saturation current I_s of 0.1 nA. It is operated at 320°K.
(a) Compute the open-circuit voltage of a single cell.
(b) Calculate the short-circuit current of a single cell.
(c) Find the extracted power from a single cell.
(d) If a system requires a power of 12 W at a voltage level of 10 V, determine the number of solar cells in series and the number of rows in parallel for this type of solar-cell array to meet the power requirement.

Solution.   (a) From Eq. (11-4-22) the open-circuit voltage of a single cell is

$$V_{oc} = \frac{kT}{q} \ell n \left(\frac{I_i}{I_s} \right) = 27.6 \times 10^{-3} \ell n \left(\frac{100 \times 10^{-3}}{0.1 \times 10^{-9}} \right) = 0.57 \text{ V}$$

(b) From Eq. (11-4-21) the short-circuit current of a single cell is

$$I_{sc} = I_i = 100 \text{ mA}$$

(c) The power extracted from a single cell is

$$P = 0.8 V_{oc} I_{sc} = 0.8 \times 0.57 \times 100 = 45.60 \text{ mW}$$

(d) The number of solar cells in series is

$$N(\text{series}) = \frac{10}{0.8 \times 0.57} = 22 \text{ cells}$$

The number of rows in parallel is

$$N(\text{parallel}) = \frac{12}{10 \times 0.1 \times 0.8} = 15 \text{ rows}$$

The silicon solar-cell array is 22×15 and the total number of solar cells is 330.

11-4-4 Phototransistors and Photoemissive Detectors

(1) Phototransistors.   *Phototransistors* are those bipolar and unipolar transistors that can detect an optical signal and convert it into an electrical current with high gain. They require a large base-collector junction as the light-collecting element and their fabrication is more complicated than in conventional transistors. Figure 11-4-10 shows the diagram for an *n-p-n* phototransistor and its equivalent circuits.

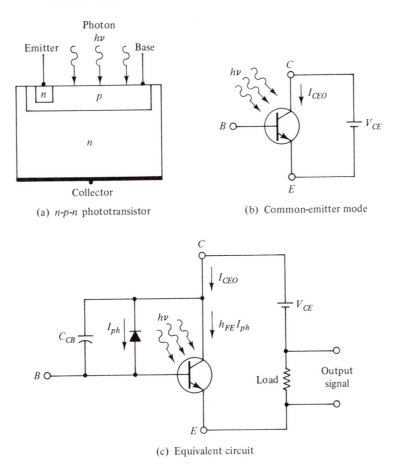

(a) *n-p-n* phototransistor

(b) Common-emitter mode

(c) Equivalent circuit

Figure 11-4-10 Schematic diagram of an *n-p-n* phototransistor with equivalent circuits

A phototransistor can operate in the common-emitter mode with the base open circuited. In the absence of photon radiation the collector current is

$$I_{CEO} = (1 + h_{FE})I_{CBO} \simeq h_{FE}I_{CBO} \qquad (11\text{-}4\text{-}24)$$

where $h_{FE} = \beta_0 = \dfrac{\Delta I_C}{\Delta I_B}$ is the dc or static common-emitter current gain

$I_{CBO} = $ reverse collector saturation leakage current with emitter open

This current is usually called the *dark current*. When the incident radiation impinges on the collector-base junction, the leakage current I_{CBO} is increased by the photocurrent I_{ph} so that the collector current becomes

$$I_{CEO} = h_{FE}(I_{CBO} + I_{ph}) = (1 + h_{FE})I_{ph} \qquad (11\text{-}4\text{-}25)$$

where $I_{ph} = 2ge$ is the photocurrent for an electron-hole pair

$g = \dfrac{\eta P_{in}}{h\nu}$ is the photon generation rate as defined in Eq. (11-4-7)

$e =$ electron charge

The base may be connected to the emitter through a resistor in order to reduce the dark current $h_{FE}I_{CBO}$ and the ratio of I_{ph} to I_{CBO} at high temperature is then greatly improved. In order to increase the current-transfer ratio of the output photocurrent to the input LED current, a photo-Darlington mode may be used. Phototransistors are particularly useful in optoisolator applications in optical-fiber communication systems. A phototransistor with an LED can provide optical coupling with electrical isolation up to 3 kV as shown in Fig. 11-4-11 [6]. Part (a) in Fig. 11-4-11 shows direct coupling, part (b) indicates coupling through a photodiode, and part (c) illustrates coupling through a logic driver with a Schottky output transistor.

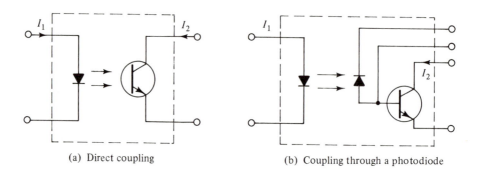

(a) Direct coupling   (b) Coupling through a photodiode

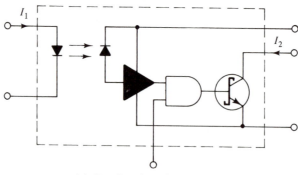

(c) Coupling through a logic driver
with Schottky output transistor

Figure 11-4-11 Optically coupled isolators with phototransistor (After J. Seymour [6]; reprinted by permission of Pitman Books Ltd., London)

Example 11-4-4   Photocurrent of a Phototransistor

A certain phototransistor has the following parameters:

Incident radiation power	$P_{in} = 10$ mW
Efficiency	$\eta = 95\%$
dc current gain	$h_{FE} = 60$
Frequency of red light	$\nu = 4.5 \times 10^{14}$ Hz
Reverse collector saturation	$I_{CBO} = 5\ \mu$A
leakage current with emitter open	

(a) Calculate the photon generation rate.
(b) Compute the photocurrent for an electron-hole pair.
(c) Determine the collector current.

Solution.   (a) Using Eq. (11-4-25), we obtain the photon generation rate as

$$g = \frac{\eta P_{in}}{h\nu} = \frac{0.95 \times 10 \times 10^{-3}}{6.625 \times 10^{-34} \times 4.5 \times 10^{14}}$$

$$= 3.19 \times 10^{16}\ \text{sec}^{-1}$$

(b) The photocurrent for an electron-hole pair is

$$I_{ph} = 2ge = 2 \times 3.19 \times 10^{16} \times 1.6 \times 10^{-19}$$

$$= 10.21\ \text{mA}$$

(c) The collector current is

$$I_{CEO} = h_{FE}(I_{CBO} + I_{ph}) = 60 \times (5 \times 10^{-6} + 10.21 \times 10^{-3})$$

$$= 613\ \text{mA}$$

(2) Photoemissive detectors.   *Photoemissive detectors* are microwave solid-state devices that operate via a photoemissive effect. Because of their high gain, low noise, and fast response, they are useful for the detection of low-intensity signals, high-speed signals, and high-resolution spatial information.

Photoemissive Effect.   The *photoemissive effect*, also known as the *external photoeffect*, is a process in which the action of the incident radiation causes electrons to be emitted from the photocathode surface into a vacuum and be collected by an anode. From modern solid-state theory the ionization energy of a solid, which is also known as the work function Φ (see Section 3-3), is the minimum energy required for an electron to escape from the solid at the Fermi level into its surface at rest (not hot). If the incident photon energy is greater than the electron bandgap energy of an intrinsic semiconductor, the electron will have kinetic energy after escaping from the intrinsic semiconductor. That is,

$$h\nu = E_g + \tfrac{1}{2}mv^2 \tag{11-4-26}$$

where $E_g = \Phi$ is the bandgap energy for an intrinsic semiconductor
$\Phi =$ work function

As described in Section 3-3, the electron affinity χ of a semiconductor is the energy level measured from the conduction energy band edge to the vacuum level E_{vac}. That is,

$$\chi = E_{vac} - E_c = \Phi + E_F - E_g \qquad (11\text{-}4\text{-}27)$$

where E_F is the Fermi energy level.

The photoemissive threshold energy E_{th} for an electron in a semiconductor is given by [14]

$$E_{th} = E_g + \chi \qquad (11\text{-}4\text{-}28)$$

This semiconductor optical ionization energy is constant regardless of doping. The photoemission threshold energy can be increased or decreased, however, depending on whether the electron affinity is positive or negative.

Photoemissive detectors can be divided into two groups. The first group, which is called the classic type, consists of metal photocathodes and semiconductor photocathodes with positive electron affinity (PEA). The second group includes the semiconductor photocathodes that have a negative electron affinity (NEA). The emitting surfaces of both groups are coated with a very thin evaporated metallic layer.

Metal Photocathodes.   Such devices are old types like the vacuum phototubes. Because the metals have a work function of several electron volts, they detect only the radiation at the wavelength range from extra violet to visible light—that is, 0.003 to 0.7 μm.

Semiconductor Photocathode with Positive Electron Affinity **(PEA).**   Because the electron affinity is positive, the photoemissive threshold energy is quite high. Even though their wavelength response to the radiation is better than the metal photocathode, they detect radiation from 0.7 to 1.0 μm at the near infrared range. Figure 11-4-12 shows the energy band diagram for a positive electron affinity (PEA) photocathode.

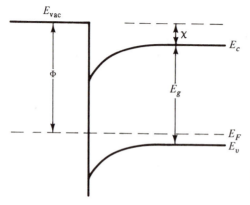

Figure 11-4-12   Energy-band diagram of a PEA semiconductor photocathode

Example 11-4-4A   Positive Electron Affinity (PEA) of a Photocathode

A typical cesium antimonide (CsSb) photocathode has the following parameters:

$$
\begin{array}{ll}
\text{Work function} & \Phi = 1.65 \text{ eV} \\
\text{Bandgap energy} & E_g = 1.60 \text{ eV} \\
\text{Fermi energy level} & E_F = 0.40 \text{ eV}
\end{array}
$$

Determine the electron affinity of the photocathode.

Solution.   Using Eq. (11-4-27), we obtain the electron affinity as

$$\chi = \Phi + E_F - E_g = 1.65 + 0.40 - 1.60 = 0.45 \text{ eV}$$

It is a PEA photocathode.

***Semiconductor Photocathode with Negative Electron Affinity* (NEA).**   This type of detector is constructed by overcoating the surface of selected *p*-type semiconductors with an evaporated layer of low work function material. Figure 11-4-13 shows the energy band diagram for a negative electron affinity (NEA) photoemitter.

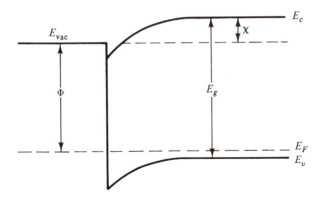

Figure 11-4-13   Energy-band diagram of a NEA semiconductor photocathode

The incident photon energy for a NEA semiconductor photoemitter must equal or exceed the bandgap energy of the semiconductor in order for photoemission to occur. The operation wavelength range for NEA semiconductor photoemitters is up to 3 μm. NEA photoemitters have five advantages over the classic ones:

1. high quantum efficiency
2. wide spectral response up to near infrared
3. high resolution
4. high uniform absolute sensitivity, and
5. low dark current at room temperature.

Still, NEA photoemitters do have some disadvantages, such as

1. high cost
2. small size
3. low durability
4. low response, and
5. difficult fabrication.

Example 11-4-4B   Negative Electron Affinity (NEA) of a Photoemitter

A typical GaAs/CsO photoemitter has the following parameters:

$$
\begin{array}{lll}
\text{Work function} & \Phi = 1.08 \text{ eV} \\
\text{Bandgap energy} & E_g = 1.43 \text{ eV} \\
\text{Fermi energy level} & E_F = 0.01 \text{ eV}
\end{array}
$$

Determine the electron affinity.

Solution.   From Eq. (11-4-27) the electron affinity is

$$ \chi = \Phi + E_F - E_g = 1.08 + 0.01 - 1.43 = -0.34 \text{ eV} $$

It is a NEA photoemitter.

11-4-5 Josephson-Effect Detectors

The *Josephson effect* refers to certain properties of electric current flow through a barrier or a weak link between two superconductors [8–12]. A superconductor is a special type of metal, such as lead, niobium, and tin, that can carry electric current without resistance at certain very low temperatures below T_c. The critical temperature T_c for a superconducting Josephson effect is close to absolute zero in degrees Kelvin. According to the Bardeen–Cooper–Schrieffer (BCS) theory, some free electrons in the metal below the critical temperature T_c are formed into what are called *Cooper pairs*. These Cooper-pair electrons can carry a supercurrent without a voltage drop across the superconductor because the resistance is zero.

All superconducting point contacts, bridges, and tunnel junctions can carry a zero-voltage current or resistanceless current; in other words, no voltage appears across the device until the current exceeds a finite critical value. The critical value of the zero-voltage current is changed when a microwave radiation is incident on the superconducting junction. Thus a superconducting junction can be used as a detector of infrared radiation and its sensitivity and speed are very high compared to other infrared detectors.

Josephson-effect detectors have three basic structures: point contact, bridge, and tunnel junction (see Fig. 11-4-14).

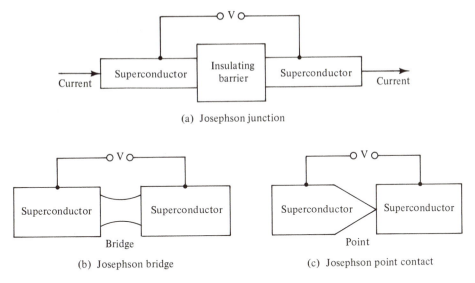

(a) Josephson junction

Bridge

(b) Josephson bridge

Point

(c) Josephson point contact

Figure 11-4-14   Basic structures of Josephson-effect detectors

(1) Josephson frequency-voltage equation.   According to Josephson's prediction, if two superconductors are separated by a very thin insulating film (10 to 30 Å) and a finite constant voltage is applied across the junction, in addition to the usual flow of dc current, an ac supercurrent will also flow at a frequency ν given by

$$\nu = \frac{2eV}{h} = 483 V \qquad \text{in GHz} \qquad (11\text{-}4\text{-}29)$$

where $e = 1.6 \times 10^{-19}$ C is the electron charge
V = applied voltage in millivolts
$h = 6.626 \times 10^{-34}$ J-s is Planck's constant

When superconductors are cooled to below the critical temperature T_c, the junction barrier resistance vanishes and the Cooper-pair electrons are going to tunnel the barrier with the emission of photons. One set of Cooper-pair electrons emits one photon, which means that $h\nu = 2eV$. This phenomenon is called *quantum mechanical tunneling*. The flow of supercurrent produces no voltage drop across the device. A Josephson junction can operate as both a local oscillator and mixer and a harmonic generator.

The upper frequency limit is in the range of 1000 GHz with several millivolts. The switching speed is approximately 10 psec.

(2) dc Josephson effect.   The dc current can be driven through a superconductor without developing any voltage across the junction. The maximum amplitude of this zero-voltage current is a quasi-periodic function of an applied magnetic field. When the Cooper-pair electrons tunnel through the superconducting junction from one end to the other, they maintain their phase coherence. Josephson has shown that when a dc voltage is applied across the junction, the phase difference θ between the two ends of a Josephson junction determines the Josephson current that may pass through the barrier as given by

$$I = I_0 \sin \theta \tag{11-4-30}$$

where I_0 is the maximum zero-voltage current that can be driven through the junction.

(3) ac Josephson effect.   When a dc bias voltage V_0 is applied to the junction of two superconductors, the phase angle is expressed as

$$\theta = \int_0^t \omega_i \, dt \tag{11-4-31}$$

where ω_i is the instantaneous angular frequency. In the absence of an RF source the instantaneous angular frequency is

$$\omega_i = \omega_0 = 2\pi\nu_0 = \frac{2eV_0}{\hbar} \tag{11-4-32}$$

where $\hbar = h/2\pi$ is the modified Planck constant. In the presence of an RF source, however, the instantaneous voltage v_i becomes

$$v_i = V_0 + V_1 \cos \omega_1 t + V_2 \cos \omega_2 t \tag{11-4-33}$$

The instantaneous angular frequency is

$$\omega_i = \frac{2ev_i}{\hbar} = \omega_0 \left(1 + \frac{V_n}{V_0} \cos \omega_n t \right) \tag{11-4-34}$$

where $\omega_n = 2\pi\nu_n$ is the angular frequency.

The phase angle can be expressed as

$$\theta = \omega_0 t + \frac{\omega_0 V_1}{\omega_1 V_0} \sin \omega_1 t + \frac{\omega_0 V_2}{\omega_2 V_0} \sin \omega_2 t \tag{11-4-35}$$

and supercurrent I becomes

$$\begin{aligned} I &= I_0 \cos \left(\omega_0 t + \frac{\omega_0 V_1}{\omega_1 V_0} \sin \omega_1 t + \frac{\omega_0 V_2}{\omega_2 V_0} \sin \omega_2 t \right) \\ &= I_0 \cos \left(\omega_0 t + \frac{2eV_1}{\hbar \omega_1} \sin \omega_1 t + \frac{2eV_2}{\hbar \omega_2} \sin \omega_2 t \right) \end{aligned} \tag{11-4-36}$$

By using standard trigonometric identities and the relations

$$\cos(X \sin \theta) = \sum_{n=-\infty}^{\infty} J_n(X) \cos n\theta \tag{11-4-37}$$

and

$$\sin(X \sin \theta) = \sum_{n=-\infty}^{\infty} J_n(X) \sin n\theta \tag{11-4-38}$$

where $J_n(X)$ is the nth-order Bessel function of the first kind, we obtain the supercurrent I of Eq. (11-4-36) as

$$I = I_0 \sum_{n=-\infty}^{\infty} \sum_{m=-\infty}^{\infty} J_n\left(\frac{2eV_1}{\hbar\omega_1}\right) J_m\left(\frac{2eV_2}{\hbar\omega_2}\right) \cos(\omega_0 t + n\omega_1 t + m\omega_2 t)$$

(11-4-39)

The dc current that flows in the step at zero voltage—namely, the zero-voltage current—is given by

$$I = I_0 J_0\left(\frac{2eV_1}{\hbar\omega_1}\right) J_0\left(\frac{2eV_2}{\hbar\omega_2}\right)$$

(11-4-40)

In conclusion, when a dc voltage V_0 is maintained across a superconducting junction, the Josephson current is oscillatory with a frequency proportional to the voltage determined by Eq. (11-4-29). Radiation emitted by these alternating Josephson currents has been observed by a number of workers. When a junction is exposed to an electromagnetic field, the dc voltage-current characteristic is modulated. Thus the Josephson junction can be used as a detector of radiation. The alternating Josephson currents are frequency modulated by RF voltages driven by the applied monochromatic radiation. Whenever one of the modulation sidebands occurs at zero frequency, a current step at constant voltage appears in the voltage-current characteristic of the junction. This voltage is given by

$$V_n = \frac{nh\nu}{2e} = \frac{n\hbar\omega_n}{2e}$$

(11-4-41)

where n is an integer.

The amplitude of the current in the nth step varies with the radiation power level of the nth-order Bessel function as shown in Eq. (11-4-39).

(4) Josephson-effect applications.

The Josephson-effect detector has a frequency response range extending into the infrared region. In addition, its sensitivity and speed are both very high. The basic detector for sensing low-frequency signals is the superconducting quantum interference device (SQUID). This type of detector uses a combination of Josephson tunneling and the magnetic flux quantization effect. There are two types of SQUID: dc SQUID and RF SQUID.

dc SQUID: Under operating conditions the dc SQUID generates an output voltage signal, which is a periodic function of the magnetic flux ϕ_0 threading a superconducting ring or loop as shown in Fig. 11-4-15.

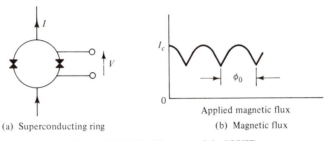

(a) Superconducting ring   (b) Magnetic flux

Figure 11-4-15 Diagrams of dc SQUID

Two Josephson junctions are mounted on the superconducting loop. The loop requires a quantized magnetic flux ϕ_0 for operation. The quantized magnetic flux is

$$\phi_0 = \frac{h}{2e} = 2.07 \times 10^{-15} \text{ weber} \tag{11-4-42}$$

The total ring flux must always equal $n\phi_0$, where n is an integer. The maximum critical current modulation is given by

$$\Delta I_c = \frac{\phi_0}{L} \tag{11-4-43}$$

where L is the loop inductance.

The maximum voltage variation across the device is then expressed by

$$\Delta V = \frac{R}{2} \frac{\phi_0}{L} \tag{11-4-44}$$

where R is the loop resistance.

The sensitivity of the dc SQUID is given by

$$\frac{\Delta V}{\Delta \phi_e} = \frac{R}{2L} \tag{11-4-45}$$

where $\Delta \phi_e$ is the small variation of the external magnetic flux.

Example 11-4-5 Sensitivity of a dc SQUID

A dc SQUID has a loop resistance R of 1 Ω and a loop inductance L of 10 nH. Find its sensitivity.

Solution. From Eq. (11-4-44) the sensitivity is

$$\Delta V = \frac{1}{2 \times 10^{-8}} \times 2.07 \times 10^{-15}$$

$$= 0.1 \ \mu V/\phi_0$$

RF SQUID: An RF SQUID is basically made by a superconducting loop with a single Josephson junction coupled to a radio-frequency-biased tank circuit as shown in Fig. 11-4-16.

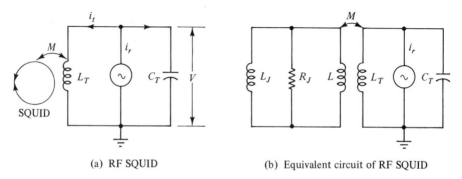

(a) RF SQUID   (b) Equivalent circuit of RF SQUID

Figure 11-4-16   Diagrams of RF SQUID

The equivalent circuit of the RF SQUID is shown in part (b) of Fig. 11-4-16. The superconducting loop is coupled to the tank-circuit inductance L_T via the mutual inductance M, where $M = k\sqrt{L_T L}$, where k is the mutual coupling coefficient. The RF current fed into the tank circuit is

$$i_r = I_r \sin \omega t \tag{11-4-46}$$

The resonant frequency of the tank circuit becomes

$$\omega_0 \simeq \frac{1}{\sqrt{L_T C_T}} \tag{11-4-47}$$

The RF current in the inductor L_T may be written

$$i_t = I_T \sin(\omega t + \theta) \tag{11-4-48}$$

The SQUID is subjected to both a dc magnetic flux ϕ_{dc} and an RF flux $MI_T \sin(\omega t + \theta)$. The total external flux is then

$$\phi_e(t) = \phi_{dc} + \phi_{rf} \sin(\omega t + \theta) \tag{11-4-49}$$

where $\phi_{rf} = MI_T$ is the magnitude of the RF magnetic flux. Peak RF voltage may be expressed as

$$V_{rf} = \omega L_T Q I_{rf} \tag{11-4-50}$$

where Q is the quality factor of the tank circuit. The voltage variation may be written

$$\Delta V = \frac{\omega L_T}{M} \frac{\phi_0}{2} = \frac{\omega \phi_0}{2k} \sqrt{\frac{L_T}{L}} \tag{11-4-51}$$

The sensitivity of the RF SQUID for a small variation of the dc magnetic flux is given by

$$\frac{\Delta V_{rf}}{\Delta \phi_{dc}} = \frac{\omega L_T}{M} = \frac{\omega}{k} \sqrt{\frac{L_T}{L}} \tag{11-4-52}$$

Example 11-4-5A   Sensitivity of an RF SQUID

An RF SQUID is operated under the following conditions:

Tank-circuit inductance	$L_T = 0.1\ \mu H$
Resonant frequency	$f = 10$ MHz
Loop inductance	$L = 2$ nH
Mutual coupling coefficient	$k = 0.3$

Determine the sensitivity of the RF SQUID.

Solution.   Using Eq. (11-4-51), we find that the sensitivity is

$$\Delta V = \frac{2\pi \times 10^7}{2 \times 0.3} \left(\frac{0.1 \times 10^{-6}}{2 \times 10^{-9}} \right)^{1/2} \times 2.07 \times 10^{-15}$$

$$= 10.47 \times 10^7 \times 7.07 \times 2.07 \times 10^{-15}$$

$$= 1.53\ \mu V/\phi_0$$

In addition, the SQUID can be used to sense very small voltages as a voltmeter. The sensitivity of the SQUID voltmeter can reach a level of 10^{-9} $\mu V/Hz^{1/2}$. Also, the SQUID may be used as a very sensitive thermometer or bolometer. Temperatures as low as a few thousandths of a Kelvin have been measured by this method. Furthermore, Josephson detectors can be used at microwave frequencies to detect signals in that frequency range. A single microwave signal coupled to a Josephson point-contact junction will give a step response with steps of constant voltage as shown in Eq. (11-4-41).

11-5 THERMAL DETECTORS

Thermal detectors are devices that operate under a thermal effect. The properties of the thermal detectors change with the temperature variations caused by the heating effect of the incident radiation. The heat effects do not depend on the photon nature of the incident energy and so they are generally wavelength independent. Thermal detectors have been used since the days of Herschel. Because most thermal detectors do not require cooling, they have found many applications in spaceborne satellites and military missiles (Falcon and Sidewinder).

The temperature dependence of a thermal detector is specified in terms of the temperature coefficient α of the element as

$$\alpha = \frac{1}{R}\frac{dR}{dT} \tag{11-5-1}$$

where R = resistance of the element
T = absolute temperature of the sample in degrees Kelvin

The temperature coefficient α could be either positive for a metal bolometer or negative for a thermistor bolometer, depending on the property of the material. Figure 11-5-1 shows a generalized diagram of a thermal detector formed on a substrate.

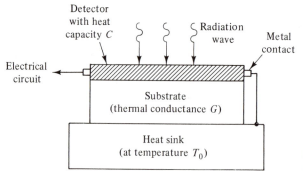

Figure 11-5-1 Generalized diagram of a thermal detector

The heat equation for thermal detectors for a sinusoidal radiation source can be expressed as

$$C\frac{d(\Delta T)}{dt} + G(\Delta T) = Pe^{j\omega t} \qquad (11\text{-}5\text{-}1a)$$

where ΔT = temperature rise of the thermal detector above the heat-sink temperature T_0 in degrees Kelvin
C = heat capacity of the thermal detector in joules per degrees Kelvin
G = thermal conductance between detector and heat sink in joules per
degrees Kelvin per second
P = peak radiation power in joules per second or in watts
ω = angular modulation frequency in radians

The solution of Eq. (11-5-1a) results in a steady-state temperature rise

$$\Delta T = \frac{P}{G}\left(1 + \omega^2\tau^2\right)^{-1/2} \qquad (11\text{-}5\text{-}1b)$$

where $\tau = C/G$ is the time constant in seconds.

When the frequency is low ($\omega\tau < 1$), the temperature rise ΔT is independent of heat capacity C. Because the time constant τ is increased with the heat capacity C, the heat capacity must be minimized in order to obtain a fast response. This condition can be achieved by decreasing the detector volume. Therefore the thickness and cross section of the detector substrate must be carefully chosen in order to obtain a minimum heat capacity.

11-5-1 Bolometers and Cryogenic Bolometer

(1) Bolometers.   *Bolometer-type detectors* are those devices that change their electrical resistance when heated by incident radiation. There are two types of bolometers: metal strip bolometer and thermistor bolometer.

Metal Strip Bolometer.   The elements of a metal strip bolometer have a positive temperature coefficient α of 0.3 to 0.4%/°C. The metal elements are either nickel or platinum. The structure may be in one arm of the Wheatstone bridge. Because the thermosensing elements of the metal strip have high reflectivity, they must be coated with a black absorber to improve absorption of incident radiation. For a platinum element coated with evaporated gold black, the standard receiver area is 7×0.3 mm^2, the resistance is 40 Ω, the flat response is from 1 to 26 μm, and the responsivity is 4 rms volts per peak-peak watt of radiation [13].

Thermistor Bolometer.   The thermistor is a thermally sensitive resistor that is constructed of an oxidic element of semiconductor material, such as manganese, cobalt, or nickel. The element exhibits a negative temperature coefficient of resistance; the resistance decreases as the temperature increases. The thermistor material is a good absorber of radiation and has a high negative temperature coefficient α of

4%/°C. So its responsivity is high. The resistance of a standard thermistor bolometer is from 1 to 5 MΩ and its noise levels are 1 mV rms or less. In response to a pulse of incident radiation energy E_i absorbed by the thermistor, the signal voltage for a thermistor bolometer in a matched bridge circuit is given by [14]

$$V_s = \frac{E_i \alpha V_b}{2C} (1 - e^{-t/\tau})$$ (11-5-2)

where E_i = incident radiation energy
  α = temperature coefficient
  V_b = bias voltage
  C = rate of heat conduction from the thermistor to the substrate
  τ = thermal response time

(2) Cryogenic bolometer.   The *cryogenic bolometer* is a detector that operates at a low temperature for far-infrared radiation. The common cryogenic bolometer is the carbon bolometer developed by Boyle and Rodgers [15]. Because the carbon composition resistor has a large negative temperature coefficient and a low specific heat, the carbon cryogenic bolometer is a highly sensitive detector at low temperatures of 4°K.

The other type of cryogenic bolometer can be constructed by a Ga-doped single crystal Ge or an In-doped Ge as the sensitive element operated at 4°K.

11-5-2 Thermocouple and Thermopile

(1) Thermocouple.   The *thermocouple* is a junction thermal detector constructed by two different metals that differ greatly in their thermoelectric power. The metal materials are bismuth-silver, copper-constantan, and bismuth-bismuth-tin alloy. Fine wires 3 or 4 mm long and 25 μm in diameter and made of two different materials are joined at one end to form a thermoelectric junction. The free ends of the wires are fastened to a metallic support to provide a reference at a constant temperature. The receiver, which is made from a blackened gold foil about 0.5 μm thick, is fastened directly to the junction. Because thermocouples are extremely fragile, they are rarely used outside the laboratory.

The time constant of a thermocouple may be expressed as

$$\tau = \frac{C}{A}$$ (11-5-3)

where C = thermal capacity of the junction and the receiver assembly
  A = rate of energy loss at the assembly

The responsivity of a thermocouple is given by

$$R = \frac{\tau}{C}$$ (11-5-4)

(2) Thermopile.   The *thermopile* is formed by several thermocouples in series and it has a higher voltage than a single thermocouple because of the addition of voltages at each junction together. Because its responsivity is higher than a single thermocouple, it can detect radiant heat from a person at a distance of 10 m. The time constant of an evaporated thermocouple can be as low as 10 ms.

11-5-3 Pyroelectric Detectors

The *pyroelectric detector* is the latest device operated under the pyroelectric effect. The pyroelectric effect is exhibited by temperature-sensitive ferroelectric crystals, including TGS (triglicine sulfate), SBN ($Sr_{1-x}Ba_xNb_2O_6$), PLZT (lanthanum-doped lead zirconate titanate), and $LiNbO_3$ [14]. Such crystals exhibit spontaneous electric polarization, which can be measured as a voltage by using electrodes attached to the sample. The voltage cannot be detected at a constant temperature, however, because the internal charge distribution is neutralized by free electrons and surface charges. If the temperature changes rapidly, the internal dipole moment will change and immediately produce a transient voltage. This pyroelectric effect can be used as a sensitive detector of modulated radiation operating at ambient temperatures.

The pyroelectric detector is a form of capacitor that has metallic electrodes applied to opposite surfaces of the temperature-sensitive ferroelectric crystals. When a modulated radiation is incident on the crystal, an alternating temperature change ΔT develops. As a result, an alternating charge ΔQ is generated on the external electrodes as

$$\Delta Q = pA\,\Delta T \tag{11-5-5}$$

where p = pyroelectric coefficient
A = area over which the incident radiation is absorbed

The alternating photocurrent is

$$i_s = pA\frac{d(\Delta T)}{dt} \tag{11-5-6}$$

The rms signal voltage is then given by

$$V_{s(\text{rms})} = \frac{I_{s(\text{rms})}R}{(1 + \omega^2R^2C^2)^{1/2}} \tag{11-5-7}$$

where R = parallel equivalent resistance
C = parallel equivalent capacitance

Example 11-5-3 Signal Voltage of a Pyroelectric Detector

A certain pyroelectric detector has the following parameters:

Photocurrent (rms)	$I_s = 10$ mA
Parallel equivalent resistance	$R = 10\ \Omega$
Parallel equivalent capacitance	$C = 10$ pF
Operating frequency	$f = 1$ GHz

Determine the signal voltage (rms).

Solution. From Eq. (11-5-7) the rms signal voltage is

$$V(\text{rms}) = \frac{10 \times 10^{-3} \times 10}{\left[1 + (2 \times 3.1416 \times 10^9 \times 10 \times 10 \times 10^{-12})^2\right]^{1/2}}$$
$$= 84.75 \text{ mV}$$

11-6 PHOTOTHERMIC DETECTORS

Photothermic detectors are devices that are based on the thermal variation of a photoemissive effect (or photoelectric effect). When a light image is focused on the target, a charge flow is induced. After the electron beam scans the target, a video signal is generated. There are two well-known types: Vidicon and Thermicon. These devices are used for commercial purposes in the TV industry and as military weapons in Walleye and Redeye missiles.

11-6-1 Vidicon

The *Vidicon* is a small television-type camera tube. It consists of an evacuated tube closed by a flat, transparent faceplate at one end and by an electron gun at the opposite end as shown in Fig. 11-6-1 [16].

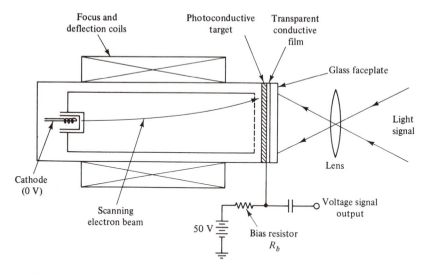

Figure 11-6-1 Schematic diagram of a Vidicon (From John Wilson and J. F. B. Hawkes [16]; reprinted by permission of Prentice-Hall, Inc.)

The faceplate, which is called the *signal electrode* or *plate*, is overcoated with a thin layer (1 to 2 μm) of photoconductive semiconductor film, such as antimony trisulfide (Sb_2S_3) for image illumination. The electron gun, which is called the *cathode electrode*, emits the electron beam that scans over the photoconductive layer (usually called the *target*). A uniform magnetic field is maintained to focus the beam. The video signal is taken from the target by connecting the amplifier to the transparent signal plate.

A fixed potential of about 20 volts positive, relative to the thermionic cathode, is applied to the transparent signal plate. The electron beam deposits electrons on the scanned surface of the photoconductor, charging it down to the thermionic cathode potential in the absence of any radiation. In effect, the conductivity of the photoconductor is sufficiently low that very little current flows in the dark.

If a light image is focused on the target, its conductivity is increased to about 10^{-12} mho in the illuminated areas, thus permitting charge to flow. In these areas the scanned surface gradually becomes charged a volt or two positive with respect to the cathode during the frame time ($1/30$ sec) between successive scans.

The electron beam deposits sufficient electrons to neutralize the charge accumulated during the frame time; and in doing so it generates the video signal in the signal plate lead. Because the target is sensitive to light throughout the entire frame time, a full storage of charge is achieved.

Many materials, such as lead oxide (SnO_2), a CdSe-based structure, and a multilayer Se(As, Te) structure, are capable of operating with a low-faceplate illumination and a lag of 10% for a 200-nA signal current. The spectral response is up to 13 μm. These three types of photoconductive materials for the surface structures of Vidicons are shown in Fig. 11-6-2 [16].

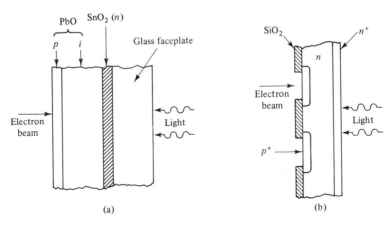

Figure 11-6-2 Surface structures of Vidicons (From John Wilson and J. F. B. Hawkes [16]; reprinted by permission of Prentice-Hall, Inc.)

11-6-2 Thermicon

Thermicon is the trade name for one of the photothermionic image converters operated under the thermal variation of photoemission. The schematic diagram of a Thermicon is shown in Fig. 11-6-3 [17].

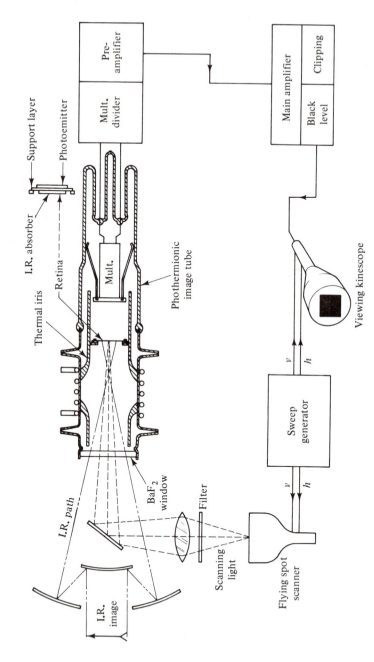

Figure 11-6-3 Schematic diagram of a Thermicon (After M. Garbuny [17]; copyright 1961, Optical Society of America.)

The target, which is called the *retina* in this tube, is a multilayer film (0.5 to 0.7 μm). An evacuated tube envelope, equipped with a sealed window of barium fluoride or sodium chloride, surrounds the retina.

The retina consists of several layers. The first serves the purpose of absorbing infrared but transmitting visible light, the second layer provides structural support, and the third is the temperature-sensitive photocathode. When a thermal radiation is incident on the retina, a temperature distribution is produced and then probed by a suitably filtered light beam from a flying spot scanner. As the light spot moves across the face of the retina, it triggers a current from the photolayer. The magnitude of the current is then modulated by the temperature distribution. As the current is received by the anode, it is amplified and impressed on the intensity control grid of a viewing monitor. If the electron beam of the viewing monitor is swept in synchronism with the flying spot scanner, it will produce, point by point, a visible reproduction of the thermal image.

REFERENCES

[1] Keyes, R. J. *Optical and Infrared Detectors*. Berlin: Springer-Verlag, 1980, p. 12.

[2] Gartner, W. W. Depletion-layer photo-effects in semiconductors. *Phys. Rev.*, **116** (1959), 84.

[3] Sze, S. M. *Physics of Semiconductor Devices*. 2nd ed. New York: John Wiley & Sons, Inc., 1981, pp. 769, 800.

[4] Long, D. Photovoltaic and photoconductive infrared detectors. In R. J. Keyes, ed., *Optical and Infrared Detectors*. Berlin: Springer-Verlag, 1980.

[6] Seymour, J. *Electronic Devices and Components*. New York: John Wiley & Sons, 1981, p. 209.

[7] Zwicker, H. R. Photoemissive detectors. In R. J. Keyes, ed., *Optical and Infrared Detectors*. Berlin: Springer-Verlag, 1980, pp. 158, 161.

[8] Josephson, Brian D. Possible new effects in superconductive tunnelling. *Phys. Lett.*, **1**, no. 7 (July 1962), pp. 251–253.

[9] Josephson, Brian D. Coupled superconductors. *Rev. Mod. Phys.*, **36** (January 1964), 216–220.

[10] Josephson, Brian D. The discovery of tunnelling supercurrents. *Rev. Mod. Phys.*, **46**, (January 1974), 251–254.

[11] Grimes, C. C., et al. Josephson-effect far-infrared detectors. *J. Appl. Phys.*, **39**, no. 8 (July 1968), 3905–3912.

[12] Grimes, C. C., and Sidney Shapiro, Millimeter-wave mixing with Josephson junctions. *Phys. Rev.*, **169**, no. 2 (May 1968), 379–406.

[13] De Waard, R., and E. M. Wormser. Description and properties of various thermal detectors. *Proc. IRE.*, **47**, no. 9 (September 1959), 1508–1513.

[14] Kruse, P. W. The photon detection process. In R. J. Keyes, ed., *Optical and Infrared Detectors*. Berlin: Springer-Verlag, 1980, p. 26.

[15] Boyle, W. S., and K. F. Rodgers, Jr. Performance characteristics of a new low-temperature bolometer. *J. Opt. Soc. Am.*, **49**, no. 1 (January 1959), 66–69.

[16] Wilson, John, and J. F. B. Hawkes. *Optoelectronics: An Introduction*. Englewood Cliffs, N.J.: Prentice-Hall, Inc., 1983, pp. 325, 326, 328.

[17] Garbuny, M., et al. Image converter for thermal radiation. *J. Opt. Soc. Am.*, **51**, no. 3 (March 1961), 261–273.

SUGGESTED READINGS

1. Arams, F. R. *Infrared-to-Millimeter Wavelength Detectors*. Dedham, Mass.: Artech House, 1973.

2. Barone, Antonio and G. Paterno. *Physics and Applications of the Josephson Effect*. New York: John Wiley & Sons, 1982.

3. Hudson, R. D., Jr., and J. W. Hudson. *Infrared Detectors*. Stroudsburg, Pa.: Dowden, Hutchinson and Ross, Inc., 1975.

4. *IEEE Transactions on Electron Devices*. Special issues on light sources and detectors.
   Vol. ED-27, No. 4, April 1980.
   Vol. ED-27, No. 10, October 1980.
   Vol. ED-28, No. 4, April 1981.

5. Keyes, R. J. ed. *Optical and Infrared Detectors*. Berlin: Springer-Verlag, 1980.

6. Kingston, R. H. *Detection of Optical and Infrared Radiation*. Berlin: Springer-Verlag, 1978.

7. Milnes, A. G. *Semiconductor Devices and Integrated Electronics*, Chapters 12 and 13. New York: Van Nostrand Reinhold Company, 1980.

8. Solymar, L. *Superconductive Tunneling and Applications*. New York: Wiley-Interscience, 1972.

9. Sze, S. M. *Physics of Semiconductor Devices*, Chapters 13 and 14. 2nd ed. New York: John Wiley & Sons, 1981.

10. Wilson, John, and J. F. B. Hawkes. *Optoelectronics: An Introduction*. Englewood Cliffs, N.J.: Prentice-Hall, Inc., 1983.

PROBLEMS

11-1   Radiation Detectors

11-1-1. A radiation detector is used to detect the infrared energy at a wavelength of 10 μm. What is the Debye temperature at that frequency?

11-1-2. A radiation detector is used to detect the infrared at a wavelength of 3 μm.
    (a) Find the frequency for the infrared.
    (b) Determine the bandgap energy in electronvolts.

(c) Compute the photon energy $h\nu$ in watt-seconds (or joules).

(d) Calculate the quantum condition temperature.

11-2 Detector Figures of Merit

11-2-1. A detector has a sensitive area of 2 cm^2 and an output signal voltage of 10 volts (rms) with an irradiance of 4 W/cm^2. Determine the responsivity of the detector.

11-2-2. A detector has a sensitive area of 3 cm^2, an output signal voltage of 8 volts (rms), and a signal-to-noise ratio of 8. The irradiance is 4 W/cm^2.
(a) Determine the noise equivalent power (NEP) of the detector.
(b) Find the detectivity D of the detector.

11-2-3. A detector has the following parameters:

$$V_s = 10 \text{ V(rms)} \qquad V_n = 0.50 \text{ V(rms)}$$
$$H = 4 \text{ W/cm}^2 \qquad A_d = 5 \text{ cm}^2$$
$$B = 10^{15} \text{ Hz}$$

(a) Determine the noise equivalent power (NEP) of the detector.
(b) Find the detectivity D.
(c) Calculate the specific detectivity D^*.

11-2-4. The current responsivity of a radiation detector is expressed by Eq. (11-2-2). Verify the equation.

11-2-5. A certain photon detector has the following parameters:

Detecting area	$A_d = 150 \times 10^{-6} \text{ m}^2$
Wavelength	$\lambda_0 = 2 \ \mu\text{m}$
Quantum efficiency	$\eta_d = 30\%$
Bandwidth	$B = 1 \text{ Hz}$
Noise current	$I_n = 10 \text{ pA}$

(a) Find the responsivity $R(f)$.
(b) Compute the noise equivalent power (NEP).
(c) Calculate the detectivity D.
(d) Determine the specific detectivity D^*.

11-3 Modes of Operation

11-3-1. Describe the principles of operation for photon-effect, thermal-effect, and photothermic-effect detectors.

11-3-2. Explain the characteristics and applications of photon, thermal, and photothermic detectors.

11-4 Photon Detectors

11-4-1. A GaAs photoconductive detector is used to detect infrared signals. Its electron concentration n is 10^{13} cm^{-3} and its hole concentration p is 10^7 cm^{-3}.
(a) Determine the conductivity of the detector.
(b) Find the current density if the applied electric field is 100 V/cm.

11-4-2. A certain *pin* photodiode has an absorption coefficient α of 5×10^3 cm^{-1} at $\lambda = 2$ μm. Its reflection coefficient R is 0.10 and its i-region width W is 10 μm. The incident signal power P_{in} is assumed to be 4 W and the detecting area A is 4 cm^2.
(a) Determine the incident photon flux density.
(b) Calculate the drift current density.

11-4-3. A certain avalanche photodiode has an i-region W of 14 μm and an absorption coefficient α $(\alpha_n = \alpha_p)$ of 7×10^3 cm^{-1}.
   (a) Determine the multiplication factor M.
   (b) If both the leakage and photocurrents are considered, calculate the combined multiplication factor M for $V_b = 40$ V, $V_r = 39$ V and $n = 2$.

11-4-4. A Si-solar cell has the following parameters:

Absorbing area	$A = 2$ cm^2
Absorption coefficient	$\alpha = 10^3$ cm^{-1} at $h\nu = 1.6$ eV
Illuminated current	$I_i = 50$ mA/cm$^2 \times A = 100$ mA
$V_{oc} = 0.7$ V	$I_s = 1$ μA
R_{sh} = very large	R_s = very large
$V = 1.5$ V	

   (a) Determine the load current I_ℓ.
   (b) Compute the maximum extracted power.
   (c) How many cells are required to generate 100 W?

11-4-5. A phototransistor has the following parameters:

Incident radiation power	$P_{in} = 5$ mW
Efficiency	$\eta = 90\%$
dc current gain	$h_{FE} = 50$
Frequency	$\nu = 4.5 \times 10^{14}$ Hz (red)
Reverse collector saturation	$I_{CBO} = 10$ μA
leakage current with emitter open	

   (a) Determine the photon generation rate.
   (b) Calculate the photocurrent for electron-hole pair.
   (c) Compute the collector current.

11-4-6. A certain PEA (positive-electron-affinity) semiconductor photocathode has a band-gap energy E_g of 1.20 eV and a work function Φ of 1.30 eV. Its Fermi level is 0.30 eV above the maximum valence band level E_v.
   (a) Determine the electron affinity χ of the material.
   (b) Find the photoemissive threshold energy E_{th} for electrons in the semiconductor.

11-4-7. A certain NEA (negative-electron-affinity) semiconductor photocathode has a band-gap energy E_g of 1.60 eV, a Fermi level 0.20 eV above the maximum valence band E_v, and a work function Φ of 1.0 eV.
   (a) Determine the electron affinity χ of the material.
   (b) Find the photoemissive threshold energy E_{th}.

11-4-8. A photocathode is illuminated with radiation of wavelength 0.60 μm. The cathode has a work function Φ of 1.10 eV. Determine the anode voltage required to produce a zero anode current.

11-4-9. If the anode voltage for a photocathode is 40 volts with respect to the cathode and the cathode is illuminated with a radiation of wavelength 0.30 μm, determine the electron velocity at the anode.

11-4-10. Describe the advantages of the NEA photocathode over the classic metal photocathode.

11-4-11. A single silicon solar cell with an area of 2 cm^2 has an illuminated current I_i of 50 mA and a diode saturation current I_s of 0.1 nA. It is operated at 325°K.
  (a) Compute the open-circuit voltage V_{oc} of a single cell.
  (b) Calculate the short-circuit current I_{sc} of a single cell.
  (c) Find the extracted power from a single cell.
  (d) If a system requires a power of 8 W at a voltage level of 8 volts, determine the number of solar cells in series and the number of rows in parallel for this array to meet the power requirement.

11-4-12. A single GaAs solar cell with an area of 4 cm^2 has an illuminated current I_i of 100 mA and a diode saturation current I_s of 0.01 $\mu\mu$A. It is operated at 328°K.
  (a) Compute the open-circuit voltage V_{oc} of a single cell.
  (b) Calculate the short-circuit current I_{sc} of a single cell.
  (c) Find the extracted power from a single cell.
  (d) If a system requires a power of 10 W at a voltage level of 12.5 volts, determine the number of GaAs solar cells in series and the number of rows in parallel for this solar array to meet the power requirement.

11-4-13. A Josephson-effect detector is operating at an upper frequency limit of 100 GHz. Determine the minimum signal voltage which the detector can detect.

11-4-14. A dc SQUID has a loop resistance R of 1.5 ohms and a loop inductance L of 8 nH. Determine its sensitivity.

11-4-15. An RF SQUID has the following parameters:

Tank-circuit inductance	$L_T = 0.2$ nH
Resonant frequency	$f = 20$ GHz
Loop inductance	$L = 3$ nH
Mutual coupling coefficient	$k = 0.4$

Determine the sensitivity of the RF SQUID.

11-5 Thermal Detectors

11-5-1. The rms signal voltage of a pyroelectric detector is expressed by Eq. (11-5-7). Verify the equation.

11-5-2. Describe the characteristics of the following detectors:
  (a) Bolometers and cryogenic bolometers
  (b) Thermocouple and thermopile
  (c) Pyroelectric detectors.

11-6 Photothermic Detectors

11-6-1. Describe the operational principles and applications of the Vidicon.

11-6-2. Describe the operational principles and applications of the Thermicon.

Chapter 12

High-Speed Switching Devices

12-0 INTRODUCTION

For microwave digital circuit applications, it is often desirable to use two-state high-speed devices to change from the binary 0 state (ON, closed, or conducting state) to the 1 state (OFF, open, or nonconducting state). When the device is driven from one extreme condition to the other, it operates much like a switch. The switching time may be in the range of nanoseconds or less. The devices should also have high sensitivity to light intensity or current injection. The metal-insulator-semi-conductor switch (MISS) diode, the microwave switching diodes, and switching transistors are capable of meeting the high-speed requirements for microwave digital circuit applications. In this chapter the electronic operations of these three devices are described.

12-1 METAL-INSULATOR-SEMICONDUCTOR SWITCH (MISS) DIODES

The *metal-insulator-semiconductor switch* (MISS) diode was proposed in 1977 by Simmons and EL-Badry [1, 2, 3]. The device has two distinct operational states: a low-impedance or conducting state and a high-impedance or nonconducting state. Moreover, it is compatible with integrated circuit (IC) fabrication techniques. Two switching mechanisms (punchthrough and avalanche) are offered to explain the operations of the devices. It should be noted that the MISS structure differs from the MIS diode, which has no p^+ layer, as described in Section 3-4.

12-1-1 Physical Structure

The MISS diode is a four-layer and three-terminal structure and is shown in Fig. 12-1-1 with its V-I characteristics.

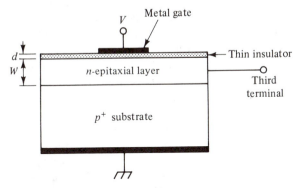

(a) Diagram of a MISS diode

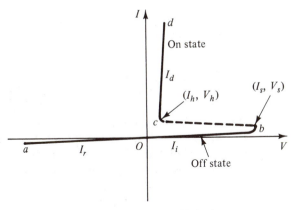

(b) V-I curves of a MISS diode

Figure 12-1-1 Diagram of a MISS diode and its V-I curves (After S. M. Sze [5]; reprinted by permission of John Wiley & Sons, Ltd.)

The metal electrode may be Al, Mb, or Pt, the insulator may be a very thin SiO_2 or Si_3N_4 ($15\ \text{Å} < d < 40\ \text{Å}$), and the thick n and p^+ layers are made of Si or GaAs, with a third terminal connected to the n layer for current injection. An ohmic contact is joined to the p^+ layer to serve as the ground electrode.

12-1-2 Switching Mechanisms

The MISS diode has two switching mechanisms, punchthrough and avalanche, depending on the doping of the epitaxial n layer in such a magnitude that the avalanche voltage V_{av} for that particular doping is less than the punchthrough voltage V_{pt} for that doping.

When the MISS diode is reverse biased, the section of the n layer under the insulator-semiconductor interface goes to deep depletion. The current through the device is limited by generation in the deeply depleted region. This is the high-imped-ance or OFF state of the device.

When the MISS diode is forward biased by a sufficiently high voltage, the switching voltage V_s, the n-p^+ junction is turned on by either avalanching in the n layer or by the deep depletion region extending through to the n-p^+ region (punchthrough). When the junction turns on, the n section goes from deep depletion toward inversion. Thus the voltage across the device decreases with a concomitant increase in the current through the device. This is the switching mode. The switch-ing voltage can be tailored by varying the doping and/or width of the n layer.

Following switching, the device comes into the steady state when the current through the insulating layer is equal to the current flowing across the n-p^+ junction. The V-I characteristic of this highly conducting (ON state) mode is determined predominantly by the V-I characteristic of the semi-insulating film. By a suitable choice of material, this portion of the characteristic can approach zero dynamic impedance—that is, a near-vertical characteristic—controlled by a low holding voltage.

(1) Punchthrough mechanism.   Figure 12-1-2 shows the energy band dia-grams of a MISS diode in the punchthrough mechanism [1].

Reverse-bias mode.   The MISS diode considered is a metal-oxide-n-p^+ structure. In order to reverse bias the n-p^+ junction, the applied gate voltage must be positive with respect to ground. Figure 12-1-2(a) is an idealized diagram of the diode at thermal equilibrium under zero bias. Figure 12-1-2(b) illustrates the diode under reverse bias across the n-p^+ junction (that is, positive applied gate voltage). It is assumed that the conductivity of the insulator oxide is much greater than that of the reverse-biased n-p^+ junction. Thus most of the applied voltage will drop across the reverse-biased n-p^+ junction and only a small portion will appear across the thin insulator layer. As a result, the tunneling current (or leakage current or dark current) flows through the diode from the p^+ layer to the n region.

Forward-bias mode.   The forward-bias mode has three states: OFF state (high-impedance state), switching state, and ON state (low-impedance state).

OFF State.   As the applied voltage increases negatively from zero to some value, the free electrons are swept out of the epitaxial n layer through the n-p^+ junction and the holes are swept to the oxide-semiconductor interface. A depletion region grows in the n layer just under the metal electrode and electron-hole pairs are generated in this depletion region. As the voltage increases more negatively, the n layer goes into deep depletion. If the oxide is thick enough, the impedance of the device is very high. The energy diagram for the OFF state is shown in Fig. 12-1-2(c).

Switching State.   When the applied voltage is increased negatively to the level of the switching voltage V_s, the depleted region of the semiconductor under the oxide

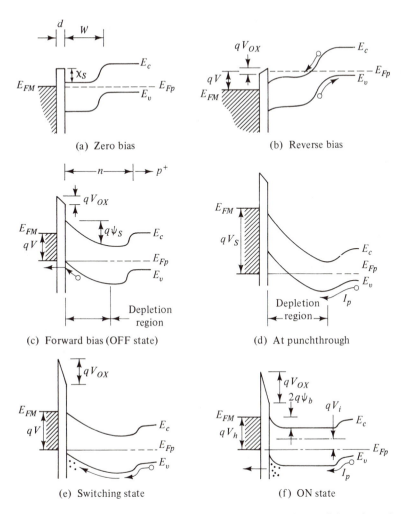

Figure 12-1-2 Energy-band diagrams of a MISS diode in punchthrough mechanism. (After J. G. Simmons and A. EL-Badry [1]; reprinted by permission of Pergamon Press, Great Britain.)

extends to the n-p^+ junction as shown in Fig. 12-1-2(d). At this point, the punchthrough effect occurs. In other words, the holes are swept directly from the p^+ layer through the n-p^+ junction toward the n layer. The accumulation of holes at the insulator-semiconductor interface results in two interacting effects:

1. The n layer begins to move from deep depletion toward inversion as shown in Fig. 12-1-2(e); consequently, the surface potential ψ_s and the voltage across the diode begin to decrease.

2. The field in the oxide begins to increase, thus allowing a larger current to pass through the n-p^+ junction. In this manner, a negative resistance region develops as shown in Fig. 12-1-1(b).

ON State. The holding voltage [Fig. 12-1-2(f)] is given by

$$V_h = V_{ox} + \psi_s + V_j \qquad (12\text{-}1\text{-}1)$$

where V_{ox} =   voltage drop across the oxide

$\psi_s = 2\psi_b$ is the surface potential

$\psi_b = \dfrac{E_F - E_i}{q}$ is the potential difference between the Fermi level E_F and the intrinsic Fermi level E_i for an n-type semiconductor

$V_j =$ voltage across the n-p^+ junction

At the holding-voltage point, the device is operating in the low-impedance state or ON state.

(2) Avalanche mechanism. If the doping of the epitaxial layer is chosen so that the avalanche voltage V_{av} is less than the punchthrough switching voltage V_s, the device will operate via the avalanche effect. Figure 12-1-3 shows the energy band diagrams for a MISS diode in the avalanche mechanism [1].

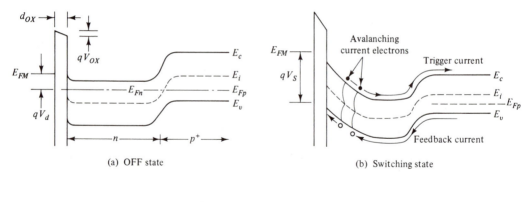

(a) OFF state   (b) Switching state

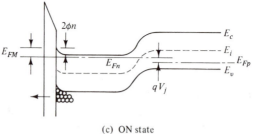

(c) ON state

Figure 12-1-3 Energy-band diagrams for a MISS diode in avalanche mechanism (After J. G. Simmons and A. EL-Badry [1]; reprinted by permission of Pergamon Press, Great Britain.)

The avalanche mechanism has three modes: OFF-state mode (high-impedance state), switching mode, and ON-state mode (low-impedance state).

OFF-State Mode.   The energy band diagram for the OFF-state mode in the avalanche mechanism as shown in Fig. 12-1-3(a) is similar to that for the punchthrough mechanism. As the avalanche voltage V_{av} is approached, however, the carriers generated in the depletion region of the epitaxial n layer are multiplied. The multiplication factor is given by

$$M = \frac{1}{1 - (V/V_{av})^{1/2}} \qquad (12\text{-}1\text{-}2)$$

where $V_{av} = 60 \left(\dfrac{E_g}{1.1} \right)^{3/2} \left(\dfrac{N_d}{10^{16}} \right)^{-3/4}$ is the avalanche voltage

$E_g =$ bandgap energy in electronvolts
$N_d =$ carrier density of the epitaxial layer and diffused well per cubic centimeter

Example 12-1-2   Multiplication Factor of a MISS Diode

A certain MISS diode has the following parameters:

Electron density of Si epitaxial layer	$N_d = 10^{17} \, cm^{-3}$
Bandgap energy	$E_g = 1.12 \, eV$
Applied voltage	$V = 10 \, V$

(a) Compute the avalanche voltage V_{av}.
(b) Calculate the multiplication factor M.

Solution.   (a) The avalanche voltage is

$$V_{av} = 60 \times \left(\frac{1.12}{1.10} \right)^{3/2} \times \left(\frac{10^{17}}{10^{16}} \right)^{-3/4}$$

$$= 11 \text{ volts}$$

(b) The multiplication factor is

$$M = \frac{1}{1 - (10/11)^{1/2}} = 20$$

Switching Mode.   When the applied voltage is equal to the avalanche voltage V_{av}, the large electron current originating in the depletion region is forced across the $n\text{-}p^+$ junction as shown in Fig. 12-1-3(b).

ON-State Mode.   When the switching process is underway, the electron current causes the $n\text{-}p^+$ junction to become forward biased and the hole current is inverted at the surface of the epitaxial n layer as shown in Fig. 12-1-3(c).

12-1-3 Voltage-Current Characteristics

As shown in Fig. 12-1-1(b), the current of a MISS diode for both the punchthrough and avalanche mechanisms has five parts: reverse current I_r at the reverse-bias mode, initial forward current I_i at the OFF-state mode, switching current I_s at the switching point, holding current I_h at the holding point, and diode current I_d at the ON-state mode.

(1) Reverse current I_r.   When the n-p^+ junction of a MISS diode is reverse biased by a positive voltage, the electrons generated in the junction tunnel through the oxide layer and the generated holes are swept toward the p^+ layer. The reverse current I_r can be expressed from Eqs. (3-4-4) and (3-4-8) as

$$I_r = \frac{q n_i A}{2 \tau_g} W_j = \frac{n_i A}{\tau_g} \left(\frac{q \epsilon_s}{2 N_d} \right)^{1/2} (V + V_b)^{1/2} \qquad \text{amperes} \qquad (12\text{-}1\text{-}3)$$

where $n_i =$ intrinsic carrier concentration per cubic centimeter
$A =$ device area in square centimeters
$W_j =$ n-p^+ junction width in centimeters
$\tau_g = (\sigma v_{th} N_t)^{-1}$ is the minority-carrier generation lifetime in seconds
$\sigma =$ trap cross section in square centimeters
$v_{th} =$ thermal velocity in centimeters per second
$N_t =$ trap density per cubic centimeter
$V =$ applied voltage in volts
$V_b =$ built-in potential of the n-p^+ junction in volts
$V_j =$ junction voltage in volts
$V_{ox} =$ oxide voltage in volts
$V_j \gg V_{ox}$ is assumed
$\epsilon_s =$ semiconductor permittivity
$N_d =$ electron concentration per cubic centimeter
$q =$ electron charge in coulombs

This reverse current I_r is illustrated by the portion a to O of the V-I curves shown in Fig. 12-1-1(b).

(2) Initial forward current I_i.   When the applied voltage increases negatively from zero to some value, a very small initial forward current I_i begins to flow. Under the deep depletion condition the depletion-layer width as shown in Eq. (3-4-5) is given by

$$W_d = \left(\frac{2 \epsilon_s \psi_s}{q N_d} \right)^{1/2} \qquad (12\text{-}1\text{-}4)$$

Because $V \gg V_{ox}$, $V \gg V_j$, and $V \gg V_b$, we obtain $V = \psi_s$. The initial forward current becomes

$$I_i = \frac{n_i A}{\tau_g} \left(\frac{q \epsilon_s V}{2 N_d} \right)^{1/2} \qquad (12\text{-}1\text{-}5)$$

This initial forward current I_i is shown by the curve O-b of the V-I curves in Fig. 12-1-1(b). The current is only weakly voltage dependent but is strongly temperature dependent through n_i.

(3) Switching current I_s.   When applied voltage V is equal to switching voltage V_s, the punchthrough effect occurs and the depletion-layer width becomes

$$W_m \simeq W_n - W_j \qquad (12\text{-}1\text{-}6)$$

where W_n = epitaxial n-layer width
   W_j = n-p^+ junction width

Then from Eq. (12-1-4) the surface potential at the punchthrough point is expressed as

$$\psi_{\text{pt}} = \frac{qN_d\left(W_n - W_j\right)^2}{2\epsilon_s} \qquad (12\text{-}1\text{-}7)$$

Because the oxide voltage V_{ox} and the junction voltage V_j are very small, the applied voltage can be written

$$V \simeq V_s \simeq \psi_{\text{pt}} = \frac{qN_d\left(W_n - W_j\right)^2}{2\epsilon_s} \qquad (12\text{-}1\text{-}8)$$

Just before punchthrough, the voltage across the oxide was sufficient to pass only a small amount of current generated in the n layer. Immediately after punchthrough, the electric field in the oxide is insufficient to allow the relatively large hole current in the n layer to pass through the insulator. In effect, the injected holes are accumulated at the insulator-semiconductor interface. The switching current is given by

$$I_s \simeq \frac{V - V_s}{R_\ell} \qquad (12\text{-}1\text{-}9)$$

where R_ℓ is the external load resistance. This switching current I_s is shown at point b in Fig. 12-1-1(b).

(4) Holding current I_h.   Immediately after the switching state takes place, the holding voltage V_h can be expressed as

$$V_h = V_{\text{ox}} + V_{jo} + \psi_s \qquad (12\text{-}1\text{-}10)$$

where V_{jo} is the turn-on voltage of the n-p^+ junction. Then the holding current in the low-impedance state is given by

$$I_h = \frac{V_s - V_h}{R_\ell} \qquad (12\text{-}1\text{-}11)$$

This holding current I_h is shown at point c in Fig. 12-1-1(b).

(5) Diode current I_d.   When the device is ON, the diode current is controlled by the load resistor as

$$I_d = \frac{V - V_h}{R_\ell} \qquad (12\text{-}1\text{-}12)$$

This diode current I_d is shown at point d in Fig. 12-1-1(b).

Example 12-1-3 Operation of a GaAs MISS Diode

A typical GaAs MISS diode has the following parameters:

Electron concentration	$N_d = 8 \times 10^{15} \, \text{cm}^{-3}$
Depletion-layer width	$W_d = 0.4 \, \mu\text{m}$
Difference between W_n and W_{di}	$W_n - W_{di} = 0.5 \, \mu\text{m}$
Difference between W_n and W_j	$W_n - W_j = 1 \, \mu\text{m}$
Relative dielectric constant	$\epsilon_r = 13.10$

(a) Calculate the surface potential ψ_s.
(b) Compute the punchthrough voltage ψ_{pt}.
(c) Estimate the holding voltage V_h.

Solution. (a) By using Eq. (12-1-4), the surface potential is obtained as

$$\psi_s = \frac{qN_dW_d^2}{2\epsilon_s} = \frac{1.6 \times 10^{-19} \times 8 \times 10^{21} \times (4 \times 10^{-7})^2}{2 \times 8.854 \times 10^{-12} \times 13.10}$$

$$= 0.88 \text{ volts}$$

(b) Then from Eq. (12-1-7) the punchthrough voltage is

$$\psi_{pt} = \frac{qN_d(W_n - W_j)^2}{2\epsilon_s} = \frac{1.6 \times 10^{-19} \times 8 \times 10^{21} \times (1 \times 10^{-6})^2}{2 \times 8.854 \times 10^{-12} \times 13.10}$$

$$= 5.52 \text{ volts}$$

(c) The holding voltage is

$$V_h \simeq \psi_s = 0.88 \text{ volts}$$

12-1-4 Switching Time and Applications

Switching time τ. The *switching time* refers to the time that it takes a MISS diode to switch from the high- to the low-impedance state. In other words, the switching time is the time for the accumulated charges to change from zero to the charges stored in the metal section of the epitaxial *n* layer [1]. That is,

$$I_h\tau \le Q_{st} \tag{12-1-13}$$

or

$$\tau \le \frac{1}{2D_p}(W_n - W_{di})^2 \tag{12-1-14}$$

where $D_p = \dfrac{\mu_p kT}{q}$ is the hole diffusion coefficient in centimeter squared per second

$W_{di} = \quad$ depletion-layer width at inversion in micrometers

Example 12-1-4 Switching Time of a Si MISS Diode

A Si MISS diode has the following parameters:

Hole mobility	$\mu_p = 600 \, \text{cm}^2/\text{V·sec}$
Temperature	$T = 300°\text{K}$
Epitaxial *n*-layer width	$W_n = 5 \, \mu\text{m}$
Depletion-layer width for inversion	$W_{di} = 3 \, \mu\text{m}$

(a) Find the hole diffusion coefficient in centimeters squared per second.
(b) Determine the switching time of the diode.

Solution.   (a) From Eq. (12-1-14) the hole diffusion coefficient is

$$D_p = \frac{\mu_p kT}{q} = \frac{600 \times 1.38 \times 10^{-23} \times 300}{1.6 \times 10^{-19}} = 15.53 \text{ cm}^2/\text{sec}$$

(b) The switching time is

$$\tau = \frac{1}{2D_p}(W_n - W_{di})^2 = \frac{(5 \times 10^{-6} - 3 \times 10^{-6})^2}{2 \times 15.53 \times 10^{-4}}$$

$$= 1.29 \text{ nsec}$$

Microwave applications.   The MISS diodes are potentially useful in microwave high-speed switching logic circuits.

12-2 FOUR-LAYER p-n-p-n THYRISTOR

Another device that finds extensive applications in switching circuits is the thyristor. The name *thyristor* is derived from *gas thyratron*, because the electronic properties of both devices are similar in many respects. Thyristors are commonly used for speed control in home appliances and for switching purposes in high-voltage transmission lines. At present, thyristors are available with current ratings from a few milliamperes to over 5 kA and voltage ratings up to 10 kV. The switching time for the forward transition from OFF to ON is less than 0.1 μsec and the time for the reverse transition is about 0.2 μsec. The basic structure of a thyristor is the four-layer *p-n-p-n* diode.

12-2-1 Physical Structure

The *p-n-p-n* diode consists of four layers of silicon doped alternately with *p*- and *n*-type impurities as shown in Fig. 12-2-1(b). The terminal *p* region is the anode, or *p* emitter, and the terminal *n* region is the cathode, or *n* emitter, as shown in Fig.

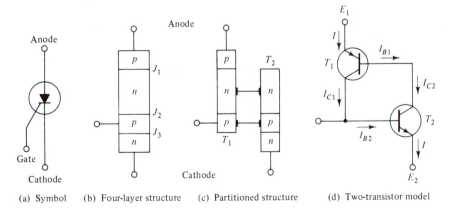

(a) Symbol   (b) Four-layer structure   (c) Partitioned structure   (d) Two-transistor model

Figure 12-2-1   Schematic diagrams of a *p-n-p-n* diode

12-2-1(b). This four-layer p-n-p-n diode can be decomposed into two transistors: one is the p-n-p transistor T1 and the other is the n-p-n transistor T2 as shown in Fig. 12-2-1(c). The equivalent circuit of the two-transistor analogy is illustrated in Fig. 12-2-1(d).

12-2-2 Operational Mechanism

Using the equivalent circuit as shown in Fig. 12-2-1(d), we can express the two collector currents, respectively, as

$$I_{C1} = \alpha_1 I + I_{CBO1} \tag{12-2-1}$$

and
$$I_{C2} = \alpha_2 I + I_{CBO2} \tag{12-2-2}$$

where $\alpha_1 = \left. \dfrac{\Delta I_{C1}}{\Delta I_{E1}} \right|_{V_{CB1}=0}$ is the common-base short-circuit forward current gain for transistor T1

$\alpha_2 = \left. \dfrac{\Delta I_{C2}}{\Delta I_{E2}} \right|_{V_{CB2}=0}$ is the common-base short-circuit forward current gain for transistor T2

$I_{CBO1} = $ reverse collector saturation current of transistor T1 with the emitter E1 open

$I_{CBO2} = $ reverse collector saturation current of transistor T2 with the emitter E2 open

The sum of the two collector currents is equal to

$$I = I_{C1} + I_{C2} \tag{12-2-3}$$

Substitution of Eqs. (12-2-1) and (12-2-2) into Eq. (12-2-3) yields

$$I = \frac{I_{CBO1} + I_{CBO2}}{1 - (\alpha_1 + \alpha_2)} \tag{12-2-4}$$

As the sum of $(\alpha_1 + \alpha_2)$ approaches unity, Eq. (12-2-4) indicates that the current I increases without limit. When condition $\alpha_1 + \alpha_2 = 1$ is attained, the switch transfers to its ON state. The voltage across the switch drops to a low value and the current becomes large, being limited by the external load resistance R_ℓ in series with the switch. When the sum of $(\alpha_1 + \alpha_2)$ is much less than unity, the current I is equal to the sum of the two reverse collector saturation currents $(I_{CBO1} + I_{CBO2})$ and the device switches OFF.

Example 12-2-2   **Anode Current of a Si p-n-p-n Thyristor**

A Si p-n-p-n thyristor has the following data:

$$I_{CBO1} = I_{CBO2} = 5 \text{ mA}$$
$$\alpha_1 = \alpha_2 = 0.40$$

Compute anode current I.

Solution.   From Eq. (12-2-4) the anode current is

$$I = \frac{5 \times 10^{-3} + 5 \times 10^{-3}}{1 - (0.40 + 0.40)} = 50 \text{ mA}$$

12-2-3   Voltage-Current Characteristics

When a positive voltage V_{dc} is applied to the anode terminal with respect to the cathode, junction J1 and junction J3 are forward biased and center junction J2 is reverse biased. The applied voltage appears principally across the reverse-biased junction J2. In effect, the current flowing through the device is very small and the device is OFF. As applied voltage is increased, the current increases slowly until a voltage called *forward-breakover voltage* V_{fb} is reached in which the current increases abruptly and voltage across the device decreases sharply. The device switches ON. The voltage-current characteristic curves are shown in Fig. 12-2-2.

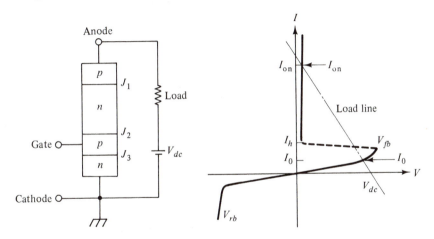

Figure 12-2-2   *V-I* curves of a *p-n-p-n* diode

In the reverse-blocking state a negative voltage is applied to the anode relative to the cathode; then junctions J1 and J3 are reverse biased and junction J2 is forward biased. Because the supply of electrons and holes to junction J2 is limited by the reverse-biased junction on either side, the device is restricted to a small saturation current arising from thermal generation near junctions J1 and J3. The reverse current remains small until negative applied voltage reaches the reverse-breakover point. At the reverse-breakover voltage the reverse current is increased rapidly as shown in Fig. 12-2-2.

12-2-4   Triggering Methods

Several methods can be used to trigger (or switch) a *p-n-p-n* diode from the OFF state (or forward-blocking state) to the ON state (or forward-conducting state).

1. Temperature excitation: An increase in the device temperature can cause triggering by increasing the carrier generation rate and lifetime sufficiently.

2. Optical excitation: Optical photons can increase the device current through the electron-hole pair generation.

3. Voltage excitation: The simple way is to raise the bias voltage to the forward-breakover level as discussed previously.

12-3 *MICROWAVE SWITCHING TRANSISTORS*

Microwave digital electronic systems often require clock speeds of gigahertz and data rates of 10 Gbits per second. All transistors can operate in two modes: ON state at the saturation region and OFF state at the cutoff region. At the microwave range, not many transistors can switch at a high speed, however.

Gallium arsenide (GaAs) is an attractive and reliable material for developing extremely high-speed operation devices. GaAs MESFETs with a short gate length of about 1 μm are better than Si devices in the microwave range. The GaAs IGFET has a small value of the gate capacitance C_{gs} and it can operate at a higher speed than a GaAs MESFET.

12-3-1 *GaAs MESFET Switches*

The GaAs MESFET (metal-semiconductor field-effect transistor) can operate in two modes: normally ON (or depletion) mode and normally OFF (or enhancement) mode. Cutoff frequency as given in Eq. (4-1-1) is

$$f_T = \frac{g_m}{2\pi C_{gs}}$$

(12-3-1)

The dual-gate GaAs MESFET may be used as a modulator for a very fast switching response in subnanosecond-pulsed amplitude modulation (PAM), phase-shift-keying (PSK), and frequency-shift-keying (FSK) carrier modulation [4]. The basic switching circuit and input and output waveforms are shown in Fig. 12-3-1.

The microwave sinusoidal output voltage from a signal generator at 8 GHz is fed directly into the first gate and the switching voltage between -2.5 V (for OFF) and $+1.5$ V (for ON) from a pulse generator is connected to the second gate of the 10-GHz GaAs MESFET as shown in Fig. 12-3-1(a). The switching pulse is synchronized with the 8-GHz signal; so the resulting microwave burst can be observed on an oscilloscope. As long as the second-gate voltage is low, the MESFET is turned OFF; no microwave signal is passed and the output voltage is at the supply-voltage level. When the second-gate voltage is pulsed to a positive potential, the MESFET is switched ON. As a result, a drain current flows, causing a voltage drop in the 50-Ω resistor and the microwave signal is transmitted.

In Fig. 12-3-1(b) the upper trace is the output-voltage waveform from the drain electrode and the lower trace is the input-voltage waveform to the first gate. The microwave burst is about 7 Hz long. The turn-on time of the switch is 100 ps and

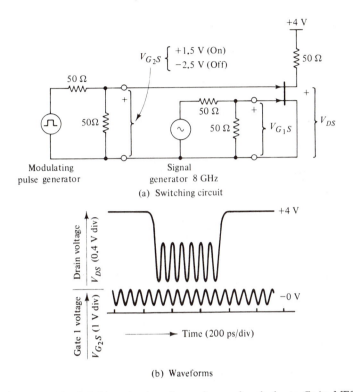

Figure 12-3-1 Switching circuit and waveforms of a dual-gate GaAs MESFET (From C. A. Liechti [4]; reprinted by permission of the IEEE, Inc.)

the turn-off time is 125 ps [4]. The conclusion is that the GaAs MESFET can be used as a switching device at X-band microwave frequencies.

Example 12-3-1   Cutoff Frequency and Switching Time of a GaAs MESFET Switch

A GaAs MESFET switch has a mutual conductance g_m of 6.283 mmho and a gate-source capacitance C_{gs} of 0.1 pF.
(a) Determine the cutoff frequency.
(b) Estimate its switching time.

Solution.   (a) Using Eq. (12-3-1), we find the cutoff frequency as

$$f_T = \frac{g_m}{2\pi C_{gs}} = \frac{6.283 \times 10^{-3}}{6.2832 \times 10^{-13}}$$

$$= 10 \text{ GHz}$$

(b) The switching time is

$$\tau = \frac{1}{2\pi f} = \frac{1}{6.2832 \times 10^{10}}$$

$$= 16 \text{ psec}$$

12-3-2 GaAs IGFET Switches

As noted, the GaAs IGFETs (insulated-gate field-effect transistors) are usually fabricated on *n*-type layer on a semi-insulating substrate as shown in Fig. 12-3-2.

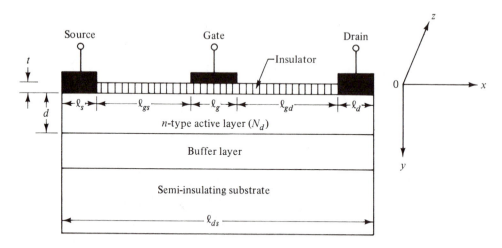

Figure 12-3-2　A schematic diagram of a GaAs IGFET

The dependence of C_{gs} on ϵ_{ox}, which is the dielectric constant of the oxide insulator, is not linear and tends to saturate in the region of very high ϵ_{ox}. This situation occurs because the gate capacitance consists of a series connection of two elements: one is the oxide capacitance and the other is the depletion-layer capacitance of the *n*-type active layer. As ϵ_{ox} increases, the oxide capacitance increases and the total input capacitance finally approaches the depletion-layer capacitance value. The deterioration in g_m is found to be small for oxide films with large dielectric constants, however. The maximum operating frequency for a GaAs IGFET as shown in Eq. (4-2-11) is

$$f_m = \frac{g_m}{2\pi C_{gs}} \tag{12-3-2}$$

To illustrate, if $g_m = 0.009$ mho and $C_{gs} = 0.09$ pF, the maximum operating frequency is 14.33 GHz.

REFERENCES

[1] Simmons, J. G., and A. EL-Badry. Theory of switching phenomena in metal/semiconductor/n-p^+ silicon devices. *Solid-State Electronics*, **20** (1977), 955–961.

[2] EL-Badry, A. and J. G. Simmons. Experimental studies of switching in metal semiconductor n-p^+ silicon devices. *Solid-State Electronics*, **20** (1977), 963–966.

[3] Habib, S. E-D., and J. G. Simmons. Theory of switching in *p-n*-insulator (Tunnel)-metal devices. *Solid-State Electronics*, **22** (1979), 181–192.

[4] Liechti, Charles A. Performance of dual-gate GaAs MESFETs as gain-controlled low-noise amplifiers and high-speed modulators. *IEEE Trans. on Microwave Theory and Techniques*, **MTT-23**, no. 6 (June 1975), 461–469.

[5] Sze, S. M. *Physics of Semiconductor Devices*. 2nd ed. New York: John Wiley & Sons, Inc., 1981, p. 550.

SUGGESTED READINGS

1. *IEEE Proceedings*, vol. 70, no. 1. Special issue on very fast solid-state technology. January 1982.

2. Moll, J. L., et al. *p-n-p-n* transistor switches. *Proc. IRE*, **44**, no. 9 (September 1956), 1174–1182.

PROBLEMS

12-1   MISS Diodes

12-1-1. A Si MISS diode has an electron concentration N_d of 10^{17} cm^{-3}. The bandgap energy E_g of Si is 1.12 eV.
   (a) Compute the avalanche voltage V_{av}.
   (b) Determine the multiplication factor M for a bias voltage of 10 volts.

12-1-2. A GaAs MISS diode has an electron concentration N_d of 10^{17} cm^{-3} and is biased by 15 volts. The bandgap energy E_g of GaAs is 1.43 eV.
   (a) Determine the avalanche voltage V_{av}.
   (b) Calculate the multiplication factor M for a bias voltage of 15 volts.

12-1-3. A Si MISS diode has an electron concentration N_d of 5×10^{17} cm^{-3} and is operating at room temperature.
   (a) Determine the punchthrough voltage ψ_{pt} for $(W_n - W_j) = 0.1$ μm.
   (b) Calculate the surface potential ψ_s at the insulator-semiconductor interface for $W_d = 0.05$ μm.
   (c) Estimate the holding voltage V_h.
   (d) Find the diode current I_d for $V = 10$ volts and $R_\ell = 10$ ohms.
   (e) Compute the switching time τ in nanoseconds for $(W_n - W_{di}) = 0.05$ μm.

12-1-4. A GaAs MISS diode has an electron concentration N_d of 5×10^{11} cm^{-3} and is operating at room temperature.
   (a) Calculate the switching time τ in nanoseconds for $(W_n - W_{di}) \simeq 1$ μm.
   (b) Compute the punchthrough voltage ψ_{pt} for $(W_n - W_j) = 100$ μm.
   (c) Find the surface potential ψ_s at the insulator-semiconductor interface for $W_d = 50$ μm.
   (d) Estimate the holding voltage V_h.
   (e) Determine the diode current for $V = 5$ volts and $R = 5$ ohms.

12-2   *p-n-p-n* Thyristors

12-2-1. A Si *p-n-p-n* thyristor has the following parameters:

$$I_{CBO1} = I_{CBO2} = 10 \text{ mA} \quad \text{and} \quad \alpha_1 = \alpha_2 = 0.45$$

Determine the anode current I.

12-2-2. Describe the operational principles and applications of thyristors.

12-3   GaAs MESFET Switches

12-3-1. A GaAs MESFET switch has a channel length L of 10 μm.
(a) Estimate the drift time τ of the carriers.
(b) Determine the cutoff frequency in gigahertz.

12-3-2. A GaAs IGFET switch has a mutual conductance g_m of 0.008 mho and a gate capacitance C_{gs} of 0.08 pF. Estimate the maximum operating frequency and switching time.

Appendixes

1. CONDUCTIVITY σ in mhos PER METER

Conductor	σ	Insulator	σ
Silver	6.17×10^7	Quartz	10^{-17}
Copper	5.80×10^7	Polystyrene	10^{-16}
Gold	4.10×10^7	Rubber (hard)	10^{-15}
Aluminum	3.82×10^7	Mica	10^{-14}
Tungsten	1.82×10^7	Porcelain	10^{-13}
Zinc	1.67×10^7	Diamond	10^{-13}
Brass	1.50×10^7	Glass	10^{-12}
Nickel	1.45×10^7	Bakelite	10^{-9}
Iron	1.03×10^7	Marble	10^{-8}
Bronze	1.00×10^7	Soil (sandy)	10^{-5}
Solder	0.70×10^7	Sands (dry)	2×10^{-4}
Steel (stainless)	0.11×10^7	Clay	10^{-4}
Nichrome	0.10×10^7	Ground (dry)	$10^{-4} - 10^{-5}$
Graphite	7.00×10^4	Ground (wet)	$10^{-2} - 10^{-3}$
Silicon	1.20×10^3	Water (distilled)	2×10^{-4}
Water (sea)	$3 - 5$	Water (fresh)	10^{-3}
		Ferrite (typical)	10^{-2}

2. DIELECTRIC CONSTANT—Relative Permittivity ϵ_r

Material	ϵ_r	Material	ϵ_r
Air	1	Sands (dry)	4
Alcohol (ethyl)	25	Silica (fused)	3.8
Bakelite	4.8	Snow	3.3
Glass	4–7	Sodium chloride	5.9
Ice	4.2	Soil (dry)	2.8
Mica (ruby)	5.4	Styrofoam	1.03
Nylon	4	Teflon	2.1
Paper	2–4	Water (distilled)	80
Plexiglass	2.6–3.5	Water (sea)	20
Polyethylene	2.25	Water (dehydrated)	1
Polystyrene	2.55	Wood (dry)	1.5–4
Porcelain (dry process)	6	Ground (wet)	5–30
Quartz (fused)	3.80	Ground (dry)	2–5
Rubber	2.5–4	water (fresh)	80

*Appendixes 1 through 5 are reprinted from Liao's *Microwave Devices and Circuits*, 2nd Ed. 1985, by permission of Prentice-Hall, Inc.

3. RELATIVE PERMEABILITY μ_r

Diamagnetic material	μ_r	Ferromagnetic material	μ_r
Bismuth	0.99999860	Nickel	50
Paraffin	0.99999942	Cast iron	60
Wood	0.99999950	Cobalt	60
Silver	0.99999981	Machine steel	300
		Ferrite (typical)	1,000
Paramagnetic material	μ_r	Transformer iron	3,000
		Silicon iron	4,000
Aluminum	1.00000065	Iron (pure)	4,000
Beryllinum	1.00000079	Mumetal	20,000
Nickel chloride	1.00004	Supermalloy	100,000
Manganese sulphate	1.0001		

4. PROPERTIES OF FREE SPACE

Velocity of light in vacuum c	2.997925×10^8 meters per second
Permittivity ϵ_0	8.854×10^{-12} farad per meter
Permeability μ_0	$4\pi \times 10^{-7}$ henry per meter
Intrinsic impedance η_0	377 or 120π ohms

5. PHYSICAL CONSTANTS

Boltzmann constant	k	1.381×10^{-23} J/°K
Electron volt	eV	1.602×10^{-19} J
Electron charge	q	1.602×10^{-19} C
Electron mass	m	9.109×10^{-31} kg
Ratio of charge to mass of an electron	e/m	1.759×10^{11} C/kg
Planck's constant	h	6.626×10^{-34} J-sec

6. ENERGY-GAP STRUCTURES OF SEMICONDUCTORS

Semi-conductor	Energy gap (eV)	Lattice constants (Å)	Energy gap structure	Expansion coefficient at 300°K ($\times 10^{-6}$°C^{-1})	Hetero-junction preferred doping	Typical dopants	Electron affinity (eV)
$Ge_{0.9}Si_{0.1}$	0.77	(5.63)	Indirect	—	n	P, As, Sb	(4.1)
Ge	0.66	5.658	Indirect	5.7	p	Al, Ga, In	4.13
GaAs	1.43	5.654	Direct	5.8	n	Se, Te	4.07
Ge	0.66	5.658	Indirect	5.7	p	Al, Ga, In	4.13
ZnSe	2.67	5.667	Direct	7.0	n	Al, Ga, In	4.09
Ge	0.66	5.658	Indirect	5.7	p	Al, Ga, In	4.13
ZnSe	2.67	5.667	Direct	7.0	n	Al, Ga, In	4.09
GaAs	1.43	5.654	Direct	5.8	p	Zn, Cd	4.07
AlAs	2.15	5.661	Indirect	5.2	p	Zn	3.5
GaAs	1.43	5.654	Direct	5.8	n	Se, Te	4.07
GaP	2.25	5.451	Indirect	5.3	n	Se, Te	4.3
Si	1.11	5.431	Indirect	2.33	p	Al, Ga, In	4.01
AlSb	1.6	6.136	Indirect	3.7	n/p	Se, Te/Zn, Cd	3.65
GaSb	0.68	6.095	Direct	6.9	p/n	Zn, Cd/Se, Te	4.06
GaSb	0.68	6.095	Direct	6.9	n	Se, Te	4.06
InAs	0.36	6.058	Direct	4.5(5.3)	p	Zn, Cd	4.9
ZnTe	2.26	6.103	Direct	8.2	p	Cu	3.5
GaSb	0.68	6.095	Direct	6.9	n	Se, Te	4.06
ZnTe	2.26	6.103	Direct	8.2	p	Cu	3.5
InAs	0.36	6.058	Direct	4.5(5.3)	n	Se, Te	4.9
ZnTe	2.26	6.103	Direct	8.2	p	Cu	3.5
AlSb	1.6	6.136	Indirect	3.7	n	Se, Te	3.65
CdTe	1.44	6.477	Direct	5	p/n	Li, Sb, P/I	4.28
PbTe	0.29	6.52	Indirect	19.8	n/p	Cl, Br/Na, K	4.6
CdTe	1.44	6.477	Direct	5	p	Li, Sb	4.28
InSb	0.17	0.479	Direct	4.9	n	Se, Te	4.59
ZnTe	2.26	6.103	Direct	8.2	p	Cu	3.5
CdSe (hex)	1.7	4.3($\sqrt{2}$) (6.05)	Direct	2.45-4.4	n	Cl, Br, I	4.95

After Milnes and Feucht, 116. From *Heterojunctions and Metal-Semiconductor Junctions*. New York: Academic Press, 1972. (Reprinted by permission of Academic Press.)

7. MOBILITIES OF Ge, GaAs, and Si

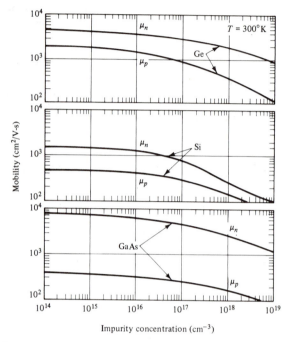

(Reprinted from Sze's *Physics of Semiconductor Devices*, 1981, by permission of John Wiley & Sons, Inc.)

8. DRIFT VELOCITIES VERSUS ELECTRIC FIELDS FOR GaAs, InP, and Si

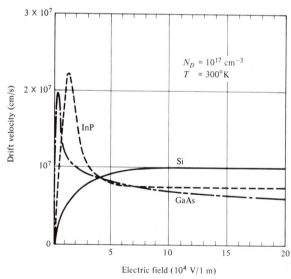

(Reprinted from Sze's *Physics of Semiconductor Devices*, 1981, by permission of John Wiley & Sons, Inc.)

9. DRIFT VELOCITIES OF GaAs, Ge, and Si

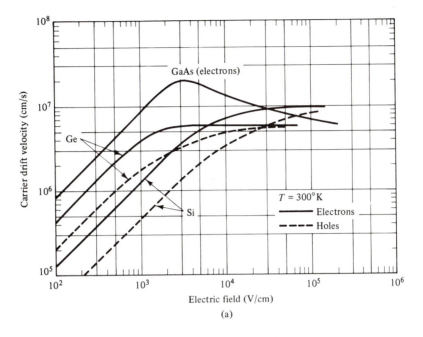

(a)

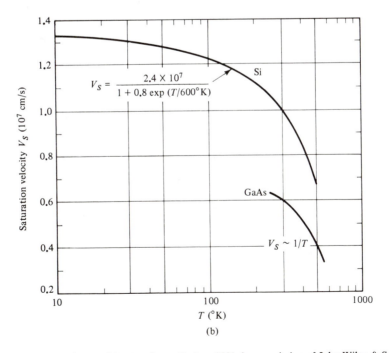

(b)

(Reprinted from Sze's *Physics of Semiconductor Devices*, 1981, by permission of John Wiley & Sons, Inc.)

10. METAL-SEMICONDUCTOR WORK FUNCTION VERSUS DOPING CONCENTRATION FOR Al-Si

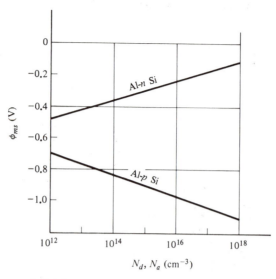

(Reprinted from Streetman's *Solid-State Electronic Devices*, 1981, by permission of Prentice-Hall, Inc.)

11. ABSORPTION COEFFICIENT VERSUS PHOTON ENERGY FOR SOME SEMICONDUCTORS

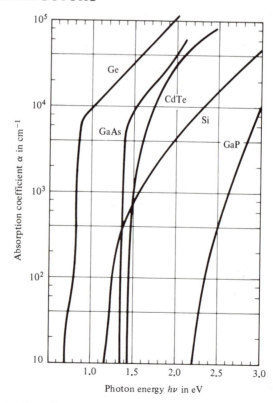

(Reprinted from Loferski's figure in *Proceedings*, vol. 63, 1963, p. 669, by permission of the IEEE, Inc.)

12. COMMERCIAL LASER AND LED SOURCES

Maker	Device	Drive current (A)	Emission wavelength $\lambda_0(\mu m)$	Spectral width (nm)	Power (mW)
			LASERs		
AEG	CQX-20	0.40	0.82	2.5	5.0
ITT	T901-L	0.35	0.84	2.0	7.5
LASER D. LABS	LCW10	0.20	0.85	2.0	14.0
RCA	C30130	0.25	0.82	2.0	6.0
			LEDs		
ASEA	1A83	0.10	0.94	40.0	10.0
ITT	T851-S	0.20	0.84	38.0	1.5
LASER D. LABS	IRE150	0.10	0.82	35.0	1.5
MERET	TL-36C	0.30	0.91	38.0	12.0
MONSANTO	ME60	0.05	0.90	40.0	0.4
PHILIPS	CQY58	0.05	0.88	45.0	0.5
RCA	SG1009	0.10	0.94	45.0	3.5
TI	TIXL472	0.05	0.91	23.0	1.0

13. HANKEL FUNCTIONS

The Hankel functions of the first and second kinds are combinations of the Bessel functions:

$$H_v^{(1)}(x) = J_v(x) + jN_v(x) \qquad \text{incoming wave} \qquad (1)$$

$$H_v^{(2)}(x) = J_v(x) - jN_v(x) \qquad \text{outgoing wave} \qquad (2)$$

For large argument, they become

$$H_v^{(1)}(x) \xrightarrow[x \to \infty]{} \sqrt{\frac{2}{j\pi x}} \, j^{-v} e^{jx} \qquad (3)$$

$$H_v^{(2)}(x) \xrightarrow[x \to \infty]{} \sqrt{\frac{2j}{\pi x}} \, j^v e^{-jx} \qquad (4)$$

When $x = ju$ is imaginary, the modified Bessel functions are

$$I_v(u) = j^v J_v(-ju) \qquad (5)$$

$$K_v(u) = \frac{\pi}{2}(-j)^{v+1} H_v^{(2)}(-ju) \qquad (6)$$

For large argument, I_v and K_v become

$$I_v(u) \xrightarrow[u \to \infty]{} \frac{e^u}{\sqrt{2\pi u}} \tag{7}$$

$$K_v(u) \xrightarrow[u \to \infty]{} \sqrt{\frac{\pi}{2u}}\, e^{-u} \tag{8}$$

Figure App-13 shows a few types.

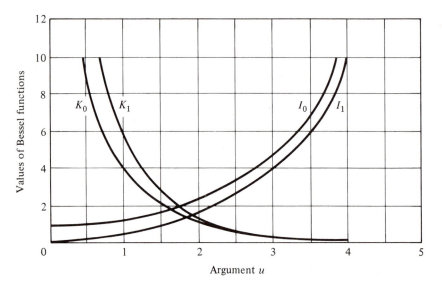

Figure App-13 Modified Bessel functions. (Reprinted from R. F. Harrington's *Time-Harmonic Electromagnetic Fields*, 1964, by permission of McGraw-Hill Book Company.)

14. MAXIMUM ELECTRIC FIELD

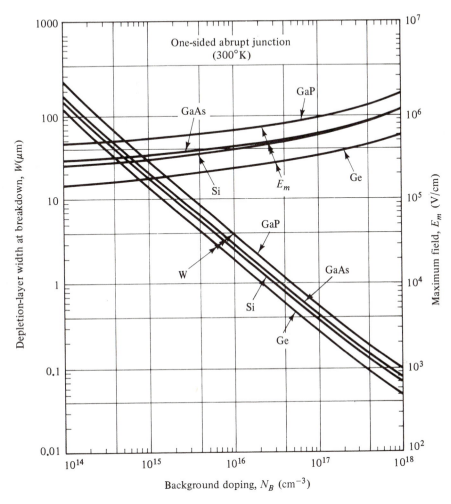

(Reprinted from Sze's *Physics of Semiconductor Devices*, 1981, by permission of John Wiley & Sons, Inc.)

Index